GSLIB
Geostatistical Software Library and User's Guide

Applied Geostatistics Series

André G. Journel, *General Editor*

Pierre Goovaerts, *Geostatistics for Natural Resources Evaluation*

Clayton V. Deutsch and André G. Journel, *GSLIB: Geostatistical Software Library and User's Guide, Second Edition*

Edward H. Isaaks and R. Mohan Srivastava, *An Introduction to Applied Geostatistics*

GSLIB

Geostatistical Software Library and User's Guide

Second Edition

Clayton V. Deutsch
Department of Petroleum Engineering
Stanford University

André G. Journel
Department of Geological and Environmental Sciences
Stanford University

New York   Oxford
OXFORD UNIVERSITY PRESS
1998

Oxford University Press

Oxford   New York
Athens   Auckland   Bangkok   Bogota   Bombay   Buenos Aires
Calcutta   Cape Town   Dar es Salaam   Delhi   Florence   Hong Kong
Istanbul   Karachi   Kuala Lumpur   Madras   Madrid   Melbourne
Mexico City   Nairobi   Paris   Singapore   Taipei   Tokyo   Toronto   Warsaw

and associated companies in
Berlin   Ibadan

Copyright © 1992, 1998 by Oxford University Press, Inc.

Published by Oxford University Press, Inc.,
198 Madison Avenue, New York, New York 10016
http://www.oup-usa.org

Oxford is a registered trademark of Oxford University Press

Library of Congress Cataloging-in-Publication Data

Deutsch, Clayton V.
    GSLIB [computer file] : geostatistical software library and user's
guide / Clayton V. Deutsch, André G. Journel. — 2nd ed.
    1 computer laser optical disc : 4 3/4 in. + 1 user's guide.
    Computer program.
    System requirements; IBM-compatible PC; FORTRAN compiler (level 77
or higher); CD-ROM drive.
    Title from disc label.
    Audience: Geoscientists.
    Summary: Thirty-seven programs that summarize data with histograms
and other graphics, calculate measures of spatial continuity,
provide smooth least-squares-type maps, and perform stochastic
spatial simulation.
    ISBN 0-19-510015-8
    1. Geology—Statistical methods—Software.   I. Journel, A. G.
II. Title.
QE33.2.S82   <1997 00505>   <MRC>   97-4525
550—DC12a   CIP

9 8 7 6 5 4 3

Printed in the United States of America
on acid-free paper

Preface to the Second Edition

The success of the first edition (1992) has proven our vision: GSLIB has received acceptance both in industry, for practical applications, and in academia, where it has allowed researchers to experiment with new ideas without tedious basic development. This success, we believe, is due to the availability and flexibility of the GSLIB source code. We also believe that communication between geostatisticians has been improved with the acceptance of the programming and notation standards proposed in the first edition.

In preparing this second edition of GSLIB we have maintained the nononsense, get-to-the-fact, guidebook approach taken in the first edition. The text has been expanded to cover new programs and the source code has been corrected, simplified, and expanded. A number of programs were combined for simplicity (e.g., the variogram programs gam2 and gam3 were combined into gam); removed due to lack of usage or the availability of better alternatives (e.g., the simulation program tb3d was removed); and, some new programs were added because they have proven useful (e.g., the smoothing program histsmth was added).

Although the programs will "look and feel" similar to the programs in the first edition, no attempt has been made to ensure backward compatibility. The organization of most parameter files has changed. As before, the GSLIB software remains unsupported, although, this second edition should alleviate any doubt regarding the commitment of the authors to maintaining the software and associated guidebook. The FORTRAN source code for the GSLIB programs are on the attached CD and are available by anonymous ftp from markov.stanford.edu

Finally, in addition to the graduate students at Stanford, we would like to acknowledge the many people who have provided insightful feedback on the first edition of GSLIB.

Stanford, California   C.V.D.
January 1997   A.G.J.

Preface to the First Edition

The primary goal of this work is to present a geostatistical software library known as GSLIB. An important prerequisite to geostatistical case studies and research is the availability of flexible and understandable software. Flexibility is achieved by providing the original FORTRAN source code. A detailed description of the theoretical background along with specific application notes allows the algorithms to be understood and used as the basis for more advanced customized programs.

The three main chapters of this guidebook are based on the three major problem areas of geostatistics: quantifying spatial variability (variograms), generalized linear regression techniques (kriging), and stochastic simulation. Additional utility programs and problem sets with partial solutions are given to allow a full exploration of GSLIB and to check new software installations.

This guidebook is aimed at graduate students and advanced practitioners of geostatistics; it is not intended to be a theoretical reference textbook. Proofs, lengthy theoretical discussions, and heavy matrix notation are omitted as much as possible. Instead, this guidebook contains many footnotes, application notes, brief warnings, and multiple cross references.

The GSLIB source code has been assembled from programs developed and used at Stanford University over the course of 12 years. These programs are constantly questioned and modified to handle new algorithms. GSLIB is not a commercial product and carries no warranties, software support, or maintenance. Undoubtedly, there will be "bugs" left in the published version of the programs, most of them introduced during the rewriting of the code and thus the sole responsibility of the authors of this guide. Even though the main avenues of these programs have been tested and used extensively there are simply too many possible combinations of input data and parameters to ensure bug-free programs.

We would like to acknowledge the many graduate students who have contributed to GSLIB during their stay at Stanford University. This text is dedicated to them and to future generations who will continue to make GSLIB a living, evolving collection of programs.

Stanford, California　　　　　　　　　　　　　　　　　　　　　　　　　C.V.D.
June 1992　　　　　　　　　　　　　　　　　　　　　　　　　　　　　　　A.G.J.

Contents

Appendices

GSLIB
Geostatistical Software Library and User's Guide

Chapter I

Introduction

GSLIB is the name of a directory containing the geostatistical software developed at Stanford. Generations of graduate students have contributed to this constantly changing collection of programs, ideas, and utilities. Some of the most widely used public-domain geostatistical software [58, 62, 72] and many more in-house programs were initiated from GSLIB. It was decided to open the GSLIB directory to a wide audience, providing the source code to seed new developments, custom-made application tools, and, it is hoped, new theoretical advances.

What started as a collection of selected GSLIB programs to be distributed on diskette with a minimal user's guide has turned into a major endeavor; a lot of code has been rewritten for the sake of uniformity, and the user's guide looks more like a textbook. Yet our intention has been to maintain the original versatility of GSLIB, where clarity of coding prevails over baroque theory or the search for ultimate speed. GSLIB programs are development tools and sketches for custom programs, advanced applications, and research; we expect them to be torn apart and parts grafted into custom software packages.

The strengths of GSLIB are also its weaknesses. The flexibility of GSLIB, its modular conception, and its machine independence may assist researchers and those beginners more eager to learn than use; on the other hand, the requirement to make decisions at each step and the lack of a slick user interface will frustrate others.

Reasonable effort has been made to provide software that is well documented and free of bugs but there is no support or telephone number for users to call. Each user of these programs assumes the responsibility for their *correct* application and careful testing as appropriate.[1]

---

[1] Reports of bugs or significant improvements in the coding should be sent to the authors at Stanford University.

Trends in Geostatistics

Most geostatistical theory [63, 118, 188, 195] was established long before the availability of digital computers allowed its wide application in earth sciences [109, 132, 133]. The development of geostatistics since the late 1960s has been accomplished mostly by application-oriented engineers: They have been ready to bend theory to match the data and their approximations have been strengthened by extended theory.

Most early applications of geostatistics were related to mapping the spatial distribution of one or more attributes, with emphasis given to characterizing the variogram model and using the kriging (error) variance as a measure of estimation accuracy. Kriging, used for mapping, is not significantly better than other deterministic interpolation techniques that are customized to account for anisotropy, data clustering, and other important spatial features of the variable being mapped [79, 92, 140]. The kriging algorithm also provides an error variance. Unfortunately, the kriging variance is independent of the data values and cannot be used, in general, as a measure of estimation accuracy.

Recent applications of geostatistics have de-emphasized the mapping application of kriging. Kriging is now used to build models of uncertainty that depend on the data values in addition to the data configuration. The attention has shifted from mapping to conditional simulation, also known as *stochastic imaging*. Conditional simulation allows drawing alternative, equally probable realizations of the spatial distribution of the attribute(s) under study. These alternative stochastic images provide a measure of uncertainty about the unsampled values taken *altogether* in space rather than one by one. In addition, these simulated images do not suffer from the characteristic smoothing effect of kriging (see frontispiece).

Following this trend, the GSLIB software and this guidebook emphasize stochastic simulation tools.

Book Content

Including this introduction, the book consists of 6 chapters, 5 appendices including a list of programs given, a bibliography, and an index. Two diskettes containing all the FORTRAN source code are provided.

Chapter II provides the novice reader with a review of basic geostatistical concepts. The concepts of random functions and stationarity are introduced as models. The basic notations, file conventions, and variogram specifications used throughout the text and in the software are also presented.

Chapter III deals with covariances, variograms, and other more robust measures of spatial variability/continuity. The "variogram" programs allow the simultaneous calculation of many such measures at little added computational cost. An essential tool, in addition to the programs in GSLIB, would be a graphical, highly interactive, variogram modeling program (such as available in other public-domain software [28, 58, 62]).

Chapter IV presents the multiple flavors of kriging, from simple and ordinary kriging to factorial and indicator kriging. As already mentioned, kriging can be used for many purposes in addition to mapping. Kriging can be applied as a filter (Wiener's filter) to separate either the low-frequency (trend) component or, conversely, the high-frequency (nugget effect) component of the spatial variability. When applied to categorical variables or binary transforms of continuous variables, kriging is used to update prior local information with neighboring information to obtain a posterior probability distribution for the unsampled value(s). Notable omissions in GSLIB include kriging with intrinsic random functions of order k (IRF-k) and disjunctive kriging (DK). From a first glance at the list of GSLIB subroutines, the reader may think that other more common algorithms are missing, e.g., kriging with an external drift, probability kriging, and multi-Gaussian kriging. These algorithms, however, are covered by programs named for other algorithms. Kriging with an external drift can be seen as a form of kriging with a trend and is an option of program kt3d. Probability kriging is a form of cokriging (program cokb3d). As for multi-Gaussian kriging, it is a mere simple kriging (program kt3d) applied to normal score transforms of the data (program nscore) followed by an appropriate back transform of the estimates.

Chapter V presents the principles and various algorithms for stochastic simulation. Major heterogeneities, possibly modeled by categorical variables, should be simulated first; then continuous properties are simulated within each reasonably homogeneous category. It is recalled that fluctuations of the realization statistics (histogram, variograms, etc.) are expected, particularly if the field being simulated is not very large with respect to the range(s) of correlation. Stochastic simulation is currently the most vibrant field of research in geostatistics, and the list of algorithms presented in this text and coded in GSLIB is not exhaustive. GSLIB includes only the most commonly used stochastic algorithms as of 1996. Physical process simulations [179], the vast variety of object-based algorithms [174], and spectral domain algorithms [18, 28, 76, 80] are not covered here, notwithstanding their theoretical and practical importance.

Chapter VI proposes a number of useful utility programs including some to generate graphics in the PostScript page description language. Seasoned users will already have their own programs. Similar utility programs are available in classical textbooks or in inexpensive statistical software. These programs have been included, however, because they are essential to the practice of geostatistics. Without a histogram and scatterplot, summary statistics, and other displays there would be no practical geostatistics; without a declustering program most geostatistical studies would be flawed from the beginning; without normal score transform and back transform programs there would be no multi-Gaussian geostatistics; an efficient and robust linear system solver is essential to any kriging exercise.

Each chapter starts with a brief summary and a first section on methodology (II.1, III.1, IV.1, and V.1). The expert geostatistician may skip this first section, although it is sometimes useful to know the philosophy underlying the software being used. The remaining sections in each chapter give commented lists of input parameters for each subroutine. A section entitled "Application Notes" is included near the end of each chapter with useful advice on using and improving the programs.

Chapters II through V end with sections suggesting one or more problem sets. These problems are designed to provide a first acquaintance with the most important GSLIB programs and a familiarity with the parameter files and notations used. All problem sets are based on a common synthetic data set provided with the GSLIB distribution diskettes. The questions asked are all open-ended; readers are advised to pursue their investigations beyond the questions asked. Limited solutions to each problem set are given in Appendix A; they are not meant to be definitive general solutions. The first time a program is encountered the input parameter file is listed along with the first and last 10 lines of the corresponding output file. These limited solutions are intended to allow readers to check their understanding of the program input and output. Working through the problem sets before referring to the solutions will provide insight into the algorithms. Attempting to apply GSLIB programs (or any geostatistical software) directly to a real-life problem without some prior experience may be extremely frustrating.

I.1   About the Source Code

The source code provided in GSLIB adheres as closely as possibly to the ANSI standard FORTRAN 77 programming language. FORTRAN was retained chiefly because of its familiarity and common usage. Some consider FORTRAN the dinosaur of programming languages when compared with newer object-oriented programming languages; nevertheless, FORTRAN remains a viable and even desirable scientific programming language due to its simplicity and the introduction of better features in new ANSI standards or as enhancements by various software companies.

GSLIB was coded to the ANSI standard to achieve machine independence. We did not want the use of GSLIB limited to any specific PC, workstation, or mainframe computer. Many geostatistical exercises may be performed on a laptop PC. Large stochastic simulations and complex modeling, however, would be better performed on a more powerful machine. GSLIB programs have been compiled on machines ranging from an XT clone to the latest multiple processor POWER Challenge. No executable programs are provided; only ASCII files containing the source code and other necessary input files are included in the enclosed diskettes. The source code can immediately be dismantled and incorporated into custom programs, or compiled and used without modification. Appendix B contains detailed information about software installation and troubleshooting.

The 1978 ANSI FORTRAN 77 standard does not allow dynamic memory allocation. This poses an inconvenience since array dimensioning and memory allocation must be *hardcoded* in all programs. To mitigate this inconvenience, the dimensioning and memory allocation have been specified in an "include" file. The include file is common to all necessary subroutines. Therefore, the array dimensioning limits, such as the maximum number of data or the maximum grid size, has to be changed in only *one* file.

Although the maximum dimensioning parameters are hardcoded in an include file, it is clearly undesirable to use this method to specify most input parameters. Given the goal of machine independence, it is also undesirable to have input parameters enter the program via a menu-driven graphical user interface. For these reasons the input parameters are specified in a prepared "parameter" file. These parameter files must follow a specified format. An advantage of this form of input is that the parameter files provide a permanent record of run parameters. With a little experience users will find the preparation and modification of parameter files faster than most other forms of parameter input. Machine-dependent graphical user interfaces would be an alternative.

Further, the source code for any specific program has been separated into a main program and then all the necessary subroutines. This was done to facilitate the isolation of key subroutines for their incorporation into custom software packages. In general, there are four computer files associated with any algorithm or program: the main program, the subroutines, an include file, and a parameter file.

Most program operations are separated into short, easily understood, program modules. Many of these shorter subroutines and funtions are then used in more than one program. Care has been taken to use a consistent format with a statement of purpose, a list of the input parameters, and a list of programmers who worked on a given piece of code. Extensive in-line comments explain the function of the code and identify possible enhancements. A detailed description of the programming conventions and a dictionary of the variable names are given in Appendix C.

Chapter II

Getting Started

This chapter provides a brief review of the geostatistical concepts underlying GSLIB and an initial presentation of notations, conventions, and computer requirements.

Section II.1 establishes the derivation of posterior probability distribution models for unsampled values as a principal goal. From such distributions, best estimates can be derived together with probability intervals for the unknowns. Stochastic simulation is the process of Monte Carlo drawing of realizations from such posterior distributions.

Section II.2 gives some of the basic notations, computer requirements, and file conventions used throughout the GSLIB text and software.

Section II.3 considers the important topic of variogram modeling. The notation and conventions established in this section are consistent throughout all GSLIB programs.

Section II.4 considers the different search strategies that are implemented in GSLIB. The super block search, spiral search, two-part search, and the use of covariance lookup tables are discussed.

Section II.5 introduces the data set that will be used throughout for demonstration purposes. A first problem set is provided in Section II.6 to acquaint the reader with the data set and the utility programs for exploratory data analysis.

II.1   Geostatistical Concepts: A Review

Geostatistics is concerned with "the study of phenomena that fluctuate in space" and/or time ([146], p. 31). Geostatistics offers a collection of deterministic and statistical tools aimed at understanding and modeling spatial variability.

The basic paradigm of predictive statistics is to characterize any unsampled (unknown) value z as a random variable (RV) Z, the probability distribution of which models the uncertainty about z. A random variable is a

variable that can take a variety of outcome values according to some probability (frequency) distribution. The random variable is traditionally denoted
by a capital letter, say Z, while its outcome values are denoted with the
corresponding lower-case letter, say z or $z^{(l)}$. The RV model Z, and more
specifically its probability distribution, is usually location-dependent; hence
the notation $Z(\mathbf{u})$, with $\mathbf{u}$ being the location coordinates vector. The RV
$Z(\mathbf{u})$ is also information-dependent in the sense that its probability distribution changes as more data about the unsampled value $z(\mathbf{u})$ become available.
Examples of continuously varying quantities that can be effectively modeled
by RVs include petrophysical properties (porosity, permeability, metal or pollutant concentrations) and geographical properties (topographic elevations,
population densities). Examples of categorical variables include geological
properties such as rock types or counts of insects or fossil species.

The cumulative distribution function (cdf) of a continuous RV $Z(\mathbf{u})$ is
denoted:

$$F(\mathbf{u}; z) = \text{Prob}\{Z(\mathbf{u}) \leq z\} \qquad (\text{II.1})$$

When the cdf is made specific to a particular information set, for example
(n) consisting of n neighboring data values $Z(\mathbf{u}_\alpha) = z(\mathbf{u}_\alpha), \alpha = 1, \ldots, n$,
the notation "conditional to (n)" is used, defining the conditional cumulative
distribution function (ccdf):

$$F(\mathbf{u}; z|(n)) = \text{Prob}\{Z(\mathbf{u}) \leq z|(n)\} \qquad (\text{II.2})$$

In the case of a categorical RV $Z(\mathbf{u})$ that can take any one of K outcome
values $k = 1, \ldots, K$, a similar notation is used:

$$F(\mathbf{u}; k|(n)) = \text{Prob}\{Z(\mathbf{u}) = k|(n)\} \qquad (\text{II.3})$$

Expression (II.1) models the uncertainty about the unsampled value $z(\mathbf{u})$
prior to using the information set (n); expression (II.2) models the posterior
uncertainty once the information set (n) has been accounted for. The goal,
implicit or explicit, of any predictive algorithm is to update prior models of
uncertainty such as (II.1) into posterior models such as (II.2). Note that the
ccdf $F(\mathbf{u}; z|(n))$ is a function of the location $\mathbf{u}$, the sample size and geometric
configuration (the data locations $\mathbf{u}_\alpha$, $\alpha = 1, \ldots, n$), and the sample values
[the n values $z(\mathbf{u}_\alpha)$s].

From the ccdf (II.2) one can derive different optimal estimates for the
unsampled value $z(\mathbf{u})$ in addition to the ccdf mean, which is the least-squares
error estimate. One can also derive various probability intervals such as the
95% interval $[q(0.025); q(0.975)]$ such that

$$\text{Prob}\{Z(\mathbf{u}) \in [q(0.025); q(0.975)]|(n)\} = 0.95$$

with $q(0.025)$ and $q(0.975)$ being the 0.025 and 0.975 quantiles of the ccdf,
for example,

$$q(0.025) \text{ is such that } F(\mathbf{u}; q(0.025)|(n)) = 0.025$$

Moreover, one can draw any number of simulated outcome values $z^{(l)}(\mathbf{u}), l = 1, \ldots, L$, from the ccdf. Derivation of posterior ccdfs and Monte Carlo drawings of outcome values are at the heart of stochastic simulation algorithms (see Chapter V).

In geostatistics most of the information related to an unsampled value $z(\mathbf{u})$ comes from sample values at neighboring locations $\mathbf{u}'$, whether defined on the same attribute z or on some related attribute y. Thus it is important to model the degree of correlation or dependence between any number of RV's $Z(\mathbf{u}), Z(\mathbf{u}_\alpha), \alpha = 1, \ldots, n$ and more generally $Z(\mathbf{u}), Z(\mathbf{u}_\alpha), \alpha = 1, \ldots, n$, $Y(\mathbf{u}'_\beta), \beta = 1, \ldots, n'$. The concept of a random function (RF) allows such modeling and updating of prior cdfs into posterior ccdfs.

II.1.1 The Random Function Concept

A random function (RF) is a set of RVs defined over some field of interest, such as $\{Z(\mathbf{u}), \mathbf{u} \in$ study area$\}$ also denoted simply as $Z(\mathbf{u})$. Usually the RF definition is restricted to RVs related to the same attribute, say z, hence another RF would be defined to model the spatial variability of a second attribute, say $\{Y(\mathbf{u}), \mathbf{u} \in$ study area$\}$.

Just as an RV $Z(\mathbf{u})$ is characterized by its cdf (II.1), an RF $Z(\mathbf{u})$ is characterized by the set of all its K-variate cdfs for any number K and any choice of the K locations $\mathbf{u}_k, k = 1, \ldots, K$:

$$F(\mathbf{u}_1, \ldots, \mathbf{u}_k; z_1, \ldots, z_K) = \text{Prob} \{Z(\mathbf{u}_1) \leq z_1, \ldots, Z(\mathbf{u}_K) \leq z_K\} \quad \text{(II.4)}$$

Just as the univariate cdf of RV $Z(\mathbf{u})$ is used to model uncertainty about the value $z(\mathbf{u})$, the multivariate cdf (II.4) is used to model joint uncertainty about the K values $z(\mathbf{u}_1), \ldots, z(\mathbf{u}_K)$.

Of particular interest is the bivariate ($K = 2$) cdf of any two RVs $Z(\mathbf{u})$, $Z(\mathbf{u}')$, or more generally $Z(\mathbf{u}), Y(\mathbf{u}')$:

$$F(\mathbf{u}, \mathbf{u}'; z, z') = \text{Prob} \{Z(\mathbf{u}) \leq z, Z(\mathbf{u}') \leq z'\} \quad \text{(II.5)}$$

Consider the particular binary transform of $Z(\mathbf{u})$ defined as

$$I(\mathbf{u}; z) = \begin{cases} 1, & \text{if } Z(\mathbf{u}) \leq z \\ 0, & \text{otherwise} \end{cases} \quad \text{(II.6)}$$

Then the previous bivariate cdf (II.5) appears as the noncentered indicator (cross) covariance:

$$F(\mathbf{u}, \mathbf{u}'; z, z') = E \{I(\mathbf{u}; z)I(\mathbf{u}'; z')\} \quad \text{(II.7)}$$

Relation (II.7) is the key to the indicator formalism as developed later in Section IV.1.9: It shows that inference of bivariate cdfs can be done through sample indicator covariances.

Rather than a series of indicator covariances, inference can be restricted to a single moment of the bivariate cdf (II.5), for example, the attribute covariance:

$$C(\mathbf{u}, \mathbf{u}') = E\left\{Z(\mathbf{u})Z(\mathbf{u}')\right\} - E\left\{Z(\mathbf{u})\right\}E\left\{Z(\mathbf{u}')\right\} \qquad (II.8)$$

The Gaussian RF model is extraordinarily convenient in the sense that any K-variate cdf of type (II.4) is fully defined from knowledge of the single covariance function (II.8). This remarkable property explains the popularity of Gaussian-related models. That property is also the main drawback: these models allow reproduction only of the data covariance when other spatial properties may also be important.

II.1.2   Inference and Stationarity

Inference of any statistic, whether a univariate cdf such as (II.1) or any of its moments (mean, variance) or a multivariate cdf such as (II.4) or (II.5) or any of its moments (covariances), requires some repetitive sampling. For example, repetitive sampling of the variable $z(\mathbf{u})$ is needed to evaluate the cdf $F(\mathbf{u}; z) = \text{Prob}\left\{Z(\mathbf{u}) \leq z\right\}$ from experimental proportions. In most applications, however, only one sample is available at any specific location $\mathbf{u}$ in which case $z(\mathbf{u})$ is known (ignoring sampling errors), and the need to consider the RV model $Z(\mathbf{u})$ vanishes. The paradigm underlying most inference processes (not only statistical inference) is to trade the unavailable replication at location $\mathbf{u}$ for another replication available somewhere else in space and/or time. For example, the cdf $F(\mathbf{u}; z)$ may be inferred from the cumulative histogram of z- samples collected at other locations, $\mathbf{u}_\alpha \neq \mathbf{u}$, within the same field, or at the same location $\mathbf{u}$ but at different times if a time series is available at $\mathbf{u}$. In the latter case the RF is actually defined in space-time and should be denoted $Z(\mathbf{u}, t)$.

This trade of replication or sample spaces corresponds to the hypothesis (rather a decision[1]) of stationarity.

The RF $\left\{Z(\mathbf{u}), \mathbf{u} \in A\right\}$ is said to be stationary within the field A if its multivariate cdf (II.4) is invariant under any translation of the K coordinate vectors $\mathbf{u}_k$, that is:

$$F(\mathbf{u}_1, \ldots, \mathbf{u}_k; z_1, \ldots, z_K) = F(\mathbf{u}_1 + \mathbf{l}, \ldots, \mathbf{u}_k + \mathbf{l}; z_1, \ldots, z_K), \qquad (II.9)$$

$$\forall \text{ translation vector } \mathbf{l}$$

Invariance of the multivariate cdf entails invariance of any lower order cdf, including the univariate and bivariate cdfs, and invariance of all their moments,

---

[1]Stationarity is a property of the RF model, not a property of the underlying spatial distribution. Thus it cannot be checked from data. In a census, the decision to provide statistics per county or state rather than per class of age is something that cannot be checked or refuted. The decision to pool data into statistics across rock types is not refutable a priori from data; however, it can be shown inappropriate a posteriori if differentiation per class of age or rock type is critical to the undergoing study. For a more extensive discussion see [89, 99, 103].

including all covariances of type (II.7) or (II.8). The decision of stationarity allows inference. For example, the unique stationary cdf

$$F(z) = F(\mathbf{u}; z), \quad \forall \, \mathbf{u} \in A$$

can be inferred from the cumulative sample histogram of the z-data values available at various locations within A. Another example concerns the inference of the stationary covariance

$$C(\mathbf{h}) = E\left\{Z(\mathbf{u} + \mathbf{h})Z(\mathbf{u})\right\} - \left[E\left\{Z(\mathbf{u})\right\}\right]^2, \qquad \text{(II.10)}$$

$$\forall \mathbf{u}, \mathbf{u} + \mathbf{h} \in A$$

from the sample covariance of all pairs of z-data values approximately separated by vector $\mathbf{h}$; see program gamv in Chapter III.

A proper decision of stationarity is critical for the representativeness and reliability of the geostatistical tools used. Pooling data across geological facies may mask important geological differences; on the other hand, splitting data into too many subcategories may lead to unreliable statistics based on too few data per category and an overall confusion. The rule in statistical inference is to pool the largest amount of *relevant* information to formulate predictive statements.

Stationarity is a property of the model; thus the decision of stationarity may change if the scale of the study changes or if more data become available. If the goal of the study is global, then local details can be averaged out; conversely, the more data available, the more statistically significant differentiations become possible.

II.1.3 Variogram

An alternative to the covariance defined in (II.8) and (II.10) is the variogram defined as the variance of the increment $[Z(\mathbf{u}) - Z(\mathbf{u} + \mathbf{h})]$. For a stationary RF:

$$
\begin{aligned}
2\gamma(\mathbf{h}) &= \operatorname{Var}\left\{Z(\mathbf{u} + \mathbf{h}) - Z(\mathbf{u})\right\} && \text{(II.11)}\\
\gamma(\mathbf{h}) &= C(0) - C(\mathbf{h}), \ \forall \, \mathbf{u}
\end{aligned}
$$

with: $C(\mathbf{h})$ being the stationary covariance, and: $C(0) = \operatorname{Var}\{Z(\mathbf{u})\}$ being the stationary variance.

Traditionally, the variogram has been used for modeling spatial variability rather than the covariance although kriging systems are more easily solved with covariance matrices [$\gamma(0)$ values are problematic when pivoting]. In accordance with tradition,[2] GSLIB programs consistently require

---

[2]In the stochastic process literature the covariance is preferred over the variogram. The historical reason for geostatisticians preferring the variogram is that its definition (II.11) requires only second-order stationarity of the RF increments, a condition also called the "intrinsic hypothesis." This lesser demand on the RF model has been recently shown to be of no consequence for most practical situations; for more detailed discussions see [88, 112, 158].

*semi*variogram models, which are then promptly converted into equivalent covariance models. The "variogram" programs of Chapter III allow the computation of covariance functions in addition to variogram functions.

A Note on Generalized Covariances

Generalized covariances of order k are defined as variances of differences of order $(k + 1)$ of the initial RF $Z(\mathbf{u})$ (see, e.g. [39, 187]). The traditional variogram $2\gamma(\mathbf{h})$, defined in relation (II.11) as the variance of the first-order difference of the RF $Z(\mathbf{u})$, is associated to a generalized covariance of order zero. The order zero stems from the variogram expression's filtering any zero-order polynomial of the coordinates $\mathbf{u}$, such as $m(\mathbf{u})$=constant, added to the RF model $Z(\mathbf{u})$. Similarly, a generalized covariance of order k would filter a polynomial trend of order k added to the RF model $Z(\mathbf{u})$.

Unfortunately, inference of generalized covariances of order $k > 0$ poses severe problems since experimental differences of order $k + 1$ are not readily available if the data are not gridded. In addition, more straightforward algorithms for handling polynomial trends exist, including ordinary kriging with moving data neighborhoods [112].

Therefore, notwithstanding the possible theoretical importance of the IRF-k formalism, it has been decided not to include generalized covariances and the related intrinsic RF models of order k in this version of GSLIB.

II.1.4   Kriging

Kriging is "a collection of generalized linear regression techniques for minimizing an estimation variance defined from a prior model for a covariance" ([146], p. 41).

Consider the estimate of an unsampled value $z(\mathbf{u})$ from neighboring data values $z(\mathbf{u}_\alpha), \alpha = 1, \ldots, n$. The RF model $Z(\mathbf{u})$ is stationary with mean m and covariance $C(\mathbf{h})$. In its simplest form, also known as simple kriging (SK), the algorithm considers the following linear estimator:

$$Z_{SK}^*(\mathbf{u}) = \sum_{\alpha=1}^{n} \lambda_\alpha(\mathbf{u})Z(\mathbf{u}_\alpha) + \left(1 - \sum_{\alpha=1}^{n} \lambda_\alpha(\mathbf{u})\right) m \qquad \text{(II.12)}$$

The weights $\lambda_\alpha(\mathbf{u})$ are determined to minimize the error variance, also called the "estimation variance." That minimization results in a set of normal equations [102, 124]:

$$\sum_{\beta=1}^{n} \lambda_\beta(\mathbf{u})C(\mathbf{u}_\beta - \mathbf{u}_\alpha) = C(\mathbf{u} - \mathbf{u}_\alpha), \qquad \text{(II.13)}$$

$$\forall \alpha = 1, \ldots, n$$

The corresponding minimized estimation variance, or kriging variance, is:

$$\sigma_{SK}^2(\mathbf{u}) = C(0) - \sum_{\alpha=1}^{n} \lambda_\alpha(\mathbf{u}) C(\mathbf{u} - \mathbf{u}_\alpha) \geq 0 \qquad (\text{II}.14)$$

Ordinary kriging (OK) is the most commonly used variant of the previous simple kriging algorithm, whereby the sum of the weights $\sum_{\alpha=1}^{n} \lambda_\alpha(\mathbf{u})$ is constrained to equal 1. This allows building an estimator $Z_{OK}^*(\mathbf{u})$ that does not require prior knowledge of the stationary mean m, yet remains unbiased in the sense that $E\{Z_{OK}^*(\mathbf{u})\} = E\{Z(\mathbf{u})\}$.

Non-linear kriging is but linear kriging performed on some non-linear transform of the z-data, e.g., the log-transform $\ln z$ provided that $z > 0$, or the indicator transform as defined in relation (II.6).

Traditionally, kriging (SK or OK) has been performed to provide a "best" linear unbiased estimate (BLUE) for unsampled values $z(\mathbf{u})$, with the kriging variance being used to define Gaussian-type confidence intervals, e.g.,

$$\text{Prob}\{Z(\mathbf{u}) \in [z_{SK}^*(\mathbf{u}) \pm 2\sigma_{SK}(\mathbf{u})]\} \cong 0.95$$

Unfortunately, kriging variances of the type (II.14), being independent of the data values, only provides a comparison of alternative geometric data configurations. Kriging variances are usually not measures of local estimation accuracy [99].

In addition, users have come to realize that kriging estimators of type (II.12) are "best" only in the least-squares error sense for a given covariance/variogram model. Minimizing an expected squared error need not be the most relevant estimation criterion for the study at hand; rather, one might prefer an algorithm that would minimize the impact (loss) of the resulting error; see [67, 167]. This decision-analysis approach to estimation requires a probability distribution of type (II.2), $\text{Prob}\{Z(\mathbf{u}) \leq z|(n)\}$, for the RV $Z(\mathbf{u})$ [15, 102].

These remarks seem to imply the limited usefulness of kriging and geostatistics as a whole. Fortunately, the kriging algorithm has two characteristic properties that allow its use in determining posterior ccdfs of type (II.2). These two characteristic properties are the basis for, respectively, the multi-Gaussian (MG) approach and the indicator kriging (IK) approach to determination of ccdfs:

(i) **The Multi-Gaussian Approach:** If the RF model $Z(\mathbf{u})$ is multivariate Gaussian,[3] then the simple kriging estimate (II.12) and variance (II.14) identify the mean and variance of the posterior ccdf. In addition, since

---

[3]If the sample histogram is not normal, a normal score-transform (a non-linear transform) can be performed on the original z-data. A multi-Gaussian model $Y(\mathbf{u})$ is then adopted for the normal score data. For example, the transform is the logarithm, $y(\mathbf{u}) = \ln z(\mathbf{u})$, if the z-sample histogram is approximately lognormal. Kriging and/or simulation are then performed on the y-data with the results appropriately back-transformed into z-values (see programs nscore and backtr in Chapter VI).

that ccdf is Gaussian, it is fully determined by these two parameters; see [10]. This remarkable result is at the basis of multi-Gaussian (MG) kriging and simulation (see program `sgsim` in Chapter V). The MG approach is said to be parametric in the sense that it determines the ccdfs through their parameters (mean and variance). The MG algorithm is remarkably fast and trouble-free; its limitation is the reliance on the very specific and sometimes inappropriate properties of the Gaussian RF model [105, 107].

(ii) **The Indicator Kriging Approach:** If the value to be estimated is the expected value (mean) of a distribution, then least-squares (LS) regression (i.e., kriging) is a priori the preferred algorithm. The reason is that the LS estimator of the variable $Z(\mathbf{u})$ is also the LS estimator of its conditional expectation $E\{Z(\mathbf{u})|(n)\}$, that is, of the expected value of the ccdf (II.2); see [109], p. 566. Instead of the variable $Z(\mathbf{u})$, consider its binary indicator transform $I(\mathbf{u}; z)$ as defined in relation (II.6). Kriging of the indicator RV $I(\mathbf{u}; z)$ provides an estimate that is also the best LS estimate of the conditional expectation of $I(\mathbf{u}; z)$. Now, the conditional expectation of $I(\mathbf{u}; z)$ is equal to the ccdf of $Z(\mathbf{u})$; indeed:

$$\begin{aligned} E\{I(\mathbf{u}; z)|(n)\} &= 1 \cdot \text{Prob}\{I(\mathbf{u}; z) = 1|(n)\} \\ &\quad +0 \cdot \text{Prob}\{I(\mathbf{u}; z) = 0|(n)\} \\ = 1 \cdot \text{Prob}\{Z(\mathbf{u}) \leq z)|(n)\} &\equiv F(\mathbf{u}; z|(n)), \text{ as defined in (II.2)} \end{aligned}$$

Thus the kriging algorithm applied to indicator data provides LS estimates of the ccdf (II.2). Note that indicator kriging (IK) is not aimed at estimating the unsampled value $z(\mathbf{u})$ or its indicator transform $i(\mathbf{u}; z)$ but at providing a ccdf model of uncertainty about $z(\mathbf{u})$. The IK algorithm is said to be non-parametric in the sense that it does not approach the ccdf through its parameters (mean and variance); rather, the ccdf values for various threshold values z are estimated directly.

A Note on Non-Parametric Models

The qualifier "non-parametric" is misleading in that it could be interpreted as the characteristic of a RF model that has no parameter. All RF models imply a full multivariate distribution as characterized by the set of distributions (II.4), but some RF models have more free parameters than others. The more free parameters that can be fitted confidently to data, the more flexible the model. Unfortunately the fewer free parameters a model has, the easier it is to handle (less inference and less computation); hence the dangerous attraction of such inflexible models.

Considering only one random variable (RV), the most inflexible models are one-free-parameter models such as the Dirac distribution corresponding to Z-constant, or the Poisson-exponential model fully determined by its mean.

Next would come two-free-parameter models such as the Gaussian distribution. The indicator approach considers a K-free-parameters model for the ccdf (II.2) if K different threshold values z are retained. Hence the terminology "parameter-rich" model should be retained for indicator based models instead of the traditional but misleading qualifier "non-parametric". Of course, a parameter-rich model is relevant only if the corresponding parameters can be confidently inferred from the available data.

The most blatant example of a parameter-poor RF model is the iid (independent identically distributed) model underlying most statistical tests and the non-parametric bootstrap [56]. An iid RF model is characterized by a single cdf (histogram), hence is particularly easy to infer and to work with. It is inappropriate for any geostatistics where the goal is to model spatial dependence. Next would come Gaussian-related RF models fully characterized by a cdf (not necessarily Gaussian) plus a covariance function or matrix. If data allow the inference of more than a single covariance function, as implicit in an indicator approach, then the parameter-poor Gaussian model is unnecessarily restrictive.

A Note on Disjunctive Kriging

Disjunctive kriging (DK) generalizes the kriging estimator to the following form:

$$Z^*_{DK}(\mathbf{u}) = \sum_{\alpha=1}^{n} f_\alpha(Z(\mathbf{u}_\alpha))$$

with $f_\alpha()$ being functions, possibly non-linear, of the data; see [109], p. 573, [136], [146], p. 18.

DK can be used, just like IK, to derive ccdf models characterizing the uncertainty about an unknown $z(\mathbf{u})$.

As opposed to the IK approach where the direct and cross indicator covariances are inferred from actual data, the DK formalism relies on parametric models for the bivariate distribution with rather restrictive assumptions. The typical bivariate distribution model used in DK is fully characterized by a transform of the original data (equivalent to the normal score transform used in the MG approach) and the covariance of those transforms. The yet-unanswered question is which transform to use. Moreover, if a normal scores transform is to be adopted, why not go all the way and adopt the well-understood multivariate Gaussian model with its unequaled analytical properties and convenience?

In the opinion of these authors, DK is conveniently replaced by the more robust MG or median IK approaches whenever only one covariance model is deemed enough to characterize the spatial continuity of the attribute under study. In cases when the model calls for reproduction of multiple indicator (cross) covariances, the indicator kriging approach allows more flexibility. Therefore, it has been decided not to include DK in this version of GSLIB.

Introducing Simulation

The simple kriging system (II.13) can be read as the conditions to identify the covariance of $Z^*_{SK}(\mathbf{u})$ with any of the n data $Z(\mathbf{u}_\alpha)$ to the covariance model. In this sense, SK is the first step towards simulation, that is, the process of reproducing a covariance model. But SK "leaves the job unfinished" in that the covariance between any two SK estimators $Z^*_{SK}(\mathbf{u})$ and $Z^*_{SK}(\mathbf{u}')$ at two different locations $\mathbf{u} \neq \mathbf{u}'$ does not identify the model value $C(\mathbf{u} - \mathbf{u}')$. Sequential simulation finishes the job by adding the value $z^*_{SK}(\mathbf{u})$, corrected for its smoothing, into the data set used at subsequent nodes $\mathbf{u}'$, see section V.1.3 and [103].

II.1.5 Stochastic Simulation

Stochastic simulation is the process of drawing alternative, equally probable, joint realizations of the component RVs from an RF model. The (usually gridded) realizations $\{z^{(l)}(\mathbf{u}), \mathbf{u} \in A\}$, $l = 1, \ldots, L$, represent L possible images of the spatial distribution of the attribute values $z(\mathbf{u})$ over the field A. Each realization, also called a "stochastic image," reflects the properties that have been imposed on the RF model $Z(\mathbf{u})$. Therefore, the more properties that are inferred from the sample data and incorporated into the RF model $Z(\mathbf{u})$ or through the simulation algorithm, the better that RF model or the simulated realizations.

Typically, the RF model $Z(\mathbf{u})$ is constrained by the sole z-covariance model inferred from the corresponding sample covariance, and the drawing is such that all realizations honor the z-data values at their locations, in which case these realizations are said to be conditional (to the data values). More advanced stochastic simulation algorithms allow reproducing more of the sample bivariate distribution by constraining the RF model to a series of indicator (cross) covariances of type (II.7).

There is another class of stochastic simulation algorithms that aim at reproducing specific geometric features of the sample data. The reproduction of bivariate statistics such as covariance(s) is then traded for the reproduction of the geometry; see literature on Boolean and object-based algorithms (e.g., [48, 77, 78, 157, 174]). In a nutshell, object-based algorithms consist of drawing shapes, e.g., elliptical shales in a clastic reservoir, from prior distributions of shape parameters, then locating these shapes at points along lines or surfaces distributed in space. GSLIB offers an elementary Boolean-type simulation program (ellipsim) that can be easily modified to handle any geometric shape. Note that the simulated annealing algorithm (anneal) in Chapter V represents an alternative for reproducing limited aspects of the geometry of categorical variables.

Conditional simulation was initially developed to correct for the smoothing effect shown on maps produced by the kriging algorithm. Indeed, kriging estimates are weighted moving averages of the original data values; thus they have less spatial variability than the data. Moreover, depending on the data configuration, the degree of smoothing varies in space, entailing possibly artifact structures. Typical conditional simulation algorithms trade the estimation variance minimization for the reproduction of a variogram/covariance seen as a model of spatial variability. A smoothed map as provided by kriging is appropriate for showing global trends. On the other hand, conditionally simulated maps are more appropriate for studies that are sensitive to patterns of local variability such as flow simulations (see, e.g., [80, 106]). A suite of conditionally simulated maps also provides a measure (model) of uncertainty about the attribute(s) spatial distribution.

Increasingly, stochastic simulations are used to provide a numerical and visual appreciation of spatial uncertainty beyond the univariate ccdf $\text{Prob}\{Z(\mathbf{u}) \leq z|(n)\}$ introduced in relation (II.2). That univariate ccdf measures only the uncertainty prevailing at a given location $\mathbf{u}$; it does not provide any indication about the joint uncertainty prevailing at several locations. For example, consider the locations within a subset B of the field A, that is, $\mathbf{u}_j \in B \subset A$. The probability distribution of the block average $1/|B| \int_B Z(\mathbf{u}) \, d\mathbf{u}$, or the probability that all values $Z(\mathbf{u}_j)$ simultaneously exceed a threshold z, cannot be derived from the set of univariate ccdfs $\text{Prob}\{Z(\mathbf{u}_j) \leq z|(n)\}, \mathbf{u}_j \in B$. They can be derived, however, from a set of stochastic images.

Equiprobability of Realizations

A set of L realizations from any specific simulation algorithm can be used to determine probabilities of occurrence of specific functions of these realizations only if the L realizations are "equiprobable", that is if any one of the L realizations had the same probability to be drawn as any other among the L. Such equiprobability is achieved if each realization can be identified to a single random number (the realization seed number) uniformly distributed in $[0, 1]$; in practice, if each realization can be reproduced exactly by re-running the simulation algorithm using its seed number. Of course, equiprobability refers to a specific simulation model including its RF model (if any), all parameter values, the implementation algorithm and computer code. Had another simulation model or another implementation of the same RF model been used, it would generate a different set of realizations leading to different probability assessments. All uncertainty assessments, even as elementary as an estimation variance, are model-dependent and are no better than the model on which they are built. In this regard no uncertainty assessment, no estimate, no statistical test can claim objectivity [104, 114].

Uncertainty About Uncertainty Models

Because all RF-based uncertainty assessments depend on the RF-model adopted, in particular on its parameter values such as cdf values, variogram type and range, there is a natural desire to factor in (add on) the uncertainty associated to inference of the model parameters. This is tantamount to access (actually define) a model for alternative models of uncertainty. The problem is that such higher order model is necessarily non-data based, and as such one should consider its own uncertainty, that is yet an even higher order model of models of models of uncertainty .... Take the example of the uncertainty associated to a covariance/variogram range (see later definition in Section II.3). Because of data sparsity all data available are used to evaluate that range value, hence none are left to calibrate a model of uncertainty about the range estimate (recall that the variogram and its range are already parameters of a model of uncertainty). Typically, the variogram range would be arbitrarily modeled as a Gaussian RV with variance known from some (unspecified) prior information or from some bootstrap assuming independence (!) then through some Bayesian formalism this arbitrary second order measure of uncertainty in channelled into an increased, say, kriging variance. Adopting higher order models of uncertainty which are not data-based does not add any objectivity to the final statement of uncertainty. We do not recommend such an open-ended procedure, although we do recommend sensitivity analysis to model parameters including, e.g., the variogram range. But as to do sensitivity analysis, we recommend varying first order model decisions such as decisions about the number of populations, the extent of each stationarity (pooling) area rather than second order model parameters such as variogram ranges. For a more extensive presentation of this point of view, see [104, 114, 115].

II.2   GSLIB Conventions

The notation adopted in this user's manual and in the GSLIB source code adheres, as much as possible, to established geostatistical conventions used in textbooks such as [89, 102, 146] and further developed in [74].

Appendix C of this manual provides the GSLIB programming conventions and a dictionary of variable names. Appendix E provides a list of acronyms and notations.

Certain typographical conventions are used to distinguish files, program names, and variable names:

- names of actual computer files, e.g., source code files, are enclosed in a box, e.g., $\boxed{\text{gam.f}}$,

- program names is written in typographic script, e.g., **gam**; and

- program variables are in bold script, e.g. **datafl**.

II.2.1   Computer Requirements

The subroutines in GSLIB are written to adhere as closely as possible to ANSI standard FORTRAN 77. Some very obvious, and easily corrected, departures from the standard have been taken. For example, although the ANSI standard specifies upper-case characters, the source code is in lower-case characters. This could be corrected if necessary by a global change. The idea is to keep the software independent of a particular compiler or computer. The software has been compiled and executed, with little modification, on a number of machines including IBM-compatible PCs, DEC, IBM, SUN, HP, and SGI workstations, and IBM and Cray mainframe computers. Therefore, the sole requirement should be the availability of a FORTRAN compiler.

For reasonably modest tasks the programs may be executed on a PC in an acceptable amount of time. Large kriging or simulation runs may require special consideration in terms of available memory (particularly RAM) and speed.

One short comment on the choice of FORTRAN: there was no need to retain FORTRAN as the programming language for this version of GSLIB; enough changes were planned to justify conversion to another language (C being a logical choice). FORTRAN was retained chiefly because of its common usage and familiarity. We also feel that structured programming is possible with FORTRAN and that extended capabilities (e.g., dynamic memory allocation, etc.) available with the next ANSI standard or as non-standard extensions of currently available compilers will keep FORTRAN a viable programming language for the near future.

II.2.2 Data Files

Although users are strongly encouraged to customize the programs, workable main programs are useful starting points. The main programs read and write data with a format similar to the menu-driven packages Geo-EAS [58] and Geostatistical Toolbox [62]. The format, described below, is a simple ASCII format with no special allowance for regular grids or data compression.

The data file format is a simplified Geo-EAS format, hence with no allowance for the user to specify explicitly the input format. The data values are always read with free format. The accessibility of the source code allows this to be easily changed.

The following conventions are used by the "simplified Geo-EAS format" used by GSLIB data files:

1. The first line in the file is taken as a title and is possibly transferred to output files.

2. The second line should be a numerical value specifying the number of numerical variables **nvar** in the data file.

3. The next **nvar** lines contain character identification labels and additional text (optional) that describe each variable.

4. The following lines, from **nvar**+3 until the end of the file, are considered as data points and must have **nvar** numerical values per line. Missing values are typically considered as large negative or positive numbers (e.g., less than -1.0e21 or greater than 1.0e21). The number of data

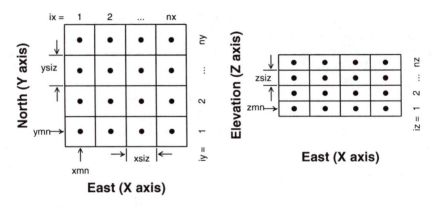

Figure II.1:   Plan and vertical cross-section views that illustrate the grid definition used in GSLIB.

will be the number of lines in the file minus **nvar**+2 minus the number of missing values. The programs read numerical values and *not* alphanumeric characters; alphanumeric variables may be transformed to integers or the source code modified.

II.2.3   Grid Definition

Regular grids of data points or block values are often considered as input or output. The conventions used throughout GSLIB are:

- The X axis is associated to the east direction. Grid node indices ix increase from 1 to nx in the positive x direction, i.e., to the east.

- The Y axis is associated to the north direction. Grid node indices iy increase from 1 to ny in the positive y direction, i.e., to the north.

- The Z axis is associated to the elevation. Grid node indices iz increase from 1 to nz in the positive z direction, i.e., upward.

The user can associate these three axes to any coordinates system that is appropriate for the problem at hand. For example, if the phenomenon being studied is a stratigraphic unit, then some type of stratigraphic coordinates relative to a marker horizon could make the most sense [48, 92, 108]. The user must perform the coordinate transformation; there is no allowance for rotation or stratigraphic grids within the existing set of subroutines.

The coordinate system is established by specifying the coordinates at the center of the first block (xmn, ymn, zmn), the number of blocks/grid nodes (nx, ny, nz), and the size/spacing of the blocks/nodes $(xsiz, ysiz, zsiz)$. Figure II.1 illustrates these parameters on two 2D sectional views.

Sometimes a special ordering is used to store a regular grid. This avoids the requirement of storing node coordinates or grid indices. The ordering is point by point to the east, then row by row to the north, and finally level by level upward, i.e., x cycles fastest, then y, and finally z. The index location of any particular node ix, iy, iz can be located by:

$$loc = (iz - 1) * nx * ny + (iy - 1) * nx + ix$$

Given the above one-dimensional index location of a node the three coordinate indices can be calculated as:

$$iz = 1 + int \left(\frac{loc - 1}{nx * ny} \right)$$

$$iy = 1 + int \left(\frac{loc - (iz - 1) * nx * ny}{nx} \right)$$

$$ix = loc - (iz - 1) * nx * ny - (iy - 1) * nx$$

The uncompressed ASCII format is convenient because of its machine independence and easy access by visual editors; however, a binary compressed format would be most efficient and even necessary if large 3D grids are being considered.

II.2.4 Program Execution and Parameter Files

The default driver programs read the name of a parameter file from standard input. If the parameter file is named for the program and has a ".par" extension, then simply keying a carriage return will be sufficient (e.g., the program gam would automatically look for $\boxed{\text{gam.par}}$). All of the program variables and names of input/output files are contained in the parameter file. A typical parameter file is illustrated in Figure II.2. The user can have as many lines of comments at the top of the file as desired, but formatted input is started immediately after the characters "START" are found at the beginning of a line. Some in-line documentation is available to supplement the detailed documentation provided in this user's manual. Example parameter files are included with the presentation of each program in Chapters III through VI and Appendix A.

An interactive and "error checking" user interface would help make GSLIB programs more user friendly. A graphical user interface (GUI) has been avoided because there is no single GUI that works on all computers. A useful addition to each program would be an *intelligent* interpretation program that would read each parameter file and create a verbose English language description of the "job" being described by the parameters.

When a serious error is encountered, an error message is written to standard output and the program is stopped. Less serious problems or apparent inconsistencies cause a warning to be written to standard output or a debugging file, and the program will continue.

```
                        Parameters for HISTPLT
                        **********************

START OF PARAMETERS:
../data/cluster.dat           \file with data
3    0                        \   columns for variable and weight
-1.0e21   1.0e21              \   trimming limits
histplt.ps                    \file for PostScript output
 0.01      100.0              \attribute minimum and maximum
-1.0                          \frequency maximum (<0 for automatic)
25                            \number of classes
1                             \0=arithmetic, 1=log scaling
0                             \0=frequency,  1=cumulative histogram
200                           \   number of cum. quantiles (<0 for all)
2                             \number of decimal places (<0 for auto.)
Clustered Data                \title
1.5                           \positioning of stats (L to R: -1 to 1)
-1.1e21                       \reference value for box plot
```

Figure II.2: An example parameter file for histplt.

II.2.5 Machine Precision

When calculations are performed on a computer, each arithmetic operation is generally affected by roundoff error. This error arises because the machine hardware can only represent a subset of the real numbers [69]. Roundoff error is not a major concern because the size of geostatistical matrices is typically small. Large problems are often separated into a large number of smaller problems by adopting local search neighborhoods.

Roundoff error is noticed most in the matrix solution of kriging systems; storing the matrix entries in single precision causes the kriging weights obtained from two different solution methods to change as soon as the third decimal place. For this reason, all matrix inversion subroutines in GSLIB have been coded in double precision. To save on storage, however, the LU simulation program lusim (V.6.2) is coded in single precision. Roundoff error becomes significant only when simulating very large grid systems.

Another source of imprecision is due to numerical approximations to certain mathematical functions. For example, the cumulative normal distribution $p = G(z)$ and inverse cumulative normal distribution $y = G^{-1}(p)$ have no closed-form expression and polynomial approximations are used [116]. Imprecision in these approximations is small relative to the error introduced by the floating-point representation of real numbers.

II.3 Variogram Model Specification

This section can be scanned quickly the first time through. The conventions for describing a variogram model are explained here rather than within all kriging and simulation programs. Most of the kriging and simulation subroutines call for covariance or pseudo-covariance values; however, a semivariogram model rather than a covariance model must be specified. This apparent inconsistency allows for the traditional practice of modeling variograms and also permits the straightforward incorporation of the power model, which has no covariance counterpart.

An acceptable semivariogram model for GSLIB consists of an isotropic nugget effect[4] and any positive linear combination of the following standard semivariogram models:

1. **Spherical** model defined by an actual range a and positive variance contribution or *sill* value c.

$$\gamma(h) = c \cdot \text{Sph}\left(\frac{h}{a}\right) = \begin{cases} c \cdot \left[1.5\frac{h}{a} - 0.5\left(\frac{h}{a}\right)^3\right], & \text{if } h \leq a \\ c, & \text{if } h \geq a \end{cases} \quad \text{(II.15)}$$

2. **Exponential** model defined by an effective range[5] a (integral range $a/3$) and positive variance contribution value c.

$$\gamma(h) = c \cdot \text{Exp}\left(\frac{h}{a}\right) = c \cdot \left[1 - \exp\left(-\frac{3h}{a}\right)\right] \quad \text{(II.16)}$$

3. **Gaussian** model defined by an effective range a and positive variance contribution value c.[6]

$$\gamma(h) = c \cdot \left[1 - exp\left(-\frac{(3h)^2}{a^2}\right)\right] \quad \text{(II.17)}$$

4. **Power** model[7] defined by a power $0 < \omega < 2$ and positive slope c.

$$\gamma(h) = c \cdot h^\omega \quad \text{(II.18)}$$

The distance h may be anisotropic in 2D or 3D.

5. **Hole effect** model defined by a length a to the first peak (size of the underlying cyclic features) and positive variance contribution value c.

$$\gamma(h) = c \cdot \left[1.0 - \cos\left(\frac{h}{a} \cdot \pi\right)\right] \quad \text{(II.19)}$$

To be a legitimate positive definite variogram model, this hole effect model may only be applied in one direction; that is, the variogram is

---

[4]Anisotropic nugget effects can be modeled by setting some of the directional ranges of the first nested structure to a very small value.

[5]For variogram models that reach their sills c asymptotically, the effective range is defined as the distance at which $\gamma(a) = 0.95 \cdot c$.

[6]Note that matrix instability problems are often encountered with a Gaussian model with no nugget effect [153, 173]. Also, a Gaussian variogram model with no nugget effect should not be used for categorical variables [137].

[7]The power model is associated to self-affine random fractals with the parameter ω related to the fractal dimension D by the relation: $D = E + 1 - \omega/2$, where E is the topological dimension of the space ($E=1$, 2, or 3) [80, 184].

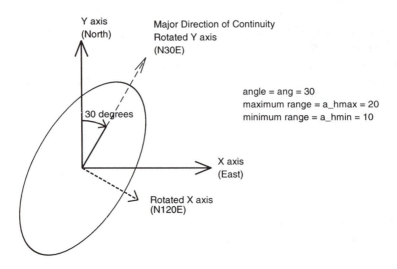

Figure II.3: An example of the two parameters needed to define the geometric anisotropy of a variogram structure in 2D.

specified by the direction of the hole effect, the range parameter in that direction, and very large range parameters in the two other directions, say, $1.0e21$.

Although not allowed for in the parameter files, a dampened hole effect model is also coded in the cova3 subroutine. The dampened hole effect model is:

$$\gamma(h) = c \cdot \left[1.0 - \exp\left(\frac{-3 \cdot h}{d}\right) \cdot \cos\left(\frac{h}{a} \cdot \pi\right) \right]$$

where d is the distance at which 95% of the hole effect is dampened out (the variance magnitude of the periodic component is then 5% of c).

The type of variogram structure is specified by an integer code, which is the order in the above list, i.e., it=1: spherical model, it=2: exponential model, it=3: Gaussian model, it=4: power model, and it=5: hole effect model. The a and c parameter values, which correspond to the description in the above list, are also needed.

Each nested structure requires an additional two or five parameters that define its own geometric anisotropy in 2D or 3D. Figure II.3 illustrates the angle and range parameters required in 2D.

The direction of maximum continuity is specified by a rotation angle ang corresponding to an azimuth angle measured in degrees clockwise from the positive Y or north direction. The range in this horizontal maximum direction is specified by a_{hmax}. The range in the perpendicular direction or the horizontal minimum direction is specified by a_{hmin}.

A very large range implies that no variance contribution will be added to that direction, a feature known as "zonal anisotropy."

Within the software the actual distance is corrected so that it accounts for the specified anisotropy. That is, the distance component along the rotated X axis (see Figure II.3) is corrected by the ratio a_{hmin}/a_{hmax}. This convention allows the anisotropy of the power model to be handled in an intuitively correct manner; an anisotropic distance is calculated and the power ω is left unchanged.

Figure II.4 illustrates the angles and ranges required in 3D. Many software packages take a shortcut and only use two angles and two ranges. The added complexity of three angles is not in programming but in documentation. It is quite straightforward to visualize and document a phenomenon that is dipping with respect to the horizontal at a dip azimuth that is not aligned with a coordinate axis. The third angle is required to account for the geological concept of a plunge or rake. One example that requires a third angle is modeling the geometric anisotropy within the limbs of a plunging syncline.

The easiest way to describe the three angles and three ranges is to imagine the rotations and squeezing that would be required to transform a sphere into an ellipsoid; see [65]. We will refer to the original Y axis as the principal direction and consider the rotations such that it ends up being the actual principal structural direction (direction of maximum continuity):

- The first rotation angle **ang1** rotates the original Y axis (principal direction) in the horizontal plane: This angle is measured in degrees clockwise.

- The second rotation angle **ang2** rotates the principal direction from the horizontal: This angle is measured in negative degrees down from horizontal.

- The third rotation angle **ang3** leaves the principal direction, defined by **ang1** and **ang2**, unchanged. The two directions orthogonal to that principal direction are rotated clockwise relative to the principal direction when looking toward the origin. The rotation of the third step in Figure II.4 appears as counterclockwise since the view is away from the origin.

Zonal anisotropy can be considered as a particular case of a geometric anisotropy, see [74] and [89], pp. 385-386. This can be handled by entering the appropriate range as a very large number; that particular variogram structure is then added only to the orthogonal directions (say, across bedding). For simulation purposes, a strong zonal anisotropy is better reproduced when the simulation grid is aligned with the direction of zonal anisotropy.

The program vmodel (see Section VI.2.8) will write out the model semivariogram in any number of arbitrary directions and lags. This will help validate the correct entry of a semivariogram model.

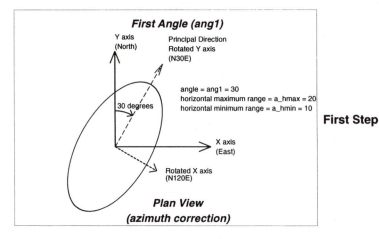

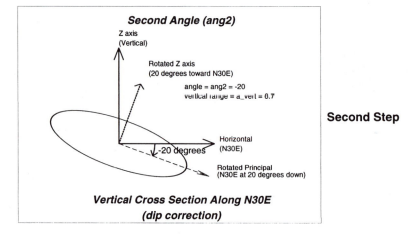

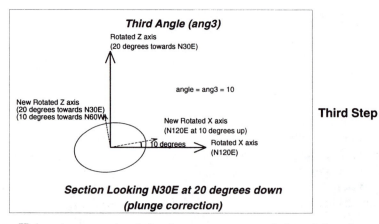

Figure II.4:   An example of the six parameters needed to define the geometric anisotropy of a variogram structure in 3D.

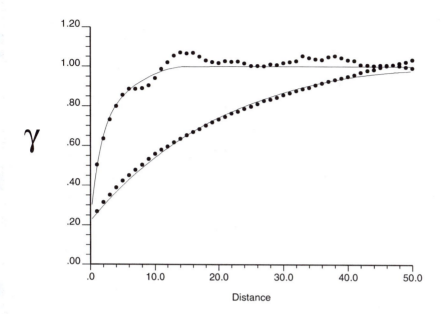

Figure II.5: Example semivariograms from a 2D data set. The variogram with the longer range is in the east-west direction. The solid line is the fitted model (as described in the text).

Note that, whether in 2D or 3D, the anisotropy directions need not be the same for each nested structure, allowing for a great flexibility in modeling experimental anisotropy.

A variogram model should not be needlessly complex. Ideally, each variogram structure should have a physical interpretation. The more complicated the variogram model, the longer it takes to construct each kriging matrix; hence, the longer the kriging or simulation program will take.

II.3.1 A Straightforward 2D Example

Consider the semivariogram shown on Figure II.5. The dots are the experimental semivariogram points in two orthogonal directions. The semivariogram that reaches the sill first (at about 10-15 distance units) is in the north-south direction (an azimuth of 0.0) and the variogram with the longer range is in the east-west direction (an azimuth of 90 degrees). The solid line in both directions is the fitted semivariogram model. For an excellent discussion on variogram modeling, refer to Chapter 16 of Isaaks and Srivastava [89] and [74].

The north-south model was fitted with a nugget effect of 0.22, an exponential structure with contribution 0.53 and range a of 4.8, and a spherical structure with contribution 0.25 and range 15.0. The east-west model was fitted with nugget effect of 0.22, an exponential structure with contribution 0.53 and range a of 48.0, and a spherical structure with contribution 0.25 and range 50.0. The semivariogram parameters required by the kriging or simulation programs would be specified as follows:

c0 = nugget = 0.22

nst = number of nested structures = 2

it (1) = type of first structure = 2 (exponential)

azimuth (1) = 90 degrees (the east-west direction)

cc (1) = contribution of first structure = 0.53

aa$_{hmax}$ (1) = maximum range of first structure = 48.0

aa$_{hmin}$ (1) = minimum range of first structure = 4.8

it (2) = type of second structure = 1 (spherical).

azimuth (2) = 90 degrees (the east-west direction).

cc (2) = contribution of second structure = 0.25.

aa$_{hmax}$ (2) = maximum range of second structure = 50.0

aa$_{hmin}$ (2) = minimum range of second structure = 15.0

II.3.2   A 2D Example with Zonal Anisotropy

Consider the semivariogram shown on Figure II.6. The dots are the experimental semivariogram points in two orthogonal and principal directions. The semivariogram that reaches the higher sill is at a direction N60E (an azimuth of 60.0) and the semivariogram with the lower sill is in the perpendicular direction (an azimuth of 150 degrees). The solid line in both directions is the fitted semivariogram model.

The semivariogram in the direction of maximum continuity (azimuth 150) was modeled with a nugget effect of 0.40, a spherical structure with a contribution of 0.40 and range of 300.0, an exponential structure with contribution 0.95 and range of 4500.0, and a spherical structure with contribution 0.90 and range 80000.0. Thus, for all practical purposes this last spherical structure does not contribute to direction N150E. The semivariogram with the higher sill was fitted with a nugget effect of 0.40, a spherical structure with a contribution of 0.40 and range of 100.0, an exponential structure with contribution 0.95 and range of 1500.0, and a spherical structure with contribution 0.90 and range 1600.0. This semivariogram model would be specified as follows:

c0 = nugget = 0.40

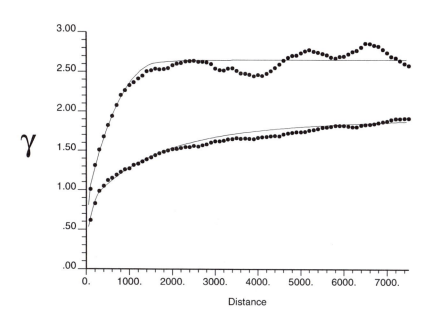

Figure II.6: Example semivariograms from a 2D data set. The variogram with the higher sill is at azimuth 60 degrees (N60E). The solid line is the fitted model (as described in the text).

nst = number of nested structures = 3

it (1) = type of first structure = 1 (spherical)

azimuth (1) = 150 degrees (direction of maximum continuity, i.e., with lower sill)

cc (1) = contribution of first structure = 0.40

aa$_{hmax}$ (1) = maximum range of first structure = 300.0

aa$_{hmin}$ (1) = minimum range of first structure = 100.0

it (2) = type of second structure = 2 (exponential)

azimuth (2) = 150 degrees (direction of maximum continuity)

cc (2) = contribution of second structure = 0.95

aa$_{hmax}$ (2) = maximum range of second structure = 4500.0

aa$_{hmin}$ (2) = minimum range of second structure = 1500.0

it (3) − type of third structure = 1 (spherical)

azimuth (3) = 150 degrees (direction of maximum continuity)

cc (3) = contribution of third structure = 0.90

aa$_{hmax}$ (3) = maximum range of third structure = 80000

aa$_{hmin}$ (3) = minimum range of third structure = 1600

II.4   Search Strategies

This section can be scanned quickly the first time through. The details of searching for nearby data are discussed here rather than with each program that calls for such a search.

Most kriging and simulation algorithms consider a limited number of nearby conditioning data. The first reason for this is to limit the CPU and memory requirements. The CPU time required to solve a kriging system increases as the number of data cubed, e.g., doubling the number of data leads to an eightfold increase in CPU time. The storage requirements for the main kriging matrix increases as the number of data squared, e.g., doubling the number of data leads to a fourfold increase in the memory requirements.

Furthermore, adopting a global search neighborhood would require knowledge of the covariance for the largest separation distance between the data. The covariance is typically poorly known for distances beyond one-half or one-third of the field size. A local search neighborhood does not call for covariance values outside the search ellipsoid.

A third reason for a limited search neighborhood is to allow local rescaling of the mean when using ordinary kriging; see Section IV.1.2 and [112]. All of the data may have been pooled together to establish a reliable histogram and variogram; however, at the time of estimation it is often better to relax the decision of stationarity locally and use only nearby data.

A number of constraints are used to establish which nearby data should be considered:

1. Only those data falling within a search ellipsoid centered at the location being estimated are considered. This anisotropic[8] search ellipsoid is specified in the same way as an anisotropic variogram (see Figures II.3 and II.4): by search radii in two or three directions.

2. The allowable data may be further restricted by a specified maximum **ndmax**. The closest data, up to **ndmax**, are retained. In all the kriging programs, and in simulation programs where original data are searched independently from simulated nodes, closeness is measured by the Euclidean distance (possibly anisotropic). In the sequential simulation algorithm (discussed in V.1.3) the previously simulated grid nodes are searched by variogram distance.[9]

3. An octant search is available as an option to ensure that data are taken on all sides of the point being estimated. This is particularly important when working in 3D with data often aligned along drillholes; an octant search ensures that data are taken from more than one drillhole. An octant search is specified by choosing the number of data **noct** to retain from each octant.

If too few data (less than **ndmin**) are found, then the node location being considered is left uninformed. This restricts estimation or simulation to areas where there are sufficient data.

Three different search algorithms have been implemented in different programs:

Exhaustive Search: The simplest approach is to check systematically all **nd** data and retain the **ndmax** closest that meet the three constraints noted above. This strategy is inefficient when there are many data and has been adopted only in the straightforward 2D kriging program.

Super Block Search: This search strategy (discussed in detail below) partitions the data into a grid network *super*imposed on the field being considered. When estimating any one location, it is then possible to limit the search to those data falling in nearby *super blocks*. This search has been adopted for non-gridded data in most kriging and simulation programs.

---

[8]Beware that an artificially anisotropic search neighborhood may create artifact anisotropic structures in the resulting kriging and/or simulated maps.

[9]A small component of the Euclidean distance is added to the variogram distance so that the nodes are ordered by Euclidean distance beyond the range of the variogram.

Spiral Search: This search strategy (discussed in detail below) is for searching gridded data. The idea is to visit the closest nearby grid nodes first and spiral away until either enough data have been found or the remaining grid nodes are beyond the search limits. This search has been adopted in all sequential simulation progams.

Super Block Search

The *super block* search strategy is an efficient algorithm to be used in cases where many points are to be estimated, using local data neighborhoods, with the same set of original data. The algorithm calls for an initial classification and ordering of the data according to a regular network of parallelipedic blocks; see [109] page 361. This grid network is independent of the grid network of points/blocks being estimated or simulated. Typically, the size of the search network is much larger than the final estimation or simulation grid node spacing.

When estimating any one point, only those data within nearby super blocks have to be checked. A large number of data are thus quickly eliminated because they have been classified in super blocks beyond the search limits. This is illustrated in 2D on Figure II.7, where an 11 by 11 super block grid network has been established over an area containing 140 data points. When estimating a point anywhere within the dark gray super block, only those data within the dark black line need be considered. Note that all search resolution less than the size of a super block has been lost. Also note that the light gray region is defined by the search ellipse (circle in this case) with its center translated to every node to be estimated within the dark gray super block. All super blocks intersected by the light gray region must be considered to ensure that all nearby data are considered for estimation of any node within the central dark gray superblock.

The first task is to build a template of super blocks, centered at the super block that contains the node being estimated. For example, the template is the relative locations of all 21 blocks enclosed by the solid line on Figure II.7. With this template, the nearby super blocks are easily established when considering any new location.

A second important computational *trick* is to sort all of the data by super block index number (the index is defined as in Section II.2.3). An array (of size equal to the total number of blocks in the super block network) is constructed that stores the cumulative number of data for each super block and all super blocks with lesser block indices, i.e.,

$$cum(i) = \sum_{j=1}^{i} nisb(j)$$

where $nisb(j)$ is the number of data in super block j, and $cum(0) = 0$. Then the number falling within any super block i is $cum(i) - cum(i-1)$ and their index location starts at $cum(i-1) + 1$. Therefore, this one array contains

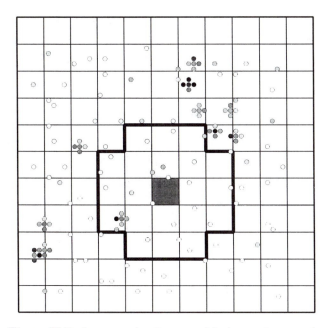

Figure II.7: An example of a super block search network.

information on the number of data in each super block and their location in memory.

Spiral Search

The *spiral* search strategy is an efficient algorithm for cases when the data are (or have been relocated) on a regular grid. The idea is to consider values at grid nodes successively farther away from the point being estimated. The ordering or template of nearby nodes to check is established on the basis of variogram distance. A spiral search template is constructed as follows:

1. Consider a *central* grid node at $ix = 0, iy = 0, iz = 0$.

2. Programs that call for a spiral search require the maximum size of the search to be input as maximum dimensioning parameters *MAXCTX*, *MAXCTY*, and *MAXCTZ* parameters, which are odd integers.[10] The maximum search distance, in terms of grid nodes, is then $nx = MAXCTX/2$, $ny = MAXCTY/2$, $nz = MAXCTZ/2$, i.e.,

$$jx = -nx, \ldots, -1, 0, 1, \ldots, nx,$$

$$jy = -ny, \ldots, -1, 0, 1, \ldots, ny,$$

$$jz = -nz, \ldots, -1, 0, 1, \ldots, nz.$$

[10]These parameters must be odd to allow n nodes to be searched on either side of a node being estimated, i.e., the search will be $2n + 1$ nodes

3. Compute the semivariogram between the fixed node ix, iy, iz and all nodes jx, jy, jz plus a small contribution of the corresponding Euclidean distance:

$$dis(jx, jy, jz) = \gamma(jx, jy, jz) + \epsilon \cdot h_{jx, jy, jz}$$

4. Sort the $dis(jx, jy, jz)$ array in ascending order.

5. Now, given a node kx, ky, kz, the closest nodes are found by spiraling away according to the ordered jx, jy, jz offsets.

When building the spiral search template, the covariance values for all offsets jx, jy, jz within the search ellipsoid are computed and stored to be later retrieved to construct the kriging matrices. This table lookup approach is much quicker than recomputing the covariance each time it is needed.

Two-part Search

Sequential simulation algorithms (see Section V.1.3) call for sequentially simulating the attribute at the grid node locations. A random path through all the nodes is followed, where the local distribution is conditioned to the original data and all previously simulated grid nodes. A common approximation is to relocate the original data to grid node locations before starting the simulation procedure. The advantage is that all conditioning data are then on a regular grid, which allows a spiral search and the use of covariance lookup tables for all covariance values needed in the kriging equations. The CPU time required is considerably reduced relative to the alternative, which is to keep the original data separate from the previously simulated values, and use a two-part search strategy, that is, search for nearby original data with a superblock strategy and search for previously simulated nodes with a spiral search strategy.

Depending on the situation the more CPU-intensive two-part search may be appropriate. Considerations include the following:

- If the grid being simulated is coarse, then original data may have to be moved a significant distance from their actual locations; the final maps will honor the data at their relocated grid node locations and not their original locations. Also, when there are more than one data near a single grid node location, only the closest is kept; thus some data may not be used.

- There may be data outside of the grid network that contribute to the simulation. These data must be relocated to blocks at the edge of the grid network or a two-part search must be implemented to use these data.

- Commonly, the number of nodes to be simulated greatly exceeds the number of data; consequently, the original data will be dominated quickly by simulated values. Special patterns in the data, that are not captured by the variogram, will be reproduced better if there is a special effort to use the original data.

Near the end of the simulation there will be so many close spaced previously simulated grid nodes that the long range variogram structure will not be used. By imposing the consideration of original data that are farther away, one can impart the long range structure by data "conditioning". Another approach to incorporate the long range structure is to consider a multiple-grid simulation procedure; see section V.1.3.

All GSLIB programs that use the sequential simulation concept allow both options: relocation of original data to grid node locations or a two-part search strategy.

II.5 Data Sets

Problem sets are proposed at the end of each chapter to allow users to check proper installation of the programs and to check their understanding of the program scope and input parameter files. Some thoughts and *partial* solutions are given in Appendix A.

All problem sets refer to the same reference data set. Results based on this particular data set should not be used to draw general conclusions about the algorithms considered.

A complete 2D gridded variable was created by simulated annealing where the first lag of a "low nugget" isotropic variogram was matched. The gridded data, defined below, provide reference data for all subsequent problem sets. It is important to note that most GSLIB programs work with 3D data. Only 2D data have been considered here to keep the problems straightforward and easy to visualize.

The reference 2D data file of 2500 values is characterized by the following geometric parameters:

xmn $= 0.5$ (origin of x axis)

nx $= 50$ (number of nodes in x direction)

xsiz $= 1.0$ (spacing of nodes in x direction)

and similarly, **ymn**=0.5, **ny**=50, and **ysiz**=1.0.

Four sample data sets were derived from this full valued 2D grid of 2500 values. The following data files may be found in the data subdirectory on the distribution diskettes:

1. The complete reference data set contained in $\boxed{\text{true.dat}}$.

2. A clustered sample of 140 values was drawn from the reference data set. The first 97 samples were taken on a pseudo-regular grid and the last

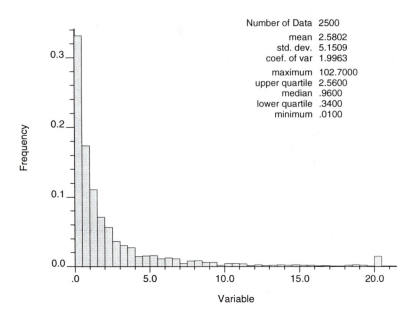

Figure II.8: Histogram of the complete 2D gridded reference data in $\boxed{\text{true.dat}}$.

43 samples were clustered in the high-valued regions (as identified from the first 97 samples). This data set, found in $\boxed{\text{cluster.dat}}$ is used for the first two problem sets on exploratory data analysis and variogram analysis.

3. A subset of 29 samples was retained in $\boxed{\text{data.dat}}$ for selected kriging and simulation exercises. This smaller number of samples is more typical of many applications and also better illustrates the use of secondary variables.

4. An exhaustively sampled secondary variable (2500 values) was created to illustrate techniques that allow the incorporation of different attributes. These 2500 secondary data *and* the 29 primary sample data are contained in $\boxed{\text{ydata.dat}}$.

The histogram of the 2500 values in $\boxed{\text{true.dat}}$ is shown on Figure II.8. This histogram was created with the program histplt described in Chapter VI. The parameter file was presented earlier (Figure II.2).

A gray-scale map of these reference values is shown on Figure II.9. The 140 clustered data shown on Figure II.10 are taken from these reference values.

A more exhaustive analysis of the sample data $\boxed{\text{cluster.dat}}$ is asked in the following problem set. A limited solution to this exploratory data analysis is given in Appendix A.

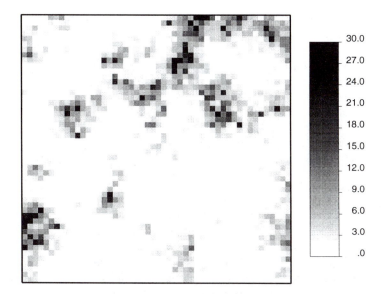

Figure II.9: The 2500 reference data values in ⃞ true.dat ⃞.

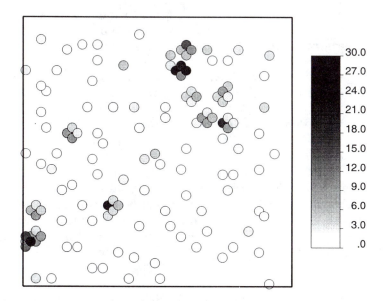

Figure II.10: Location map of the sample data in ⃞ cluster.dat ⃞.

II.6   Problem Set One: Data Analysis

The goal of this first problem set is to learn the basics of GSLIB through experimentation with data analysis techniques. Also, the user will get familiar with the data sets used extensively in later chapters. For the sole purpose of exploratory data analysis, there is no need nor advantage in using GSLIB programs; any graphical-based software would be appropriate, e.g., Geo-EAS [58], Geostatistical Toolbox [62], SAS, S, your own custom programs, etc.

The GSLIB software must first be installed and compiled (see Appendix B). The specific programs called for in this problem set are documented in Chapter VI.

The Data

The data can be interpreted as representing the sampling of a hazardous waste site. There are 97 initial samples taken on a pseudo-regular grid over a 2D area 50 miles by 50 miles. An additional 43 samples were taken around those original sample locations with high concentration values. The total 140 samples are used to infer global statistics representative of the entire 2500 square-mile area. The location map of these samples is shown in Figure II.10; the corresponding data are stored in cluster.dat .

Questions

1. Plot the location of the 140 data using locmap. Comment on the sample data and the effect of clustering on the sample histogram and statistics (mean, variance, and high quantiles).

2. Create an equal-weighted sample histogram, a normal and lognormal probability plot. Comment on the results.

3. Can you think of a quick and easy way to correct the sample histogram for spatial clustering (in geostatistical jargon to "decluster" the sample histogram)? Try it.

4. Perform cell declustering using the original 140 data. Choose a reasonable range of cell sizes. What would be a "natural" cell size? Plot a scatterplot of the declustered cell mean versus the cell size. Check out the cell size providing the smallest mean and calculate the corresponding declustered histogram. Comment on the results. For more details about declustering algorithms, see [89].

5. Generate the histogram and probability plot of the 2500 reference data stored in true.dat (see Figure II.9). Comment on the effectiveness of the declustering algorithms you have tried.

6. Plot Q-Q (quantile-quantile) and P-P (cumulative probabilities) plots of the sample data (both equal-weighted and declustered) versus the reference data. Comment on the results.

7. Smooth the histogram of the 140 data extending the data range to a range of $[0.01, 100]$ assumed known. Use program `histsmth` (documented in Section VI.2.3) with 100 classes evenly spaced (logarithmic scaling option).

8. Smooth the histogram of the 140 secondary data in $\boxed{\text{cluster.dat}}$. Consider the data range $[0.01, 100]$ and use 100 evenly spaced classes (logarithmic scaling option).

9. Use `scatsmth` to model the bivariate relationship between the primary and secondary data and program `bivplt` to view the results. Consider different options and weighting in program `scatsmth` and comment on the results.

10. Draw 2500 independent realizations from the previous smoothed and declustered sample histograms. Use program `draw` (documented in Section VI.2.4). Compare the histograms of these simulated values to the target model and to the histogram of the reference data set.

Chapter III

Variograms

This chapter presents the variogram programs of GSLIB. These facilities extend well beyond the calculation of traditional variograms and cross variograms. A single subroutine call allows alternative measures of spatial continuity such as the correlogram or pairwise relative semivariogram to be computed in addition to the semivariogram. These alternative measures are known to be more resistant to data sparsity, outliers values, clustering, and sampling error.

Section III.1 presents the ten measures of spatial variability/continuity considered by GSLIB with a brief discussion of their respective advantages and disadvantages. The principles underlying the two variogram subroutines are presented in Section III.2. The detailed parameter specification of the programs for regularly and irregularly spaced data are given in Sections III.3 and III.4 respectively.

Section III.5 introduces the concept of a variogram map in 2D and variogram volume in 3D. Variogram maps (volumes) are useful to detect anisotropy.

Some useful application notes are presented in Section III.6. Finally, Section III.7 proposes a problem set that explores the capability of the variogram programs.

III.1 Measures of Spatial Variability

A variogram as defined in relation (II.11) or, more generally, a measure of spatial variability, is the key to any geostatistical study. In essence, the variogram replaces the Euclidean distance $\mathbf{h}$ by a structural distance $2\gamma(\mathbf{h})$ that is specific to the attribute and the field under study. The variogram distance measures the average degree of dissimilarity between an unsampled value $z(\mathbf{u})$ and a nearby data value. For example, given only two data values $z(\mathbf{u}+\mathbf{h})$ and $z(\mathbf{u}+\mathbf{h}')$ at two different locations the more "dissimilar" sample value should receive lesser weight in the estimation of $z(\mathbf{u})$.

The distance measure need not be inferred from the sample semivariogram

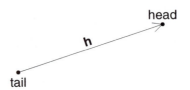

Figure III.1: An example of the *tail* and *head* naming convention.

(III.1); the sole constraint is that the measure of distance be modeled by a positive definite model[1] (to ensure existence and uniqueness of solutions to the kriging equations; see [26]).

The following ten experimental measures of spatial variability/continuity can be computed with GSLIB subroutines:

1. *Semivariogram:* this traditional measure is defined as half of the average squared difference between two attribute values approximately separated by vector **h**:

$$\gamma(\mathbf{h}) = \frac{1}{2N(\mathbf{h})} \sum_{i=1}^{N(\mathbf{h})} (x_i - y_i)^2 \qquad (\text{III.1})$$

   where $N(\mathbf{h})$ is the number of pairs, x_i is the value at the start or *tail* of the pair i, and y_i is the corresponding end or *head* value. Figure III.1 illustrates this naming convention relative to the separation vector **h**. In general, the separation vector **h** is specified with some direction and distance (lag) tolerance. It is not recommended to use expression (III.1) with tail and head values corresponding to two different variables. The resulting cross "variogram" is not equal to the sample cross semivariogram [see definition (III.2) and the related discussion in Section III.6]. A better measure of cross dependence is the cross covariance [see definition (III.3)].

2. *Cross semivariogram:* this measure of cross variability is defined as half of the average product of **h**-increments relative to two different attributes:

$$\gamma_{ZY}(\mathbf{h}) = \frac{1}{2N(\mathbf{h})} \sum_{i=1}^{N(\mathbf{h})} (z_i - z_i')(y_i - y_i') \qquad (\text{III.2})$$

   where z_i is the value of attribute z at the tail of pair i and z_i' is the corresponding head value; the locations of the two values z_i and z_i' are separated by vector **h** with specified direction(s) and distance tolerance. $(y_i - y_i')$ is the corresponding **h**-increment of the other attribute y.

---

[1]In fact, it is the covariance [constant minus $\gamma(\mathbf{h})$] that must be positive definite. All variogram models considered by GSLIB correspond to allowable, positive definite, covariance models. The only exception is the power variogram model, which has no covariance counterpart and hence cannot be used with simple kriging.

3. *Covariance:* this measure, also called the *non-ergodic* covariance [88], is the traditional covariance commonly used in statistics. The covariance does not implicitly assume that the mean of the tail values is the same as the mean of the head values ([89], p. 59).

$$C(\mathbf{h}) = \frac{1}{N(\mathbf{h})} \sum_{i=1}^{N(\mathbf{h})} x_i y_i - m_{-\mathbf{h}} m_{+\mathbf{h}} \qquad (\text{III.3})$$

where $m_{-\mathbf{h}}$ is the mean of the tail values, i.e., $m_{-\mathbf{h}} = \frac{1}{N(\mathbf{h})} \sum_1^{N(\mathbf{h})} x_i$, and $m_{+\mathbf{h}}$ is the mean of the head values, i.e., $m_{+\mathbf{h}} = \frac{1}{N(\mathbf{h})} \sum_1^{N(\mathbf{h})} y_i$.

If x and y refer to two different attributes, expression (III.3) identifies the sample cross covariance.

4. *Correlogram:* the previous measure (III.3) is standardized by the respective tail and head standard deviations:

$$\rho(\mathbf{h}) = \frac{C(\mathbf{h})}{\sigma_{-\mathbf{h}} \sigma_{+\mathbf{h}}} \qquad (\text{III.4})$$

where $\sigma_{-\mathbf{h}}$ and $\sigma_{+\mathbf{h}}$ are the standard deviation of the tail and head values, i.e.,

$$\sigma_{-\mathbf{h}}^2 = \frac{1}{N(\mathbf{h})} \sum_{i=1}^{N(\mathbf{h})} x_i^2 - m_{-\mathbf{h}}^2, \text{ and } \sigma_{+\mathbf{h}}^2 = \frac{1}{N(\mathbf{h})} \sum_{i=1}^{N(\mathbf{h})} y_i^2 - m_{+\mathbf{h}}^2$$

When x and y refer to two different attributes, expression (III.4) identifies the sample cross correlogram. Moreover, $\rho(\mathbf{h} = 0)$ identifies the (linear) correlation coefficient between the two attributes.

5. *General relative semivariogram:* the semivariogram as defined in (III.1) is standardized by the squared mean of the data used for each lag:

$$\gamma_{GR}(\mathbf{h}) = \frac{\gamma(\mathbf{h})}{\left(\frac{m_{-\mathbf{h}} + m_{+\mathbf{h}}}{2}\right)^2} \qquad (\text{III.5})$$

6. *Pairwise relative semivariogram:* each pair is normalized by the squared average of the tail and head values:

$$\gamma_{PR}(\mathbf{h}) = \frac{1}{2N(\mathbf{h})} \sum_{i=1}^{N(\mathbf{h})} \frac{(x_i - y_i)^2}{(\frac{(x_i + y_i)}{2})^2} \qquad (\text{III.6})$$

Practical experience has shown that the general relative (III.5) and pairwise relative (III.6) sample variograms are resistant to data sparsity and outliers when applied to positively skewed sample distributions.

They sometimes reveal spatial structure and anisotropy that could not be detected otherwise. Because of the divisors in expressions (III.5) and (III.6), the general and relative pairwise variograms should be limited to strictly positive variables.

7. *Semivariogram of logarithms:* the semivariogram is computed on the natural logarithms of the original variables (provided that they are positive):

$$\gamma_L(\mathbf{h}) = \frac{1}{2N(\mathbf{h})} \sum_1^{N(\mathbf{h})} [\ln(x_i) - \ln(y_i)]^2 \qquad (\text{III.7})$$

8. *Semimadogram:* this measure is similar to the traditional variogram; instead of squaring the difference between x_i and y_i, the absolute difference is taken:

$$\gamma_M(\mathbf{h}) = \frac{1}{2N(\mathbf{h})} \sum_1^{N(\mathbf{h})} |x_i - y_i| \qquad (\text{III.8})$$

Madograms are particularly useful for establishing large-scale structures (range and anisotropy). They should not be used for modeling the nugget variance of semivariograms.

9. *Indicator semivariogram (continuous variable):*[2] the semivariogram is computed on an internally constructed indicator variable. This requires the specification of a continuous variable and cutoff to create the indicator transform. For the cutoff cut_k and datum value x_i the indicator transform ind_i is defined as:

$$ind_i = \begin{cases} 1, & \text{if } x_i \leq cut_k \\ 0, & \text{otherwise} \end{cases} \qquad (\text{III.9})$$

where the subscript i refers to a particular location.

10. *Indicator semivariogram (categorical variable):* the semivariogram is computed on an internally constructed indicator variable. This requires the specification of a categorical variable and category to create the indicator transform. For the category s_k and datum value x_i the indicator transform ind_i is defined as:

$$ind_i = \begin{cases} 1, & \text{if } x_i = s_k \\ 0, & \text{otherwise} \end{cases} \qquad (\text{III.10})$$

where the subscript i refers to a particular location.

---

[2] An evident alternative is to define the indicator variable outside of the variogram program, then call for variogram type 1 as in (III.1). It is often more convenient to define the indicator variable as 1 if the indicator variable exceeds the cutoff. The indicator variogram is unchanged by this transform, i.e., $\gamma_I = \gamma_J$, with $J = 1 - I$.

Within the GSLIB source code, the variogram type is specified by the integer number in the list given above, that is, 1 for the traditional semivariogram, 2 for the covariance, etc.

Any of the previous experimental distance/correlation measures can be used qualitatively to infer structures of spatial continuity. More than one type of measure can be computed at the same time. For example, modeling of the sample variogram can be facilitated by inferring anisotropy directions and range ratios from the more robust relative variograms or madogram. In general, the features observed along the abscissa axis (distance $|\mathbf{h}|$) are common to all measures of variability/continuity (perhaps more apparent on some). The actual variogram distance value, however, as read from the ordinate axis, is specific to the variogram type chosen.

Since the actual variogram value depends on the type chosen, the resulting kriging variance should be interpreted only in relative terms. Many excellent discussions on variogram inference, interpretation, and modeling are available in the literature; see [11, 31, 34, 74, 88, 89, 100, 101, 109, 147, 168, 172, 173]; these discussions will not be reproduced here. Practitioners should expect to spend more time on data analysis and variogram modeling than on all kriging and simulation runs combined.

The MG Case

When a multivariate Gaussian model is being used, as in the MG approach to kriging or simulation, a covariance model for the normal score transforms of the data is required. The corresponding semivariogram model should have a sill of 1.0, which excludes the power model unless used only for small distances $\mathbf{h}$. Any of the three distance measures (III.1), (III.3), or (III.4), may be used since they provide the correct ordinate axis scaling. The other measures, however, can help in detecting nested structures, anisotropy, and evaluating ranges.

III.2   GSLIB Variogram Programs

The two variogram subroutines provided with GSLIB differ in their capability to handle regular versus irregular data layout. Both can handle 1D, 2D or 3D data. The first variogram subroutine gam is for gridded data. The other subroutine gamv is for irregularly spaced data.

The subroutines can handle many different directions, variables, and variogram types in a single pass. For example, it is possible to consider four different variables, eight directions, all four direct (auto-)variograms, two cross variograms, and the correlogram in one subroutine call. This flexibility has been achieved by making users explicitly state the tail variable, head variable, and variogram type for every "variogram"[3] they want to compute.

---

[3]The generic term variogram will be used hereafter to designate any measure of spatial variability/continuity. Whenever appropriate the specific distance measure will be speci-

All directions called for are computed for every variogram.

Warning: When specifying the tail variable, head variable, and variogram type the user should understand what is being requested. For example, it would be difficult to interpret the cross relative variogram of porosity and permeability. Rather the cross covariance of porosity-permeability should be inferred and modeled through expression (III.3).

Gridded Data

When data are on a regular grid, the directions are specified by giving the number of grid nodes that must be shifted to move from a node on the grid to the next nearest node on the grid that lies along the directional vector. Some examples for a cubic grid:

x shift	y shift	z shift	
1	0	0	aligned along the x axis
0	1	0	aligned along the y axis
0	0	1	aligned along the z axis
1	1	0	horizontal at 45° from y
1	-1	0	horizontal at 135° from y.
1	1	1	dipping at -45°, 225° clockwise from y.

Within the **gam** subroutine the directions are specified by offsets of the form shown above. No direction or lag tolerances are allowed. In some cases this direction definition is too restrictive and may result in not enough pairs being combined. The solution is either to average multiple directions from **gam** with weighting by the number of pairs, or to store the grid as irregularly spaced data and use **gamv** to compute the variograms (at substantially more computational cost).

Irregularly Spaced Data

The angles and lag distances are entered explicitly for irregularly spaced data. The azimuth angle is measured clockwise from north-south, and the dip angle measured in negative degrees down from horizontal (see Figure III.2). Angular half-window tolerances are required for both the azimuth and the dip. These tolerances may overlap, causing pairs to report to more than one direction. The unit lag distance and the lag tolerance are also required (see Figure III.2).

Note that the first lag reported in the output file (*lag0*) corresponds to $|h| \in [0, \epsilon)$. For a cross covariance and cross correlogram this value corresponds to the standard covariance or correlation coefficient of collocated points. The second lag reported to the output file (*lag1*) corresponds to $|h| \in [\epsilon, xlag - xtol)$.

---

fied, e.g., semivariogram (III.1) or madogram (III.8).

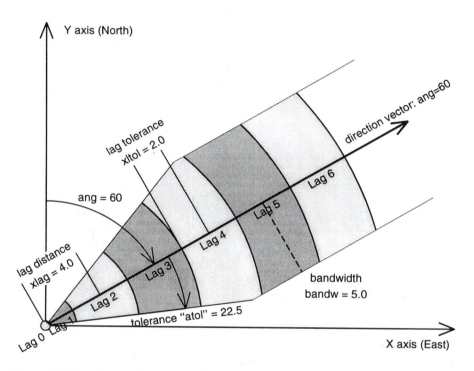

Figure III.2: Some of the parameters required by gamv. The azimuth angle is measured clockwise from North; the angle tolerance (half window) is restricted once the deviation from the direction vector exceeds the bandwidth; the shaded areas represent the different lags.

General Comments

A number of general comments are useful before documenting each program in more detail:

- The pairwise relative variogram and the variogram of logarithms require that the data be strictly positive. When the pairwise relative variogram is calculated, the sum of each pair is checked and the pair is discarded if that sum is less than a small positive minimum. When computing the variogram of logarithms, both the tail and head value must be greater than a small positive minimum. The user may wish to perform the necessary checks before passing the data to the calculation subroutine.

- There are no default variograms computed. For all subroutines and main programs the user must specify the number of variograms *nvarg* and their corresponding types. Variograms are not automatically computed for all variables in the data file.

- It may be convenient to perform some data transformation outside the subroutines. For example, normal scores, indicator, or logarithm transformations are better performed within a preprocessing program or a customized main program.

- The present allowance for indicator variograms is not especially efficient. For most applications it will not really matter, but the execution time could be considerably reduced with special coding (i.e., using Boolean logic). The current approach is to create a new variable (a 1/0 transform) internally within the main program. A more efficient coding would make use of the particular features of indicator variables, e.g., there must be one 1 and one 0 before anything is added to the experimental variogram, and a prior ordering of the indicator cutoffs would be beneficial.

- The general nature of these variogram programs makes them slightly inefficient. Checking for which type of variogram to compute adds about 3% to the overall execution time. This part of the code could be removed if one is interested in a single variogram type.

- In many cases the covariance (III.3) is the most appropriate measure of spatial distance; however, for convenience of plotting and modeling it should be inverted, i.e., $\gamma_C(\mathbf{h}) = C(0) - C(\mathbf{h})$. Variogram type (*ivtype*) option 3 specifies the covariance; setting the variogram type to -3 automatically calculates $C(0) - C(\mathbf{h})$.

III.3   Regularly Spaced Data gam

The GSLIB library and gam program should be compiled following the instructions in Appendix B. The parameters required for the main program

```
                    Parameters for GAM
                    *******************

START OF PARAMETERS:
../data/true.dat           \file with data
2    1    2                 \   number of variables, column numbers
-1.0e21      1.0e21         \   trimming limits
gam.out                    \file for variogram output
1                          \grid or realization number
50    0.5    1.0           \nx, xmn, xsiz
50    0.5    1.0           \ny, ymn, ysiz
 1    0.5    1.0           \nz, zmn, zsiz
2   10                     \number of directions, number of lags
 1   0   0                 \ixd(1),iyd(1),izd(1)
 0   1   0                 \ixd(2),iyd(2),izd(2)
1                          \standardize sill? (0=no, 1=yes)
5                          \number of variograms
1    1    1                \tail variable, head variable, variogram type
1    1    3                \tail variable, head variable, variogram type
2    2    1                \tail variable, head variable, variogram type
2    2    3                \tail variable, head variable, variogram type
1    1    9   2.5          \tail variable, head variable, variogram type
```

Figure III.3: An example parameter file for gam.

gam are listed below and shown on Figure III.3:

- **datafl:** the input data in a simplified Geo-EAS formatted file. The data are ordered rowwise (X cycles fastest, then Y, then Z).

- **nvar** and **ivar(1) ... ivar(nvar):** the number of variables and their columns in the data file.

- **tmin** and **tmax:** all values, regardless of which variable, strictly less than **tmin** and greater than or equal to **tmax** are ignored.

- **outfl:** the output variograms are written to a single output file named **outfl**. The output file contains the variograms ordered by direction and then variogram type specified in the parameter file (the directions cycle fastest then the variogram number). For each variogram there is a one-line description and then **nlag** lines each with the following:

  1. lag number (increasing from 1 to **nlag**).

  2. average separation distance for the lag.

  3. the *semivariogram* value (whatever type was specified).

  4. number of pairs for the lag.

  5. mean of the data contributing to the tail.

  6. mean of the data contributing to the head.

  7. the tail and head variances (for the correlogram).

The `vargplt` program documented in Section VI.1.8 may be used to create PostScript displays of multiple variograms.

- **igrid:** the grid or realization number. Recall that realizations or grids are written one after another; therefore, if **igrid=2** the input file must contain at least $2 \cdot$ **nx** $\cdot$ **ny** $\cdot$ **nz** values and the second set of **nx** $\cdot$ **ny** $\cdot$ **nz** values will be taken as the second grid.

- **nx, xmn, xsiz:** definition of the grid system (x axis).

- **ny, ymn, ysiz:** definition of the grid system (y axis).

- **nz, zmn, zsiz:** definition of the grid system (z axis).

  One- or two-dimensional data may be considered by setting the number of nodes in some directions to 1. Often, **gam** is used to check the variogram reproduction of realizations from a simulation program.

- **ndir** and **nlag:** the number of directions and lags to consider. The same number of lags are considered for all directions and all directions are considered for all of the **nvarg** variograms specified below.

- **ixd, iyd** and **izd:** these three arrays specify the unit offsets that define each of the **ndir** directions (see Section III.2).

- **standardize:** if set to 1, the semivariogram values will be divided by the variance.

- **nvarg:** the number of variograms to compute.

- **ivtail, ivhead**, and **ivtype:** for each of the **nvarg** variograms one must specify which variables should be used for the tail and head and which type of variogram is to be computed. For direct variograms the **ivtail** array is identical to the **ivhead** array. Cross variograms are computed by having the tail variable different from the head variable, e.g., if **ivtail(i)** is set to 1, **ivhead(i)** is set to 2, and **ivtype(i)** is set to 2, then distance measure **i** will be a cross semivariogram between variable 1 and variable 2. Note that **ivtype(i)** should be set to something that makes sense (e.g., types 1, 2, or 3); a cross relative variogram would be difficult to interpret. Further, note that for the cross semivariogram (**ivtype=2**) the two variables **ivtail** and **ivhead** are used at both the tail and head locations. The **ivtype** variable corresponds to the integer code in the list given in Section III.1.

- **cut:** whenever the **ivtype** is set to 9 or 10, i.e., asking for an indicator variogram, then a cutoff must be specified immediately after the **ivtype** parameter on the same line in the input file (see Figure III.3). Note that if an indicator variogram is being computed, then the cutoff/category applies to variable **ivtail(i)** in the input file [although the **ivhead(i)** variable is not used, it must be present in the file to maintain consistency with the other variogram types].

```
                        Parameters for GAMV
                        *******************
START OF PARAMETERS:
../data/cluster.dat                \file with data
1    2    0                        \   columns for X, Y, Z coordinates
2    3    4                        \   number of varables,column numbers
-1.0e21        1.0e21              \   trimming limits
gamv.out                          \file for variogram output
10                                \number of lags
5.0                               \lag separation distance
3.0                               \lag tolerance
3                                 \number of directions
0.0   90.0 50.0    0.0  90.0  50.0  \azm,atol,bandh,dip,dtol,bandv
0.0   22.5 25.0    0.0  22.5  25.0  \azm,atol,bandh,dip,dtol,bandv
90.   22.5 25.0    0.0  22.5  25.0  \azm,atol,bandh,dip,dtol,bandv
1                                 \standardize sills? (0=no, 1=yes)
3                                 \number of variograms
1    1    1                       \tail var., head var., variogram type
1    1    3                       \tail var., head var., variogram type
2    2    1                       \tail var., head var., variogram type
```

Figure III.4: An example parameter file for **gamv**.

Regularly spaced data in 1D can be handled by setting ny, nz to one and iyd, izd to zero.

The maximum size of the input array, the maximum number of variograms, and other maximum dimensioning parameters are specified in the file | gam.inc |, which is included in the **gam** program.

III.4 Irregularly Spaced Data gamv

The GSLIB library and **gamv** program should be compiled following the instructions in Appendix B. The parameters required for the main program **gamv** are listed below and shown in Figure III.4. Figure III.2 graphically illustrates some of the parameters that specify the distance and direction parameters. The parameters required by **gamv**:

- **datafl:** the input data in a simplified Geo-EAS formatted file.

- **icolx, icoly** and **icolz:** the columns for the x, y, and z coordinates (any of the column numbers may be set to zero if that coordinate is not present in the data, e.g., 2D data may be handled by setting **icolz** to 0).

- **nvar** and **ivar(1)** ... **ivar(nvar):** the number of variables and their column order in the data file.

- **tmin** and **tmax:** all values, regardless of which variable, strictly less than **tmin** and greater than or equal to **tmax** are ignored.

- **outfl:** the output variograms are written to a single output file named **outfl**. The output file contains the variograms ordered by direction and then variogram type specified in the parameter file (the directions cycle fastest then the variogram number). For each variogram there is a one-line description and then **nlag** lines each with the following:

1. lag number (increasing from 1 to **nlag**).
2. average separation distance for the lag.
3. the *semivariogram* value (whatever type was specified).
4. number of pairs for the lag.
5. mean of the data contributing to the tail.
6. mean of the data contributing to the head.
7. the tail and head variances (for the correlogram).

- **nlag:** the number of lags to compute (same for all directions).

- **xlag:** the unit lag separation distance.

- **xltol:** the lag tolerance. This could be one-half of **xlag** or smaller to allow for data on a pseudoregular grid. If **xltol** is entered as negative or zero, it will be reset to **xlag**/2. A pair will report to multiple lags if **xltol** is greater than one half of **xlag**.

- **ndir:** the number of directions to consider. All these directions are considered for all the **nvarg** variograms specified below.

- **azm, atol, bandwh, dip, dtol,** and **bandwd:** the azimuth angle, the half-window azimuth tolerance, the azimuth bandwidth, the dip angle, the half-window dip tolerance, and the dip bandwidth. The azimuth is measured in degrees clockwise from north, e.g., **azm=0** is north, **azm=90** is east, and **azm=135** is southeast. The dip angle is measured in negative degrees down from horizontal; i.e., **dip=0** is horizontal, **dip=-90** is vertical downward, and **dip=-45** is dipping down at 45 degrees. **bandwh** is the horizontal "bandwidth" or maximum acceptable horizontal deviation from the direction vector; see Figure III.2. **bandwd** is the vertical "bandwidth" or maximum acceptable deviation perpendicular to the dip direction in the vertical plane.

- **standardize:** if set to 1, the semivariogram values will be divided by the variance.

- **nvarg:** the number of variograms to compute.

- **ivtail, ivhead,** and **ivtype:** for each of the **nvarg** variograms one must specify which variables should be used for the tail and head and which type of variogram is to be computed. For direct variograms the **ivtail** array is identical to the **ivhead** array. Cross variograms are computed by having the tail variable different from the head variable. The **ivtype** variable corresponds to the integer code in the list given in Section III.1.

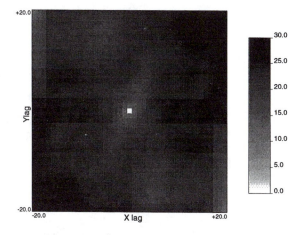

Figure III.5: An example variogram map.

- **cut:** whenever the **ivtype** is set to 9 or 10, i.e., asking for an indicator variogram, then a cutoff must be specified immediately after the **ivtype** parameter on the same line in the input file. Note that if an indicator variogram is being computed, then the cutoff/category applies to variable **ivtail(i)** in the input file [although the **ivhead(i)** variable is not used, it must be present in the file to maintain consistency with the other variogram types].

The maximum number of data, the maximum number of variograms, and other maximum dimensioning parameters are specified in the file $\boxed{\text{gamv.inc}}$ which is included in the **gamv** program.

III.5 Variogram Maps varmap

Variograms are traditionally presented as 1D curves: $\gamma(\mathbf{h})$ as a function of the distance $\mathbf{h}$ along a particular direction. It is often useful to have a global view of the variogram values in all directions. The variogram map [28, 89] is a 2D plot of $\gamma(\mathbf{h}_1, \mathbf{h}_2)$ of the sample semivariogram for all experimentally available separation vectors $\mathbf{h} = (\mathbf{h}_1, \mathbf{h}_2)$); see Figure III.5. The value $\gamma(0) = 0$ plots at the center of the figures. The value $\gamma(\mathbf{h}_1, \mathbf{h}_2)$ plots as a grayscale or colorscale at offset $(\mathbf{h}_1, \mathbf{h}_2)$ from that center. Any pixel not filled by an experimental value is left uninformed (it could be filled in with some type of interpolation). The variogram volume is the 3D plot of $\gamma(\mathbf{h}_1, \mathbf{h}_2, \mathbf{h}_3)$.

Directions of anisotropy are usually evident from a variogram map; see Figure III.5. Visualization of a variogram volume is best done using a graphical routine that allows continuous sectioning of the volume along any of the three axes $\mathbf{h}_1, \mathbf{h}_2, \mathbf{h}_3$, [162].

The GSLIB library and **varmap** program should be compiled following the

```
                         Parameters for VARMAP
                         *********************

START OF PARAMETERS:
../data/cluster.dat              \file with data
1   3                            \   number of variables: column numbers
-1.0e21        1.0e21            \   trimming limits
0                                \1=regular grid, 0=scattered values
  50    50     1                 \if =1: nx,       ny,    nz
1.0   1.0   1.0                  \       xsiz, ysiz, zsiz
1    2    0                      \if =0: columns for x,y, z coordinates
varmap.out                       \file for variogram output
  10     10      0               \nxlag, nylag, nzlag
5.0    5.0    1.0                \dxlag, dylag, dzlag
5                                \minimum number of pairs
1                                \standardize sill? (0=no, 1=yes)
1                                \number of variograms
1   1   1                        \tail, head, variogram type
```

Figure III.6: An example parameter file for **varmap**.

instructions in Appendix B. The parameters required for the main program varmap are listed below and shown on Figure III.6:

- **datafl:** the input data in a simplified Geo-EAS formatted file.

- **nvar** and **ivar(1) ... ivar(nvar):** the number of variables and their column order in the data file.

- **tmin** and **tmax:** all values, regardless of which variable, strictly less than **tmin** and greater than or equal to **tmax** are ignored.

- **igrid:** set to 1 if the data are on a regular grid and set to 0 if the data are irregularly spaced.

- **nx, ny**, and **nz:** the number of nodes in the x, y, and z coordinate directions (used if **igrid** is set to 1). Any of the numbers may be set to one if that coordinate is not present in the data; e.g., 2D data may be handled by setting **nz** to 1.

- **xsiz, ysiz**, and **zsiz:** the separation distance between the nodes in the x, y, and z coordinate directions (used if **igrid** is set to 1).

- **icolx, icoly**, and **icolz:** the columns for the x, y, and z coordinates (used if **igrid** is set to 0). Any of the column numbers may be set to zero if that coordinate is not present in the data; e.g., 2D data may be handled by setting **icolz** to 0.

- **outfl:** the output file for the variogram maps/volumes are written to a single output file named **outfl**. This file contains each variogram map/volume written sequentially as a grid with the x direction cycling fastest, then y, then z.

- **nxlag, nylag**, and **nzlag:** the number of lags to compute in the X, Y and Z directions.

- **dxlag, dylag** and **dzlag:** the lag tolerances or "cell sizes" in the X, Y and Z directions.

- **minpairs:** the minimum number of pairs needed to define a variogram value (set to missing if fewer than **minpairs** is found).

- **standardize:** if set to 1, the semivariogram values will be divided by the variance.

- **nvarg:** the number of variograms to compute.

- **ivtail, ivhead**, and **ivtype:** for each of the *nvarg* variograms one must specify which variables should be used for the tail and head and which type of variogram is to be computed. For direct variograms the **ivtail** array is identical to the **ivhead** array. Cross variograms are computed by having the tail variable different from the head variable. The **ivtype** variable corresponds to the integer code in the list given in Section III.1.

- **cut:** whenever the **ivtype** is set to 9 or 10, i.e., asking for an indicator variogram, then a cutoff must be specified immediately after the **ivtype** parameter on the same line in the input file (see Figure III.3). Note that if an indicator variogram is being computed, then the cutoff/category applies to variable **ivtail(i)** in the input file [although the **ivhead(i)** variable is not used, it must be present in the file to maintain consistency with the other variogram types].

The maximum number of data, the maximum number of variograms, and other maximum dimensioning parameters are specified in the file $\boxed{\text{varmap.inc}}$, which is included in the `varmap` program.

III.6 Application Notes

- A good understanding of the spatial arrangement of the data is essential to make intelligent decisions about variogram computation parameters like lag spacing, directions, and direction tolerances. It is good practice to plot the data locations and contour the data values with any off-the-shelf program prior to choosing the variogram computation parameters; this allows prior detection of data clusters, trends, discontinuities, and other features.

- The coordinate values should be rescaled so that they have similar ranges and their squared values do not exceed the computer's precision. Similarly, data values may have to be rescaled, particularly when working with different attribute values expressed in widely different unit scales. A common rescaling formula uses the minimum and maximum data values:

$$z' = \frac{z - z_{min}}{z_{max} - z_{min}}$$

- When the data are clustered (e.g., along drillholes or in specific areas), it is common to modify the variogram program to have a smaller lag separation distance **xlag** for the first few lags. The lags for longer distances are made larger with a corresponding increase in the lag tolerance.

- Certain geometrically complicated situations may require special attention. For example, a 2D isotropic variogram within the plane of a 2D dipping tabular layer will require either a prior geometric (stratigraphic) transformation of coordinates, or some special coding. Variograms along a meandering direction or undulating surface also require specific coding (see [32, 48, 108, 113]); **gamv** does not handle these situations. A worthwhile addition to GSLIB would be a program that would provide 3D graphic visualization of an ellipsoid defined with all three angles (azimuth, dip, and rake); see Figure II.4 and [65]. Such a program is not included here because it would be machine dependent.

- Care must be taken when attempting to compute omnidirection variograms with a stratigraphic coordinate system. The stratigraphic coordinate must be scaled to the same units as the other directional coordinates.

- It is good practice to run at least two alternate measures of spatial variability/continuity, for example, the semivariogram (III.1) and the correlogram (III.4). The additional computing cost is negligible if they are run simultaneously. If these alternate measures appear similar, everything is fine. If not, their differences must be understood, e.g., by plotting the lag means m_{-h} and m_{+h} or the lag variances σ^2_{-h} and σ^2_{+h} versus the distance $|h|$; see definitions (III.3) and (III.4). Sometimes, the structure read on the semivariogram is an artifact of the preferential clustering of the sample data. These artifacts can be reduced by the measures (III.3 and III.4) that filter the lag means and lag variances; see [88]. Similarly, clustering and outlier values usually affect the madogram or relative semivariogram measures much less than the semivariogram.

- Within reason, it is justifiable to adopt the measure of spatial continuity that gives the most interpretable and cleanest results. It is easy to mask spatial continuity by a poor choice of lag spacing, direction angles, or a poor handling of outlier data values. It is rare to generate spatial continuity that does not exist. There are two notable exceptions to this statement: (1) clustered data may cause certain measures of spatial continuity to show an artificial structure (see above), and (2)

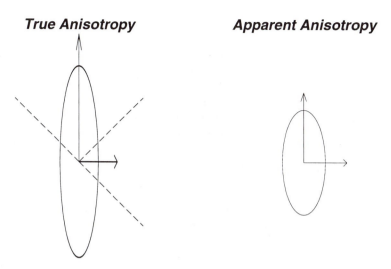

True Anisotropy **Apparent Anisotropy**

Figure III.7: An illustration of how the apparent anisotropy inferred from experimental variograms, computed with a large angular tolerance, can be considerably less than the true anisotropy. The elliptical outline on the left represents the locus of points that are equally close in terms of variogram distance to the central point. If a 45° tolerance is used to compute variograms, the apparent anisotropy is reduced to that shown on the right.

the combination of a severe anisotropy and large angular tolerance can artificially increase the range of correlation in the direction of minimum continuity (see below).

- In the presence of few, widely spaced, data (almost always the case in practice), it is common to use a fairly large angular tolerance to obtain enough pairs for a stable "variogram." This is not a problem when the underlying phenomenon is not strongly anisotropic. If the phenomenon is truly anisotropic, however, then directional variograms will always yield an apparent anisotropy that is less than the true anisotropy; i.e., the most continuous direction will appear less continuous and the least continuous direction will appear more continuous by combining pairs that are in different directions. This is illustrated in Figure III.7, where a true 5:1 anisotropy appears as a 2.4:1 anisotropy if the variogram measure is computed with a 45° tolerance. One should use the smallest tolerance possible (in the previous example the apparent anisotropy would be 3.6:1 with a 22.5^o tolerance) and possibly increase the anisotropy ratios after modeling the experimental variograms.

- Considering pairs at relatively long distances, to infer the short scale structure, often leads to experimental semivariograms with too high nugget effects.

- As in any inference endeavor, it is not the properties of the limited sample nor the properties of the elusive RF model that are of prime interest. It is the underlying properties of the (one and only one) physical population that are of interest. In the process of variogram inference and that of variogram modeling (not covered by GSLIB), the practitioner should allow some initiative in cleaning or departing from the data, as long as any such departure is clearly documented. It is the subjective interpretation, possibly based on prior experience or ancillary information, that makes a *good* model; the data, by themselves, are rarely enough. An excellent discussion on variogram analysis is given in [74, 185].

- When modeling a variogram for the purpose of stochastic simulation, one should try to model part or all of the apparent nugget variance with a nonzero range. Indeed, simulation is most likely to be performed on a grid much denser than the average data spacing. Most likely, at such short scales there exist structures not revealed by the available data.

- When used as a cross variogram (with tail and head values corresponding to different variables Z and Y), the semivariogram measure (III.1) corresponds to the moment:

$$\gamma_{ZY}^{(1)}(\mathbf{h}) = \frac{1}{2} E\left\{[Z(\mathbf{u}) - Y(\mathbf{u} + \mathbf{h})]^2\right\} \qquad \text{(III.11)}$$

which differs from the traditional cross semivariogram:

$$\gamma_{ZY}^{(2)}(\mathbf{h}) = \frac{1}{2} E\left\{[Z(\mathbf{u}) - Z(\mathbf{u} + \mathbf{h})]\,[Y(\mathbf{u}) - Y(\mathbf{u} + \mathbf{h})]\right\} \qquad \text{(III.12)}$$

The cross covariance, as required by cokriging systems, is:

$$\begin{aligned}
C_{ZY}(\mathbf{h}) &= Cov\{Z(\mathbf{u}), Y(\mathbf{u} + \mathbf{h})\} \\
&= E\{Z(\mathbf{u})Y(\mathbf{u} + \mathbf{h})\} - m_Z m_Y
\end{aligned}$$

with, in general, $C_{ZY}(\mathbf{h}) \neq C_{YZ}(\mathbf{h})$; see "lag effect" in [109], p. 41.

Development of the two alternative definitions (III.11) and (III.12) leads to the two relations:

$$\gamma_{ZY}^{(1)}(\mathbf{h}) = \frac{1}{2}\left[C_Z(0) + C_Y(0) + (m_Z - m_Y)^2 - 2C_{ZY}(\mathbf{h})\right]$$

and

$$\gamma_{ZY}^{(2)}(\mathbf{h}) = C_{ZY}(0) - \frac{1}{2}\left[C_{ZY}(\mathbf{h}) + C_{YZ}(\mathbf{h})\right]$$

with m_Z, m_Y and $C_Z(0)$, $C_Y(0)$ being the means and variances of variables Z and Y. If the two variables are standardized, then $m_Z = m_Y = 0$ and $C_Z(0) = C_Y(0) = 1$. Introducing the correlogram $\rho_{ZY}(\mathbf{h})$ [equal to $C_{ZY}(\mathbf{h})$ if Z and Y are standardized], the two previous relations become:

$$\gamma_{ZY}^{(1)}(\mathbf{h}) = 1 - \rho_{ZY}(\mathbf{h}) \qquad (\text{III.13})$$

and

$$\gamma_{ZY}^{(2)}(\mathbf{h}) = \rho_{ZY}(0) - \frac{1}{2}\left[\rho_{ZY}(\mathbf{h}) + \rho_{YZ}(\mathbf{h})\right] \qquad (\text{III.14})$$

Remark: Variograms and cross variograms can be used in cokriging systems only under specific constraints on the cokriging weights. It is safe practice to transform sample cross semivariograms of either type (III.11) and (III.12) into the corresponding cross covariance $C_{ZY}(\mathbf{h})$ and then model the covariance/cross covariance matrix with a linear model of coregionalization before using it in cokriging systems.

When modeling spatial cross dependence, there are clear advantages in directly calculating cross covariances, $C_{ZY}(\mathbf{h})$, using the covariance measure (III.3) with tail and head values corresponding to different variables Z and Y; see [88, 172]. If cross semivariograms are inferred, then the traditional expressions (III.2) and (III.12) should be preferred.

- Experimental indicator semivariograms will reach their sills (if any) at the nondeclustered indicator variances $\hat{F}(z_c)[1 - \hat{F}(z_c)]$, where $\hat{F}(z_c)$ is the sample mean of the corresponding indicator data, not at the declustered variances $F(z_c)[1 - F(z_c)]$ corresponding to the declustered cdf values $F(z_c)$, z_c being the threshold values at which the indicator data are defined; see definition (II.6). Hence, if these experimental indicator semivariograms are to be standardized to a common unit sill, the sample variances $\hat{F}(z_c)[1 - \hat{F}(z_c)]$ should be used. Also in the presence of zonal anisotropy, certain experimental indicator semivariograms may not reach the variance defined on either $F(z_c)$ or $\hat{F}(z_c)$.

- **Variogram modeling** requires interactive and graphical programs not within the scope of GSLIB because they would require a machine-dependent graphics library. The output from GSLIB variogram subroutines could be reformatted to be used in interactive variogram modeling programs such as the one provided in the Geostatistical Toolbox [62], program xgam [28], or by any versatile spreadsheet.

The user is warned to exercise caution when using software that automatically fits variograms and cross variograms without user interaction. Good fitting algorithms should require a prior choice of the number of variogram structures, their types, anisotropies, and, only then, the parameters (sill, range) would be chosen to match the experimental variogram values. For a discussion on variogram fitting, see [74].

III.7   Problem Set Two: Variograms

The goal of this second problem set is to experiment with variogram calculation and model fitting techniques. Once again, compiled GSLIB programs are called for. The `vargplt` utility given in Chapter VI may be useful.

The 2D irregularly spaced data in $\boxed{\text{cluster.dat}}$ and the 2D grid of reference data in $\boxed{\text{true.dat}}$ will be used for this problem set.

Questions

1. Perform a variogram study on the irregularly spaced data (with `gamv`). Try all variogram types and at least three directions (omnidirectional, NS, and EW). Compute and plot (with `vargplt`) the variograms using all 140 data and, then, using only the first 97 data. Comment on each variogram type, the presence or absence of anisotropy, the impact of clustering, etc.

2. Perform a variogram study on the exhaustively sampled grid (with `gam`). Just consider a few representative directions - `gam` does not compute omnidirectional variograms. Comment on the results and compare to the results of question one. Generate a variogram map using, in sequence, programs `varmap` and `pixelplt` (documented in Section VI.1.2).

3. Model one directional variogram calculated from the sample data and one directional variogram calculated from the exhaustive data. The program `vmodel` may be used in conjunction with `vargplt` to overlay your model on the experimental variogram points. An interactive variogram modeling program would be helpful. Comment on the results.

4. Compute the normal score transform of the data with `nscore` (documented in Section VI.2.5), compute the normal score variogram, and model it. Keep this model in preparation for Problem Set 6 dealing with Gaussian simulations.

5. Establish the theoretical (given a bi-Gaussian hypothesis) indicator variograms for the three quartile cutoffs using `bigaus` (documented in Section VI.2.9). Compute the actual indicator variograms, compare them to the previously obtained theoretical models, and comment on the appropriateness of a multivariate Gaussian RF model; see Section V.2.2.

Chapter IV

Kriging

This chapter presents the kriging programs of GSLIB. These programs allow for simple kriging (SK), ordinary kriging (OK), and kriging with various trend models (KT). A cokriging program provides the ability to use secondary variables and an indicator kriging program allows the direct estimation of posterior probability distributions.

Section IV.1 describes the various kriging algorithms and their underlying principles. Section IV.2 presents straightforward 2D ordinary kriging programs for grid kriging (points or blocks) and cross validation.

Section IV.3 presents a more elaborate kriging program for point/block kriging and cross validation in 2D and 3D. This program allows many alternate kriging algorithms and options such as kriging the trend and filtering certain nested covariance structures.

Section IV.4 presents a 3D cokriging program that will accept a number of covariates with either simple or ordinary cokriging. The indicator kriging (IK) program is presented in Section IV.5.

Some useful application notes are presented in Section IV.6. Finally, a problem set is proposed in Section IV.7 that allows the kriging paradigm to be tested and understood. Special problem sets on cokriging and indicator kriging are proposed in Sections IV.8 and IV.9.

IV.1   Kriging with GSLIB

Although kriging was initially introduced to provide estimates for unsampled values [121], it is being used increasingly to build probabilistic models of uncertainty about these unknown values; see Section II.1.4 and Lesson 4 in [102]. In a nutshell, the kriging algorithm provides a minimum error-variance estimate of any unsampled value. Contouring a grid of kriging estimates is the traditional mapping application of kriging. Kriging used as a mapping algorithm is a low-pass filter that tends to smooth out details and extreme values of the original data set.

Since kriging is a minimum error-variance estimation algorithm, it approximates, and in some cases, identifies the conditional expectation of the variable being estimated. Thus kriging can be used to estimate a series of posterior conditional probability distributions from which unsmoothed images of the attribute spatial distribution can be drawn. In the multi-Gaussian (MG) case the conditional distribution is identified by the mean and variance obtained from simple kriging. In the indicator kriging (IK) approach a series of conditional cumulative distribution function (ccdf) values are estimated directly.

The kriging principle, applied both as a mapping algorithm and as a tool to obtain conditional probability distributions, has been presented in numerous papers and textbooks [22, 31, 34, 68, 74, 79, 82, 89, 102, 109, 124, 134, 160]. In electrical engineering circles, kriging as a time series interpolation algorithm is known as the Wiener filter [19, 188]. Only a brief summary and information specific to GSLIB implementations are given in this section.

IV.1.1   Simple Kriging

All versions of kriging are elaborations on the basic generalized linear regression algorithm and corresponding estimator:

$$[Z_{SK}^*(\mathbf{u}) - m(\mathbf{u})] = \sum_{\alpha=1}^{n} \lambda_\alpha(\mathbf{u}) [Z(\mathbf{u}_\alpha) - m(\mathbf{u}_\alpha)] \qquad \text{(IV.1)}$$

where $Z(\mathbf{u})$ is the RV model at location $\mathbf{u}$, the $\mathbf{u}_\alpha$'s are the n data locations, $m(\mathbf{u}) = E\{Z(\mathbf{u})\}$ is the location-dependent expected value of RV $Z(\mathbf{u})$, and $Z_{SK}^*(\mathbf{u})$ is the linear regression estimator, also called the "simple kriging" (SK) estimator.

The SK weights $\lambda_\alpha(\mathbf{u})$ are given by the system (II.13) of normal equations written in their more general nonstationary form as follows:

$$\sum_{\beta=1}^{n} \lambda_\beta(\mathbf{u}) C(\mathbf{u}_\beta, \mathbf{u}_\alpha) = C(\mathbf{u}, \mathbf{u}_\alpha), \quad \alpha = 1, \ldots, n \qquad \text{(IV.2)}$$

The SK algorithm requires prior knowledge of the $(n + 1)$ means $m(\mathbf{u})$, $m(\mathbf{u}_\alpha), \alpha = 1, \ldots, n$, and the $(n+1)$ by $(n+1)$ covariance matrix $[C(\mathbf{u}_\alpha, \mathbf{u}_\beta), \alpha, \beta = 0, 1, \ldots, n]$ with $\mathbf{u}_0 = \mathbf{u}$. In most practical situations, inference of these means and covariance values requires a prior hypothesis (rather a *decision*) of stationarity of the random function $Z(\mathbf{u})$; see the discussion in Section II.1.2. If the RF $Z(\mathbf{u})$ is stationary with constant mean m, and covariance function $C(\mathbf{h}) = C(\mathbf{u}, \mathbf{u} + \mathbf{h}), \forall \mathbf{u}$, the SK estimator reduces to its stationary version:

$$Z_{SK}^*(\mathbf{u}) = \sum_{\alpha=1}^{n} \lambda_\alpha(\mathbf{u}) Z(\mathbf{u}_\alpha) + \left[1 - \sum_{\alpha=1}^{n} \lambda_\alpha(\mathbf{u})\right] m \qquad \text{(IV.3)}$$

with the traditional stationary SK system:

$$\sum_{\beta=1}^{n} \lambda_\beta(\mathbf{u}) C(\mathbf{u}_\beta - \mathbf{u}_\alpha) = C(\mathbf{u} - \mathbf{u}_\alpha), \ \alpha = 1, \ldots, n \qquad \text{(IV.4)}$$

Stationary SK does not adapt to local trends since it assumes that the mean value m is constant (and known) throughout the area. In some situations secondary information allows a prior determination of the locally varying means $m(\mathbf{u})$, the SK estimator (IV.1) can then be used with the system (IV.4) assuming a stationary residual covariance.

In the system (IV.4) the covariance values $C(\mathbf{h})$ cannot be replaced by semivariogram values $\gamma(\mathbf{h}) = C(0) - C(\mathbf{h})$ unless $\sum_\beta \lambda_\beta(\mathbf{u}) = 1$, which is the ordinary kriging constraint.

According to strict stationary theory, it is SK that should be applied to the normal score transforms in the MG approach. The OK algorithm, however, might be considered if enough data are available to re-estimate locally the normal score mean; see the related discussion in Section V.2.3.

IV.1.2 Ordinary Kriging

Ordinary kriging (OK) filters the mean from the SK estimator (IV.3) by requiring that the kriging weights sum to one. This results in the following ordinary kriging estimator:

$$Z^*_{OK}(\mathbf{u}) = \sum_{\alpha=1}^{n} \lambda_\alpha^{(OK)}(\mathbf{u}) Z(\mathbf{u}_\alpha) \qquad \text{(IV.5)}$$

and the stationary OK system:

$$\begin{cases} \sum_{\beta=1}^{n} \lambda_\beta^{(OK)}(\mathbf{u}) C(\mathbf{u}_\beta - \mathbf{u}_\alpha) + \mu(\mathbf{u}) = C(\mathbf{u} - \mathbf{u}_\alpha), \ \alpha = 1, \ldots, n \\ \sum_{\beta=1}^{n} \lambda_\beta^{(OK)}(\mathbf{u}) = 1 \end{cases} \qquad \text{(IV.6)}$$

where the $\lambda_\alpha^{(OK)}(\mathbf{u})$'s are the OK weights and $\mu(\mathbf{u})$ is the Lagrange parameter associated with the constraint $\sum_{\beta=1}^{n} \lambda_\beta^{(OK)}(\mathbf{u}) = 1$. Comparing systems (IV.4) and (IV.6), note that the SK weights are different from the OK weights.

It can be shown that ordinary kriging amounts to re-estimating, at *each* new location $\mathbf{u}$, the mean m as used in the SK expression; see [74, 112, 134]. Since ordinary kriging is most often applied within moving search neighborhoods, i.e., using different data sets for different locations $\mathbf{u}$, the implicit re-estimated mean denoted $m^*(\mathbf{u})$ depends on the location $\mathbf{u}$. Thus the OK estimator (IV.5) is, in fact, a simple kriging of type (IV.3), where the constant mean value m is replaced by the location-dependent estimate $m^*(\mathbf{u})$:

$$Z_{OK}^*(\mathbf{u}) \;=\; \sum_{\alpha=1}^{n} \lambda_\alpha^{(OK)}(\mathbf{u}) Z(\mathbf{u}_\alpha) \tag{IV.7}$$

$$\equiv \sum_{\alpha=1}^{n} \lambda_\alpha^{(SK)}(\mathbf{u}) Z(\mathbf{u}_\alpha) + \left[1 - \sum_{\alpha=1}^{n} \lambda_\alpha^{(SK)}(\mathbf{u})\right] m^*(\mathbf{u})$$

where $\lambda_\alpha^{(OK)}$ are the OK weights given by system (IV.6) and $\lambda_\alpha^{(SK)}$ are the SK weights given by system (IV.4).

Hence, ordinary kriging as applied within moving data neighborhoods is already a nonstationary algorithm, in the sense that it corresponds to a nonstationary RF model with varying mean but stationary covariance. This ability to rescale locally the RF model $Z(\mathbf{u})$ to a different mean value $m^*(\mathbf{u})$ explains the extreme robustness of the OK algorithm. Ordinary kriging has been and will remain the anchor algorithm of geostatistics.

The subroutine kb2d provides a basic 2D simple and ordinary kriging program stripped of fancy data classification and search strategies.

The more complex kt3d subroutine allows SK, OK, and kriging with various trend models in either 1, 2 or 3D.

IV.1.3   Kriging with a Trend Model

The term universal kriging has been traditionally used to denote what is, in fact, kriging with a prior trend model [68, 74, 85, 109, 112, 145, 187]. The terminology kriging with a trend model (KT) is more appropriate since the underlying RF model is the sum of a trend component plus a residual:

$$Z(\mathbf{u}) = m(\mathbf{u}) + R(\mathbf{u}) \tag{IV.8}$$

The trend component, defined as $m(\mathbf{u}) = E\{Z(\mathbf{u})\}$, is usually modeled as a smoothly varying deterministic function of the coordinates vector $\mathbf{u}$ whose unknown parameters are fitted from the data:

$$m(\mathbf{u}) = \sum_{k=0}^{K} a_k f_k(\mathbf{u}) \tag{IV.9}$$

The $f_k(\mathbf{u})$'s are known functions of the location coordinates and the a_k's are unknown parameters. The trend value $m(\mathbf{u})$ is itself unknown since the parameters a_k are unknown.

The residual component $R(\mathbf{u})$ is usually modeled as a stationary RF with zero mean and covariance $C_R(\mathbf{h})$.

Kriging with the trend model (IV.9) results in the so-called "universal" kriging estimator and system of equations. This system, which is actually a system of constrained normal equations [68, 124], would be better named the

KT system. The KT estimator is written:

$$Z_{KT}^*(\mathbf{u}) = \sum_{\alpha=1}^{n} \lambda_{\alpha}^{(KT)}(\mathbf{u}) Z(\mathbf{u}_\alpha) \tag{IV.10}$$

and the KT system is:

$$\begin{cases} \sum_{\beta=1}^{n} \lambda_{\beta}^{(KT)}(\mathbf{u}) C_R(\mathbf{u}_\beta - \mathbf{u}_\alpha) + \sum_{k=0}^{K} \mu_k(\mathbf{u}) f_k(\mathbf{u}_\alpha) = C_R(\mathbf{u} - \mathbf{u}_\alpha), \\ \quad \alpha = 1, \ldots, n \\ \sum_{\beta=1}^{n} \lambda_{\beta}^{(KT)}(\mathbf{u}) f_k(\mathbf{u}_\beta) = f_k(\mathbf{u}), \quad k = 0, \ldots, K \end{cases}$$

$$\tag{IV.11}$$

where the $\lambda_{\beta}^{(KT)}(\mathbf{u})$'s are the KT weights and the $\mu_k(\mathbf{u})$'s are the $(K+1)$

Lagrange parameters associated with the $(K+1)$ constraints on the weights.

Remarks

Ideally, the functions $f_k(\mathbf{u})$ that define the trend should be specified by the physics of the problem. For example, a sine function $f_k(\mathbf{u})$ with specific period and phase would be considered if this periodic component is known to contribute to the spatial or temporal variability of $z(\mathbf{u})$; the amplitude of the periodic component, i.e., the parameter a_k, is then implicitly estimated from the z data through the KT system.

In the absence of any information about the shape of the trend, the dichotomization (IV.8) of the z data into trend and residual components is somewhat arbitrary: What is regarded as stochastic fluctuations $[R(\mathbf{u})]$ at large scale may later be modeled as a trend if additional data allow focusing on the smaller-scale variability. In the absence of a physical interpretation, the trend is usually modeled as a low-order (≤ 2) polynomial of the coordinates $\mathbf{u}$, e.g., with $\mathbf{u} = (x, y)$:

- a linear trend in 1D: $m(\mathbf{u}) = a_0 + a_1 x$

- a linear trend in 2D limited to the 45° direction: $m(\mathbf{u}) = a_0 + a_1(x + y)$

- a quadratic trend in 2D: $m(\mathbf{u}) = a_0 + a_1 x + a_2 y + a_3 x^2 + a_4 y^2 + a_5 xy$

By convention, $f_0(\mathbf{u}) = 1$, $\forall \mathbf{u}$. Hence the case $K = 0$ corresponds to ordinary kriging with a constant but unknown mean: $m(\mathbf{u}) = a_0$.

Trend models using higher-order polynomials ($n > 2$) or arbitrary non-monotonic functions of the coordinates $\mathbf{u}$ are better replaced by a random function component with a large-range variogram; see hereafter and Section IV.1.6.

When only z data are available, the residual covariance $C_R(\mathbf{h})$ is inferred from linear combinations of z data that filter the trend $m(\mathbf{u})$. For example, differences of order 1 such as $z(\mathbf{u} + \mathbf{h})$ - $z(\mathbf{u})$ would filter any trend of order zero $m(\mathbf{u}) = a_0$; differences of order 2 such as $z(\mathbf{u} + 2\mathbf{h}) - 2z(\mathbf{u} + \mathbf{h}) + z(\mathbf{u})$

would filter any trend of order 1 such as $m(\mathbf{u}) = a_0 + a_1\mathbf{u}$; see the related discussion in Section II.1.3.

In most practical situations it is possible to locate subareas or directions along which the trend can be ignored, in which case $Z(\mathbf{u}) \simeq R(\mathbf{u})$, and the residual covariance can be directly inferred from the local z data.

Exactitude of Kriging

If the location $\mathbf{u}$ to be estimated coincides with a datum location $\mathbf{u}_\alpha$, the normal system (SK, OK, or KT) returns the datum value for the estimate. Thus kriging is an exact interpolator in the sense that it honors the (hard) data values at their locations.

However, in the presence of a nonzero nugget effect, this exactitude is obtained through discontinuities at the data locations. In mapping applications a solution is to move the data locations so that they do not coincide with any of the kriging nodes, this is equivalent to filtering the nugget effect; see Section IV.1.6. Another solution is to consider using stochastic simulation, but then there are several realizations possible.

Random Trend Model

A model similar to (IV.8) consists of interpreting the trend as a random component, denoted $M(\mathbf{u})$, added to a residual $R(\mathbf{u})$ independent from it:

$$Z(\mathbf{u}) = M(\mathbf{u}) + R(\mathbf{u}), \quad \text{where} \;\; E\{Z(\mathbf{u})\} = E\{M(\mathbf{u})\}$$

Prior trend-data are assumed available, for example they can be prior guesses $m(\mathbf{u})$ about the local z-values. These trend data would allow inference of the M-covariance $C_M(\mathbf{h})$, and the corresponding residual data can be used to infer the R-covariance, then per the independence assumption:

$$C_Z(\mathbf{h}) = C_M(\mathbf{h}) + C_R(\mathbf{h})$$

Kriging can then be performed using the z-data and the covariance model $C_Z(\mathbf{h})$ with the results (z-kriging estimates and variances) depending on the statistics $E\{M(\mathbf{u})\}$ and $C_M(\mathbf{h})$. The variance $Var\{M(\mathbf{u})\}$ can be made non-stationary and used to measure the reliability of the prior "guess" $m(\mathbf{u})$, in which case the M-covariance is not anymore stationary and its inference becomes problematic.

The random trend model is equivalent (yields the same results) yet is much simpler to establish than the Bayesian kriging model [148]. The weakness of either model lies in the inference of the statistics of $M(\mathbf{u})$, whether interpreted as a random trend or as prior guess on z-values, and on the key hypothesis of independence of M and R-values. The only physical reality is z, not M or R. A similar and better alternative is provided by factorial kriging, see Section IV.1.6, where $C_M(\mathbf{h})$ is obtained by decomposition of the experimentally available Z-covariance.

IV.1.4 Kriging the Trend

Rather than estimating the sum $Z(\mathbf{u}) = m(\mathbf{u}) + R(\mathbf{u})$ one could estimate only the trend component $m(\mathbf{u})$. Starting directly from the original z data the KT system (IV.11) is easily modified to yield a KT estimate for $m(\mathbf{u})$,

$$m_{KT}^*(\mathbf{u}) = \sum_{\alpha=1}^{n} \lambda_\alpha^{(m)}(\mathbf{u}) Z(\mathbf{u}_\alpha) \qquad (IV.12)$$

and the KT system:

$$\begin{cases} \sum_{\beta=1}^{n} \lambda_\beta^{(m)}(\mathbf{u}) C_R(\mathbf{u}_\beta - \mathbf{u}_\alpha) + \sum_{k=0}^{K} \mu_k^{(m)}(\mathbf{u}) f_k(\mathbf{u}_\alpha) = 0, \quad \alpha = 1, \dots, n \\ \sum_{\beta=1}^{n} \lambda_\beta^{(m)}(\mathbf{u}) f_k(\mathbf{u}_\beta) = f_k(\mathbf{u}), \quad k = 0, \dots, K \end{cases}$$

$$\qquad (IV.13)$$

where the $\lambda_\alpha^{(m)}$'s are the KT weights and the $\mu_k^{(m)}$'s are Lagrange parameters. Note that this system differs from the KT system (IV.11) of the variable $Z(\mathbf{u})$.

This algorithm identifies the least-squares fit of the trend model (IV.9) when the residual model $R(\mathbf{u})$ is assumed to have no correlation: $C_R(\mathbf{h}) = 0, \forall \mathbf{h} \neq 0$; see [37], p. 405.

The program kt3d allows, as an option, kriging the trend as in expression (IV.12) and system (IV.13).

The direct KT estimation of the trend component can also be interpreted as a low-pass filter that removes the random (high-frequency) component $R(\mathbf{u})$. The same principle underlies the algorithm of "factorial kriging" (see Section IV.1.6 and [161]) and that of the Wiener-Kalman filter [19].

The traditional notation (IV.9) for the trend does not reflect the general practice of kriging with moving data neighborhoods. Because the data used for estimation change from one location $\mathbf{u}$ to another, the resulting implicit estimates of the parameters a_l's are different. Hence the following notation for the trend is more appropriate:

$$m(\mathbf{u}) = \sum_{k=0}^{K} a_k(\mathbf{u}) f_k(\mathbf{u}) \qquad (IV.14)$$

Ordinary kriging (OK) corresponds to the case $K = 0$ and $m(\mathbf{u}) = a_0(\mathbf{u})$, which is not fundamentally different from the more general expression (IV.14). In the case of OK, the varying trend $m(\mathbf{u})$ is specified as a single unknown value $a_0(\mathbf{u})$, whereas in the case of KT the trend is split (often arbitrarily) into $(K + 1)$ components $f_k(\mathbf{u})$.

When the trend functions $f_k(\mathbf{u})$ are not based on physical considerations (often the case in practice), *and* in interpolation conditions, it can be shown [74, 112] that the choice of the specific functions $f_k(\mathbf{u})$ does not change the estimated values $z_{KT}^*(\mathbf{u})$ or $m_{KT}^*(\mathbf{u})$. When working with moving neighborhoods the important aspect is the residual covariance $C_R(\mathbf{h})$, not the choice of the trend model. The trend model, however, is important in extrapolation conditions, i.e., when the data locations $\mathbf{u}_\alpha$ do not surround within the covariance range the location $\mathbf{u}$ being estimated. Extrapolating a constant yields significantly different results for either $z^*_{KT}(\mathbf{u})$ or $m^*_{KT}(\mathbf{u})$ than extrapolating either a line or a parabola (nonconstant trend).

The practitioner is warned against overzealous modeling of the trend and the unnecessary usage of "universal kriging" (KT) or intrinsic random functions of order k (IRF-k): In most interpolation situations the simpler and well-proven OK algorithm within moving search neighborhoods will suffice. In extrapolation situations, almost by definition, the sample z data alone cannot justify the trend model chosen.

IV.1.5   Kriging with an External Drift

Kriging with an external drift variable [74, 130] is an extension of KT. The trend model is limited to two terms $m(\mathbf{u}) = a_0 + a_1 f_1(\mathbf{u})$, with the term $f_1(\mathbf{u})$ set equal to a secondary (external) variable. The *smooth* variability of the second variable is deemed related to that of the primary variable $Z(\mathbf{u})$ being estimated.

Let $y(\mathbf{u})$ be the secondary variable; the trend model is then:

$$E\{Z(\mathbf{u})\} = m(\mathbf{u}) = a_0 + a_1 y(\mathbf{u}) \tag{IV.15}$$

$y(\mathbf{u})$ is assumed to reflect the spatial trends of the z variability up to a linear rescaling of units (corresponding to the two parameters a_0 and a_1).

The estimate of the z variable and the corresponding system of equations are identical to the KT estimate (IV.10) and system (IV.11) with $K = 1$, and $f_1(\mathbf{u}) = y(\mathbf{u})$, i.e.:

$$Z^*_{KT}(\mathbf{u}) = \sum_{\alpha=1}^{n} \lambda^{(KT)}_\alpha(\mathbf{u}) Z(\mathbf{u}_\alpha) \ , \ \ \text{and}$$

$$\begin{cases} \sum_{\beta=1}^{n} \lambda^{(KT)}_\beta(\mathbf{u}) C_R(\mathbf{u}_\beta - \mathbf{u}_\alpha) + \mu_0(\mathbf{u}) + \mu_1(\mathbf{u}) y(\mathbf{u}_\alpha) = \ \ C_R(\mathbf{u} - \mathbf{u}_\alpha), \\ \hspace{7cm} \alpha = 1, \ldots, n \\ \sum_{\beta=1}^{n} \lambda^{(KT)}_\beta(\mathbf{u}) = 1 \\ \sum_{\beta=1}^{n} \lambda^{(KT)}_\beta(\mathbf{u}) y(\mathbf{u}_\beta) = y(\mathbf{u}) \end{cases}$$

$$\tag{IV.16}$$

where the $\lambda^{(KT)}_\alpha$'s are the kriging (KT) weights and the μ's are Lagrange parameters.

Kriging with an external drift is a simple and efficient algorithm to incorporate a secondary variable in the estimation of the primary variable $z(\mathbf{u})$. The fundamental (hypothesis) relation (IV.15) must make physical sense. For example, if the secondary variable $y(\mathbf{u})$ represents the travel time to a seismic reflective horizon, assuming a constant velocity, the depth $z(\mathbf{u})$ of that horizon should be (in average) proportional to the travel time $y(\mathbf{u})$. Hence a relation of type (IV.15) makes sense. But, if the primary variable is permeability, it is not clear that the general trends of the spatial variability of permeability is revealed by the variability of seismic data. Alternatives to kriging with an external drift are cokriging or soft kriging; see Sections IV.1.7 and IV.1.12.

Two conditions must be met before applying the external drift algorithm: (1) The external variable must vary smoothly in space, otherwise the resulting KT system (IV.16) may be unstable; and (2) the external variable must be known at *all* locations $\mathbf{u}_\alpha$ of the primary data values and at all locations $\mathbf{u}$ to be estimated.

Note that the residual covariance rather than the covariance of the original variable $Z(\mathbf{u})$ must be used in the KT system. Both covariances are equal in areas or along directions where the trend $m(\mathbf{u})$ can be ignored (can be set to zero). Note also that the cross covariance between variables $Z(\mathbf{u})$ and $Y(\mathbf{u})$ plays no role in system (IV.16); this is different from cokriging.

Kriging with an external drift yields maps whose trends reflect the y variability. This is a result of the decision (IV.15). It is not proof that those trends necessarily pertain to the z variable.

An option of program kt3d allows kriging with an external drift.

Bayesian Kriging

Under a different presentation, Bayesian kriging [148] amounts to kriging with an external drift of type (IV.15) where the coefficient a_1 is set to 1, and the $y(\mathbf{u})$'s represent prior (known) guess values related to the z-spatial trend: $E\{Z(\mathbf{u})\} = a_0 + y(\mathbf{u})$.

The corresponding KT estimator is written as a residual (SK-type) estimator:

$$Z^*(\mathbf{u}) - y(\mathbf{u}) = \sum_{\alpha=1}^{n} \lambda_\alpha(\mathbf{u}) \left[z(\mathbf{u}_\alpha) - y(\mathbf{u}_\alpha)\right]$$

Unbiasedness then calls for the only condition, $\sum_\alpha \lambda_\alpha(\mathbf{u}) = 1$. The KED system (IV.16) is then reduced by one unknown and one equation:

$$\begin{cases} \sum_{\beta=1}^{n} \lambda_\beta(\mathbf{u}) C_R(\mathbf{u}_\beta - \mathbf{u}_\alpha) + \mu_0(\mathbf{u}) = C_R(\mathbf{u} - \mathbf{u}_\alpha), \alpha = 1, \ldots, n \\ \sum_{\beta=1}^{n} \lambda_\beta(\mathbf{u}) = 1 \end{cases} \quad \text{(IV.17)}$$

This is exactly an OK system except for the Z-covariance being replaced by the covariance of the residual $Z(\mathbf{u}) - y(\mathbf{u})$.

IV.1.6   Factorial Kriging

Rather than splitting the RF model $Z(\mathbf{u})$ into a deterministic trend $m(\mathbf{u})$ plus a stochastic component $R(\mathbf{u})$, one could consider a model consisting of two or more *independent* stochastic components (also called "factors" in relation to factor analysis):

$$Z(\mathbf{u}) = Z_0(\mathbf{u}) + Z_1(\mathbf{u}) + \ldots + Z_L(\mathbf{u}) \quad \text{(IV.18)}$$

The Z covariance is then the sum of the $(L+1)$ component covariances:

$$C_Z(\mathbf{h}) = \sum_{l=0}^{L} C_l(\mathbf{h}) \qquad (IV.19)$$

For example, the $(L+1)$ RFs $Z_l(\mathbf{u})$ can be modeled from the $(L+1)$ nested covariance structures $C_l(\mathbf{h})$ used to model the overall z sample covariance.

Filtering

A kriging estimator of the partial sum of any number of the previous $(L+1)$ components $Z_l(\mathbf{u})$ can be obtained by filtering out the covariance contribution of the nonselected components [19, 126, 161, 185]. For example, starting from the OK system (IV.6), filtering of the first l_0 component leads to the estimator:

$$\left[\sum_{l=l_0}^{L} Z_l(\mathbf{u})\right]^*_{OK} = \sum_{\alpha=1}^{n} d_\alpha(\mathbf{u}) Z(\mathbf{u}_\alpha) \qquad (IV.20)$$

with the kriging system:

$$\begin{cases} \sum_{\beta=1}^{n} d_\beta(\mathbf{u}) C_Z(\mathbf{u}_\beta - \mathbf{u}_\alpha) + \mu_0(\mathbf{u}) & = \sum_{l=l_0}^{L} C_l(\mathbf{u} - \mathbf{u}_\alpha), \alpha = 1, \ldots, n \\ & = C_Z(\mathbf{u} - \mathbf{u}_\alpha) - \sum_{l=0}^{l_0-1} C_l(\mathbf{u} - \mathbf{u}_\alpha) \\ \sum_{\beta=1}^{n} d_\beta(\mathbf{u}) = 1 \end{cases}$$

$$(IV.21)$$

The last equation of system (IV.21) ensures unbiasedness of the estimator (IV.20) in the case $E\{Z_l(\mathbf{u})\} = 0, l = 0, \ldots, l_0 - 1$. Other unbiasedness conditions should be considered otherwise. Alternatively, no unbiasedness condition are needed if all factors have a mean of zero.

Kriging the trend (IV.12) and filtering the high frequency components (corresponding to, say, $l \leq l_0$) are two examples of low-pass filters: In one case, a deterministic trend is fitted to the z data; in the other case, a smooth covariance-type interpolator [the right-hand side covariance of system (IV.21)] is fitted to the same z data. Note that in both cases the resulting z map need not honor the z data at their locations. Note that strict nugget effect filtering only changes the kriging estimate at the data locations.

Both programs kb2d and kt3d can be easily modified to handle any such type of filtering.

Warning

Be aware that the result of any factorial kriging depends heavily on the (usually arbitrary) decomposition (IV.18). If that decomposition is based solely on the z sample covariance, then it is an artifact of the model (IV.19) rather than a proof of physical significance of the "factors" $Z_l(\mathbf{u})$.

IV.1.7   Cokriging

The term kriging is traditionally reserved for linear regression using data on the same attribute as that being estimated. For example, an unsampled porosity value $z(u)$ is estimated from neighboring porosity sample values defined on the same volume support.

The term cokriging is reserved for linear regression that also uses data defined on different attributes. For example, the porosity value $z(\mathbf{u})$ may be estimated from a combination of porosity samples and related acoustic data values.

In the case of a single secondary variable (Y), the ordinary cokriging estimator of $Z(\mathbf{u})$ is written:

$$Z^*_{COK}(\mathbf{u}) = \sum_{\alpha_1=1}^{n_1} \lambda_{\alpha_1}(\mathbf{u}) Z(\mathbf{u}_{\alpha_1}) + \sum_{\alpha_2=1}^{n_2} \lambda'_{\alpha_2}(\mathbf{u}) Y(\mathbf{u}'_{\alpha_2}) \qquad (IV.22)$$

where the λ_{α_1}'s are the weights applied to the n_1 z samples and the λ'_{α_2}'s are the weights applied to the n_2 y samples.

Kriging requires a model for the Z covariance. Cokriging requires a *joint* model for the matrix of covariance functions including the Z covariance $C_Z(\mathbf{h})$, the Y covariance $C_Y(\mathbf{h})$, the cross Z-Y covariance $C_{ZY}(\mathbf{h}) = \mathrm{Cov}\{Z(\mathbf{u}), Y(\mathbf{u}+\mathbf{h})\}$, and the cross Y-Z covariance $C_{YZ}(\mathbf{h})$.

The covariance matrix requires K^2 covariance functions when K different variables are considered in a cokriging exercise. The inference becomes extremely demanding in terms of data and the subsequent joint modeling is particularly tedious [74]. This is the main reason why cokriging has not been extensively used in practice. Algorithms such as kriging with an external drift (see Section IV.1.5) and collocated cokriging (see hereafter) have been developed to shortcut the tedious inference and modeling process required by cokriging.

Another reason that cokriging is not used extensively in practice is the screen effect of the better correlated data (usually the z samples) over the data less correlated with the z unknown (the y samples). Unless the primary variable, that which is being estimated, is undersampled with respect to the secondary variable, the weights given to the secondary data tend to be small, and the reduction in estimation variance brought by cokriging is not worth the additional inference and modeling effort ([73, 74, 109], p. 324).

Other than tedious inference and matrix notations, cokriging is the same as kriging. Cokriging with trend models could be developed, and a cokriging that filters specific components of the spatial variability of either Z or Y can be designed. These notation-heavy developments will not be given here: the reader is referred to [25, 51, 74, 109, 144, 143, 185].

GSLIB gives only one cokriging program, cokb3d. This simple program does not do justice to the full theoretical potential of cokriging. The three most commonly applied types of cokriging, however, are included:

1. **Traditional ordinary cokriging:** the sum of the weights applied to the primary variable is set to one, and the sum of the weights applied to any other variable is set to zero. In the case of two variables as in expression (IV.22), these two conditions are:

$$\sum_{\alpha_1} \lambda_{\alpha_1}(\mathbf{u}) = 1 \text{ and } \sum_{\alpha_2} \lambda'_{\alpha_2}(\mathbf{u}) = 0 \qquad (IV.23)$$

The problem with this traditional formalism is that the second condition (IV.23) tends to limit severely the influence of the secondary variable(s).

2. **Standardized ordinary cokriging:** often, a better approach (see [89], pp. 400-16) consists of creating new secondary variables with the same mean as the primary variable. Then all the weights are constrained to sum to one.

In this case expression (IV.22) could be rewritten as:

$$Z^*_{COK}(\mathbf{u}) = \sum_{\alpha_1=1}^{n_1} \lambda_{\alpha_1}(\mathbf{u}) Z(\mathbf{u}_{\alpha_1}) + \sum_{\alpha_2=1}^{n_2} \lambda'_{\alpha_2}(\mathbf{u}) \left[Y(\mathbf{u}'_{\alpha_2}) + m_Z - m_Y \right]$$

$$(IV.24)$$

with the single condition: $\sum_{\alpha_1=1}^{n_1} \lambda_{\alpha_1}(\mathbf{u}) + \sum_{\alpha_2=1}^{n_2} \lambda'_{\alpha_2}(\mathbf{u}) = 1$, where $m_Z = E\{Z(\mathbf{u})\}$ and $m_Y = E\{Y(\mathbf{u})\}$ are the stationary means of Z and Y.

3. **Simple cokriging:** there is no constraint on the weights. Just like simple kriging (IV.1), this version of cokriging requires working on data residuals or, equivalently, on variables whose means have all been standardized to zero. This is the case when applying simple cokriging in an MG approach (the normal score transforms of each variable have a stationary mean of zero).

Except when using traditional ordinary cokriging, covariance measures should be inferred, modeled, and used in the cokriging system rather than variograms or cross variograms; see related discussion in Section IV.1.1.

Collocated Cokriging

A reduced form of cokriging consists of retaining only the collocated secondary variable $y(\mathbf{u})$, provided that it is available at all locations $\mathbf{u}$ being estimated [9, 53, 194]. The cokriging estimator (IV.22) is written:

$$Z^*_{COK}(\mathbf{u}) = \sum_{\alpha_1}^{n_1} \lambda_{\alpha_1}(\mathbf{u}) Z(\mathbf{u}_{\alpha_1}) + \lambda'(\mathbf{u}) Y(\mathbf{u}) \qquad (IV.25)$$

The corresponding cokriging system requires knowledge of only the Z covariance $C_Z(\mathbf{h})$ and the Z-Y cross covariance $C_{ZY}(\mathbf{h})$. The latter can be approximated through the following model [9]:

$$C_{ZY}(\mathbf{h}) = B \cdot C_Z(\mathbf{h}), \quad \forall \mathbf{h} \qquad (IV.26)$$

where $B = \sqrt{C_Y(0)/C_Z(0)} \cdot \rho_{ZY}(0)$, $C_Z(0)$, $C_Y(0)$ are the variances of Z and Y, and $\rho_{ZY}(0)$ is the linear coefficient of correlation of collocated $z - y$ data.

If the secondary variable $y(\mathbf{u})$ is densely sampled but not available at all locations being estimated, it may be simulated at those locations conditional to the y data [9, 194]. The various realizations of the simulated value differ little if y is densely sampled; also keep in mind that y is but a secondary variable.

Warning

The approximation consisting of retaining only the collocated secondary datum does not affect the estimate (close-by secondary data are typically very similar in values), but it may affect the resulting cokriging estimation variance: that variance is overestimated, sometimes significantly. In a kriging (estimation) context this is not a problem, because kriging variances are of little use. In a simulation context, see later Section V.2, where the kriging variance defines the spread of the conditional distribution from which simulated values are drawn, this may be a problem. The collocated cokriging variance should then be reduced by a factor (assumed constant $\forall \mathbf{u}$) to be determined by trial and error.

IV.1.8 Nonlinear Kriging

All nonlinear kriging algorithms are actually linear kriging (SK or OK) applied to specific nonlinear transforms of the original data. The nonlinear transform used specifies the nonlinear kriging algorithm considered. Examples are:

- **lognormal kriging,** or kriging applied to logarithms of the data; see [94, 152, 156].

- **multi-Gaussian kriging,** or kriging applied to the normal score transforms of the data; see [109, 182]. Multi-Gaussian kriging is actually a generalization of lognormal kriging.

- **rank kriging,** or kriging of the rank or uniform transform of the data; see Section IV.1.11 and [115, 175].

- **indicator kriging,** or kriging of indicator transforms of the data; see Section IV.1.9 and [74, 95, 96].

- **disjunctive kriging,** or kriging of specific polynomial transforms of the data [136].

For example, consider the log transform $y(\mathbf{u}) = \ln z(\mathbf{u})$ of the original strictly positive variable $z(\mathbf{u})$. Simple or ordinary kriging of the log data yields an estimate $y^*(\mathbf{u})$ for $\ln z(\mathbf{u})$. Unfortunately, a "good" estimate of ln $z(\mathbf{u})$ need not lead straightforwardly to a good estimate of $z(\mathbf{u})$; in particular the antilog back-transform $e^{y^*(\mathbf{u})}$ is a biased estimator of $Z(\mathbf{u})$. The unbiased back-transform of the simple lognormal kriging estimate $y^*(\mathbf{u})$ is actually

$$z^*(\mathbf{u}) = \exp\left[y^*(\mathbf{u}) + \frac{\sigma^2_{SK}(\mathbf{u})}{2}\right]$$

where $\sigma^2_{SK}(\mathbf{u})$ is the simple lognormal kriging variance. The exponentiation involved in that back-transform is particularly delicate since it exponentiates any error in the process of estimating either the lognormal estimate $y^*(\mathbf{u})$ or its SK variance $\sigma^2_{SK}(\mathbf{u})$ [94]. This extreme sensitivity to the back-transform explains why direct nonlinear kriging of unsampled values has fallen into disuse. Instead, nonlinear kriging algorithms, essentially the MG and IK approaches, have been developed for stochastic simulations.

In the MG approach, the ccdf mean and variance obtained by SK applied to the normal score transform data are *not* back-transformed. Instead, it is the simulated normal score values, as drawn from the normal ccdf, which are back-transformed. That back-transform is the straight inverse normal score transform without any need for a bias correction.

Similarly, the IK estimates of the ccdf values are *not* back-transformed. Instead, realizations of the original z attribute are drawn directly from the ccdf (see Sections IV.1.9 and V.3.2).

Multi-Gaussian kriging is straightforward: It consists of SK or OK of the normal score data using the covariance modeled from the sample normal score covariance. Such kriging does not require any specific program (kb2d or kt3d could be used); multi-Gaussian kriging has been coded in the corresponding Gaussian simulation program sgsim (see Section V.7.2). The normal score transform itself and the corresponding straight back-transform are presented in Sections V.2.1, VI.2.5, and VI.2.6 (programs nscore and backtr).

As for indicator kriging (IK), its theoretical background is equally straightforward [74, 95, 102]; however, its implementation is delicate enough to justify the next Section IV.1.9.

IV.1.9   Indicator Kriging

Indicator kriging of a continuous variable is *not* aimed at estimating the indicator transform $i(\mathbf{u}; z_k)$ set to 1 if $z(\mathbf{u}) \leq z_k$, and to 0 if not. Indicator kriging provides a least-squares estimate of the conditional cumulative distribution function (ccdf) at cutoff z_k:

$$\begin{aligned}[i(\mathbf{u}; z_k)]^* &= E\left\{I(\mathbf{u}; z_k|(n)\right\}^* \qquad\qquad\text{(IV.27)}\\ &= \text{Prob}^*\left\{Z(\mathbf{u}) \leq z_k|(n)\right\}\end{aligned}$$

where (n) represents the conditioning information available in the neighborhood of location $\mathbf{u}$.

The IK process is repeated for a series of K cutoff values $z_k, k = 1, \ldots, K$, which discretize the interval of variability of the *continuous* attribute z. The ccdf, built from assembling the K indicator kriging estimates of type (IV.27), represents a probabilistic model for the uncertainty about the unsampled value $z(\mathbf{u})$.

If $z(\mathbf{u})$ is itself a binary categorical variable, e.g., set to 1 if a specific rock type prevails at $\mathbf{u}$, to 0 if not, then the direct kriging of $z(\mathbf{u})$ provides a model for the probability that $z(\mathbf{u})$ be one, i.e., for that rock type to prevail at location $\mathbf{u}$.

If $z(\mathbf{u})$ is a continuous variable, then the correct selection of the cutoff values z_k at which indicator kriging takes place is essential: Too many cutoff values and the inference and computation becomes needlessly tedious and expensive; too few, and the details of the distribution are lost.

Simple IK

The stationary mean of the binary indicator RF $I(\mathbf{u}; z)$ is the cumulative distribution function (cdf) of the RF $Z(\mathbf{u})$ itself; indeed:

$$
\begin{aligned}
E\{I(\mathbf{u}; z)\} &= 1 \cdot \text{Prob}\{Z(\mathbf{u}) \le z\} + 0 \cdot \text{Prob}\{Z(\mathbf{u}) > z\} \\
&= \text{Prob}\{Z(\mathbf{u}) \le z\} = F(z)
\end{aligned}
\tag{IV.28}
$$

The SK estimate of the indicator transform $i(\mathbf{u}; z)$ is thus written, according to expression (IV.3):

$$
[i(\mathbf{u}; z)]^*_{SK} = [\text{Prob}\{Z(\mathbf{u}) \le z|(n)\}]^*_{SK}
\tag{IV.29}
$$

$$
= \sum_{\alpha=1}^{n} \lambda_\alpha(\mathbf{u}; z) i(\mathbf{u}_\alpha; z) + \left[1 - \sum_{\alpha=1}^{n} \lambda_\alpha(\mathbf{u}; z)\right] F(z)
$$

where the $\lambda_\alpha(\mathbf{u}; z)$'s are the SK weights corresponding to cutoff z. These weights are given by a SK system of type (IV.4):

$$
\sum_{\beta=1}^{n} \lambda_\beta(\mathbf{u}; z) C_I(\mathbf{u}_\beta - \mathbf{u}_\alpha; z) = C_I(\mathbf{u} - \mathbf{u}_\alpha; z), \quad \alpha = 1, \ldots, n
\tag{IV.30}
$$

where $C_I(\mathbf{h}; z) = \text{Cov}\{I(\mathbf{u}; z), I(\mathbf{u} + \mathbf{h}; z)\}$ is the indicator covariance at cutoff z. If K cutoff values z_k are retained, simple IK requires K indicator covariances $C_I(\mathbf{h}; z_k)$ in addition to the K cdf values $F(z_k)$. If the z data are preferentially clustered, the sample cdf values $\hat{F}(z)$ should be declustered before being used (see program `declus` in Section VI.2.2).

Simple IK with Prior Means

There are cases when the prior means can be considered non-stationary: $F(\mathbf{u}; z)$, and can be inferred from a secondary variable. In such cases one can use the general non-stationary expression (IV.1) of simple kriging. For example, if $I(\mathbf{u}; s_k)$ represents the indicator of presence/absence of category

k at location $\mathbf{u}$, and $p(\mathbf{u}; s_k)$ is a prior probability of presence of category k at $\mathbf{u}$, the updated probability is given by a simple IK of type (IV.1):

$$[Prob\{\mathbf{u} \in k|(n)\}]^*_{IK} = p(\mathbf{u}; s_k) + \sum_{\alpha=1}^{n} \lambda_\alpha(\mathbf{u}; s_k)\,[i(\mathbf{u}_\alpha; s_k) - p(\mathbf{u}; s_k)]$$

(IV.31)

where $p(\mathbf{u}; s_k)$ is the prior indicator mean at location $\mathbf{u}$. The SK system remains as above assuming a stationary residual indicator covariance model.

Ordinary IK

Just like any simple kriging, simple IK is dependent on the stationarity decision and on the cdf values $F(z)$ interpreted as mean indicator values. When data are abundant, ordinary indicator kriging within moving data neighborhoods may be considered; this amounts to re-estimating locally the prior cdf values $F(z)$; see Section IV.1.2.

Both simple and ordinary kriging are implemented in the GSLIB program ik3d.

Median IK

In indicator kriging the K cutoff values z_k are usually chosen so that the corresponding indicator covariances $C_I(\mathbf{h}; z_k)$ are significantly different one from another. There are cases, however, when the sample indicator covariances/variograms appear proportional to each other; i.e., the sample indicator correlograms are all similar. The corresponding continuous RF model $Z(\mathbf{u})$ is the so-called "mosaic"[1] model [96] such that:

$$\rho_Z(\mathbf{h}) = \rho_I(\mathbf{h}; z_k) = \rho_I(\mathbf{h}; z_k, z_{k'}), \ \forall z_k, z_{k'}$$

(IV.32)

where $\rho_Z(\mathbf{h})$ and $\rho_I(\mathbf{h}; z_k, z_{k'})$ are the correlograms and indicator cross correlograms of the continuous RF $Z(\mathbf{u})$.

Then the single correlogram function is better estimated either directly from the sample z correlogram or from the sample indicator correlogram at the median cutoff $z_k = M$, such that $F(M) = 0.5$. Indeed, at the median cutoff, the indicator data are evenly distributed as 0 and 1 values with, by definition, no outlier values.

Indicator kriging under the model (IV.32) is called "median indicator kriging" [95]. It is a particularly simple and fast procedure since it calls for a single easy-to-infer median indicator variogram that is used for all K cutoffs. Moreover, if the indicator data configuration is the same for all cutoffs,[2] one

---

[1] The term mosaic model is unfortunate, because the realizations of this continuous RF model do not appear as pavings of subareas with constant z values (mosaic pieces).

[2] Unless inequality constraint-type data $z(\mathbf{u}_\alpha) \in (a_\alpha, b_\alpha]$ are considered, the indicator data configuration is the same for all cutoffs z_k's as long as the same data locations $\mathbf{u}_\alpha$'s are retained for all cutoffs. For more details see Section IV.1.12.

single IK system needs to be solved with the resulting weights being used for all cutoffs. For example, in the case of simple IK,

$$[i(\mathbf{u}; z_k)]^*_{SK} = \sum_{\alpha=1}^{n} \lambda_\alpha(\mathbf{u}) i(\mathbf{u}_\alpha; z_k) + \left[1 - \sum_{\alpha=1}^{n} \lambda_\alpha(\mathbf{u})\right] F(z_k) \qquad \text{(IV.33)}$$

where the $\lambda_\alpha(\mathbf{u})$'s are the SK weights common to all cutoffs z_k and are given by the single SK system:

$$\sum_{\beta=1}^{n} \lambda_\beta(\mathbf{u}) C(\mathbf{u}_\beta - \mathbf{u}_\alpha) = C(\mathbf{u} - \mathbf{u}_\alpha), \ \alpha = 1, \ldots, n \qquad \text{(IV.34)}$$

The covariance $C(\mathbf{h})$ is modeled from either the z sample covariance or, better, the sample median indicator covariance.

Note that the weights $\lambda_\alpha(\mathbf{u})$ are also the SK weights of the simple kriging estimate of $z(\mathbf{u})$ using the $z(\mathbf{u}_\alpha)$ data.

Median indicator kriging can be performed with program ik3d.

Using Inequality Data

In the IK expression (IV.29), the indicator data $i(\mathbf{u}_\alpha; z)$ originate from data $z(\mathbf{u}_\alpha)$ that are deemed perfectly known; thus the indicator data $i(\mathbf{u}_\alpha; z)$ are "hard" in the sense that they are valued either 0 or 1 and are available at any cutoff value z.

There are applications where some of the z information takes the form of inequalities such as:

$$z(\mathbf{u}_\alpha) \in (a_\alpha, b_\alpha] \qquad \text{(IV.35)}$$

or $z(\mathbf{u}_\alpha) \leq b_\alpha$ equivalent to $z(\mathbf{u}_\alpha) \in (-\infty, b_\alpha]$, or $z(\mathbf{u}_\alpha) > a_\alpha$ equivalent to $z(\mathbf{u}_\alpha) \in (a_\alpha, +\infty]$. The indicator data corresponding to the constraint interval (IV.35) are available only outside that interval:

$$i(\mathbf{u}_\alpha; z) = \begin{cases} 0 \text{ for } z \leq a_\alpha \\ \text{undefined for } z \in (a_\alpha, b_\alpha] \\ 1 \text{ for } z > b_\alpha \end{cases} \qquad \text{(IV.36)}$$

The undefined (missing) indicator data in the interval $(a_\alpha, b_\alpha]$ are ignored, and the IK algorithm applies identically; the constraint interval information (IV.35) is honored by the resulting ccdf; see hereafter and [98]. The program ik3d allows simple and ordinary indicator kriging in 2D or 3D with, possibly, constraint intervals of the type (IV.35).

Exactitude of IK

If the location $\mathbf{u}$ to be estimated coincides with a datum location $\mathbf{u}_\alpha$, whether a hard datum or a constraint interval of type (IV.35), the exactitude of kriging entails that the ccdf returned is either a zero variance cdf identifying the class to which the datum value $z(\mathbf{u}_\alpha)$ belongs or a cdf honoring the constraint interval up to the cutoff interval amplitude, in the sense that:

$$[i(\mathbf{u}_\alpha; z)]^* = \text{Prob}^*\{Z(\mathbf{u}_\alpha) \leq z|(n)\}$$

$$= \left\{ \begin{array}{ll} 0, & \text{if } z \leq a_\alpha \\ 1, & \text{if } z > b_\alpha \end{array} \right. , \text{if } z(\mathbf{u}_\alpha) \in (a_\alpha, b_\alpha]$$

This is true whether simple, ordinary, or median IK is used.

Programs ik3d **and** postik

Indicator kriging is primarily used to generate conditional probabilities within the sisim stochastic simulation program. GSLIB offers one independent IK program, ik3d, for simple and ordinary indicator kriging of categorical variables or cumulative indicators defined on continuous variables [see definition (II.6)]. IK of categorical variables can also be done by directly applying kb2d or kt3d to the categorical data.

In the case of continuous variables, the IK algorithm provides independent discrete models for the probabilities at various cutoffs; in the case of categorical variables, IK provides discrete probabilities. For these "independent" probability values to constitute a legitimate distribution, they must be corrected to verify the following order relations,[3]

- for ccdf's of continuous variables $z(\mathbf{u})$:

$$\text{Prob}\{Z(\mathbf{u}) \leq z|(n)\} = F(\mathbf{u}; z|(n)) \in [0, 1] \qquad (\text{IV.37})$$

$$\text{and } F(\mathbf{u}; z_{k'}|(n)) \geq F(\mathbf{u}; z_k|(n)), \ \forall \ z_{k'} > z_k$$

- for conditional probabilities of an exhaustive set of mutually exclusive categorical variables $I(\mathbf{u}; s_k), k = 1, \ldots, K$:

$$\text{Prob}\{I(\mathbf{u}; s_k) = 1|(n)\} = F(\mathbf{u}; k|(n)) \in [0, 1] \qquad (\text{IV.38})$$

$$\text{and } \sum_{k=1}^{K} F(\mathbf{u}; s_k|(n)) = 1$$

The indicator $i(\mathbf{u}; s_k)$ is set to 1 if category s_k prevails at location $\mathbf{u}$, to zero if not.

Regardless of the estimation algorithm used it is imperative to correct for order relations. ik3d performs these corrections and provides a detailed report of the number and magnitude of corrections at each cutoff.

---

[3] Order relation deviations are not specific to IK. Most methods that attempt to estimate a ccdf from partial information, including IK and DK, will incur such problems. The exceptions to this rule are methods that rely on a prior multivariate distribution model to determine the ccdf as a whole; for example, the MG approach assumes that all ccdf's are Gaussian, exactly determined by a mean and variance.

The result of `ik3d`, i.e., the conditional probability values, can also be smoothed into a higher resolution and smoother probability distribution; see program `histsmth` presented in Section VI.2.3.

The program `postik` (Section VI.2.12) performs within-class interpolation (see Section V.1.6) to provide any required quantile value or probabilities of exceeding any given threshold value. `postik` also returns the mean of the ccdf, called the "E-type" estimate of $z(\mathbf{u})$, and defined as:

$$
[z(\mathbf{u})]_E^* \;=\; \int_{-\infty}^{+\infty} z \, dF(\mathbf{u}; z|(n)) \tag{IV.39}
$$

$$
\approx \sum_{k=1}^{K+1} z_k' \left[F(\mathbf{u}; z_k|(n)) - F(\mathbf{u}; z_{k-1}|(n)) \right]
$$

where $z_k, k = 1, \ldots, K$ are the K cutoffs retained, and $z_0 = z_{min}$, $z_{K+1} = z_{max}$ are the minimum and maximum of the z range, to be entered as input parameters. The conditional mean value z_k' within each class, $(z_{k-1}, z_k]$, is obtained by the interpolation procedure specified as input to `postik` (see Section V.1.6).

Exactitude of the E-type Estimate

Because the ccdf returned by IK honors both hard z data and constraint intervals of the type (IV.35), the corresponding E-type estimate also honors that information. More precisely, at a datum location $\mathbf{u}_\alpha$, $[z(\mathbf{u}_\alpha)]_E^* = z(\mathbf{u}_\alpha)$, if the z datum is hard, and $[z(\mathbf{u}_\alpha)]_E^* \in (a_\alpha, b_\alpha]$, if the information at $\mathbf{u}_\alpha$ is the constraint interval $z(\mathbf{u}_\alpha) \in (a_\alpha, b_\alpha]$. In practice, the exactitude of the E-type estimate is limited by the finite discretization into K cutoff values z_k. For example, in the case of a hard z datum, the estimate is: $[z(\mathbf{u}_\alpha)]_E^* \in (z_{k-1}, z_k]$, with z_k being the upper bound of the interval containing the datum value $z(\mathbf{u}_\alpha)$. Thus the E-type estimate attached to IK provides a straightforward solution to the difficult problem of constrained interpolation. The quadratic programming solution would limit the estimate $z^*(\mathbf{u}_\alpha)$ to either bound a_α or b_α if the constraint interval is active [55, 127].

The IK solution is particularly fast if median IK is used. However, the data configuration may change if constraint intervals of type (IV.35) are considered; in such case, one may have to solve a different IK system for each cutoff.

Correcting for Order Relation Deviations

The flexibility of the IK approach comes from modeling the various probabilities with different variogram distances. As a consequence, the set of IK-derived conditional probabilities may not verify the order relations conditions (IV.37) or (IV.38). In any particular study one would expect to meet order relation deviation for up to one-half or two-thirds of the IK-derived ccdf's.

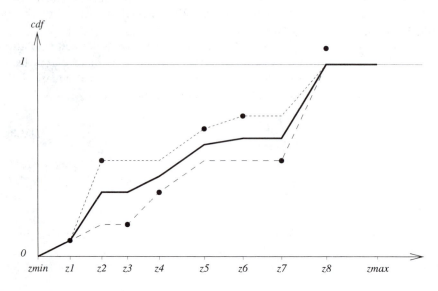

Figure IV.1: Order relation problems and their correction. The dots are the "independent" ccdf values returned by IK. The corrected ccdf is obtained by averaging the forward and downward corrections.

Fortunately, the average magnitude of the probability corrections is usually on the order of 0.01, much smaller than shown in Figure IV.1. Program `ik3d` provides statistics of the order relation problems encountered.

In the case of categorical probabilities, the first constraint (IV.38) is easily met by resetting the estimated value $F^*(\mathbf{u}; s_k|(n))$ to the nearest bound, 0 or 1, if originally valued outside the interval $[0,1]$. This resetting corresponds exactly to the solution provided by quadratic programming [127].

The second constraint (IV.38) is tougher because it involves K separate krigings. One solution consists of kriging only $(K-1)$ probabilities leaving aside one category s_{k_o}, chosen with a large enough prior probability p_{k_o}, so that:

$$F^*(\mathbf{u}; s_{k_0}|(n)) = 1 - \sum_{k \neq k_o} F^*(\mathbf{u}; s_k|(n)) \quad \in [0,1]$$

Another solution, applied after the first constraint (IV.38) has been met, is to restandardize each estimated probability $F^*(\mathbf{u}; s_k|(n)) \in [0,1]$ by the sum $\sum_k F^*(\mathbf{u}; s_k|(n)) < 1$.

Correcting for order relations of continuous variable ccdf's is more delicate, because of the ordering of the cumulative indicators.

Figure IV.1 shows an example with the following order relation problems:

$$F(\mathbf{u}; z_3|(n)) \;<\; F(\mathbf{u}; z_2|(n))$$
$$F(\mathbf{u}; z_7|(n)) \;<\; F(\mathbf{u}; z_6|(n))$$
$$F(\mathbf{u}; z_8|(n)) \;>\; 1$$

There are two sources of order relation problems:

1. Negative indicator kriging weights. One solution is to constrain the IK system to deliver only nonnegative weights, [13, 155]. One would have to forfeit, however, the sometimes beneficial properties of having a nonconvex kriging estimate [99].

2. Lack of data in some classes; see hereafter.

Practice has shown that the majority of order relation problems are due to a lack of data, more precisely, to cases when IK is attempted at a cutoff z_k which is the upper bound of a class $(z_{k-1}, z_k]$ that contains no z data. In such case the indicator data set is the same for both cutoffs z_{k-1} and z_k and yet the corresponding indicator variogram models are different; consequently, the resulting ccdf values will likely be different with a good chance for order relation problems.

There are many implementations, specific to each data set considered, that will reduce the occurrence of order relation problems. Examples include:

- Smooth the original discrete distribution of z data into a more continuous distribution by replacing each set of binary indicator data $\{i(\mathbf{u}_\alpha; z_k),\ k = 1, \ldots, K\}$ with a prior parametric kernel cdf $F(\mathbf{u}_\alpha; z)$ that varies continuously in $[0, 1]$; see [123, 165]. The IK procedure is unchanged; the indicator data $i(\mathbf{u}_\alpha; z_k)$ in expression (IV.29) are simply replaced by the kernel values $F(\mathbf{u}_\alpha; z_k)$; see Figure IV.2. The price to pay for getting a more continuous ccdf is the loss of exactitude due to the spread of the kernel function $F(\mathbf{u}_\alpha; z)$ around the hard datum value $z(\mathbf{u}_\alpha)$. The indicator kriging subroutine ik3d could be used with a preprocessor that generates such kernel cdf values.

- Define a continuum in the variability of the indicator variogram parameters with increasing (or decreasing) cutoff values [28, 47]. There should not be sudden large changes in these variogram parameters from one cutoff to the next. It is good practice to model all indicator variograms with the same nested sum of basic structures, e.g.,

$$\gamma_I(\mathbf{h}; z_k) = C_0(z_k) + C_1(z_k)Sph(|\mathbf{h}|/a(z_k)) \qquad (\text{IV.40})$$

where, e.g., $Sph(\cdot)$ is a spherical variogram model with unit sill and range $a(z_k)$. The variability of each parameter (e.g., C_0, C_1, a) is plotted versus the cutoff value z_k to ensure smoothly changing parameter values. If the decision of stationarity is reasonable, then one would

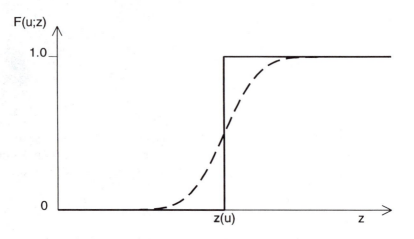

Figure IV.2: The binary step function $i(\mathbf{u}; z)$ (solid line) is replaced by the continuous kernel cdf $F(\mathbf{u}; z)$ (dashed line).

expect smoothly changing parameters. Moreover, these variogram parameter plots allow interpolation or extrapolation beyond the cutoff values z_k initially retained. An extreme case would be to retain the same indicator variogram for all cutoffs. Choosing the robust median indicator variogram would lead to the "median IK" approach presented in relation (IV.33). Median IK drastically reduces the number of order relation deviations at the expense of less flexibility: The single indicator correlogram used imparts a dependence between the probability values estimated at different cutoffs.

- Identify the cutoff values z_k to the z data values retained in the neighborhood of the location $\mathbf{u}$ being estimated. The corresponding indicator variogram parameters are then interpolated from the parameters of the available indicator variogram models; see relation (IV.40) and [28, 47].

- Retain only those prior cutoff values z_k such that the class $(z_{k-1}, z_k]$ has at least one datum; see Figure IV.3.

Such implementation procedures reduce, but do not eliminate, order relation problems. A final correction of the IK-returned ccdf values is necessary. The following correction algorithm,[4] implemented in the program ik3d, considers the average of an upward and downward correction:

1. Upward correction resulting in the upper line of Figure IV.1:

   - Start with the lowest cutoff z_1.

---

[4]This correction corresponds to a sequence of quadratic optimizations aimed at ensuring each of the inequality constraints (IV.37).

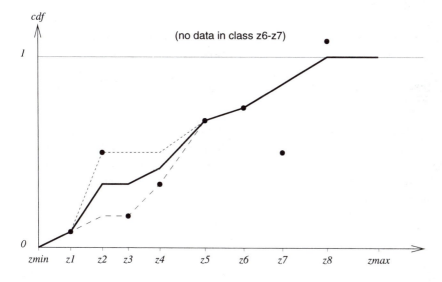

Figure IV.3: Order relation problems and their correction ignoring the class $(z_6, z_7]$ that did not contain any z data.

- If the IK-returned ccdf value $F(\mathbf{u}_\alpha; z_1|(n))$ is not within $[0, 1]$, reset it to the closest bound.

- Proceed to the next cutoff z_2. If the IK-returned ccdf value $F(\mathbf{u}_\alpha; z_2|(n))$ is not within $[F(\mathbf{u}_\alpha; z_1|(n)), 1]$, reset it to the closest bound.

- Loop through all remaining cutoffs $z_k, k = 3, \ldots, K$.

2. Downward correction resulting in the lower line of Figure IV.1:

- Start with the largest cutoff z_K.

- If the IK-returned ccdf value $F(\mathbf{u}_\alpha; z_K|(n))$ is not within $[0, 1]$, reset it to the closest bound.

- Proceed to the next lower cutoff z_{K-1}. If the IK-returned ccdf value $F(\mathbf{u}_\alpha; z_{K-1}|(n))$ is not within $[0, F(\mathbf{u}_\alpha; z_K|(n))]$, reset it to the closest bound.

- Loop downward through all remaining cutoffs $z_k, k = K - 2, \ldots, 1$.

3. Average the two sets of corrected ccdf's resulting in the thick middle line of Figure IV.1.

As mentioned above, one option is to ignore ccdf values for cutoffs that are upper bounds of classes with no z data. Consider the example of Figure IV.1 and suppose there are no z data in class $(z_6, z_7]$, yet all other classes contain at least one z datum. The ccdf value $F(\mathbf{u}_\alpha; z_7|(n))$ is ignored and the correction algorithm given above is applied to the remaining cutoff values. This correction is illustrated in Figure IV.3. Note that this latter correction is implemented within the indicator simulation programs, but not within ik3d.

Another more straightforward option to correct all order relations is to perturb the original IK-derived probability values until they constitute a licit probability distribution. The perturbation tries to mininimize departure from original quantile values while ensuring all order relations. Program histsmth presented in Section VI.2.3 can be used for this purpose by switching off the constraint of reproduction of the target mean and variance.

Order relation correction entails departure from the reproduction of the indicator covariance models as ensured from the theory of sequential indicator simulation ([102], p. 35). They may also entail departure from honoring constraint intervals if the correction is done within that interval. Order relation problems represent the most severe drawback of the indicator approach. They are the price to pay for trying to reproduce (even approximately) more than a single sample covariance.

IV.1.10   Indicator Cokriging

Indicator kriging does not make full use of the information contained in the original z data. In the case of a continuous variable, IK considers only whether the z data values exceed the cutoff value z. In the case of several categorical variables, e.g., defining the presence or absence of several rock types, IK retains only information about one rock type at a time.

One solution is to consider all K indicator data $i(\mathbf{u}_\alpha; z_k)$, $k = 1, \ldots, K$, for the estimation of each ccdf value. The corresponding estimate would be a coindicator kriging (coIK) estimate defined as:

$$[i(\mathbf{u}; z_{k_0})]^*_{coIK} = \sum_{k=1}^{K} \sum_{\alpha=1}^{n} \lambda_{\alpha,k}(\mathbf{u}; z_{k_0}) \cdot i(\mathbf{u}; z_k) \qquad \text{(IV.41)}$$

The corresponding cokriging system would call for a matrix of K^2 direct and cross indicator covariances of the type

$$C_I(\mathbf{h}; z_k, z_{k'}) = \text{Cov}\{I(\mathbf{u}; z_k), I(\mathbf{u} + \mathbf{h}; z_{k'})\}$$

The direct inference and joint modeling of the K^2 covariances [74, 185] is not practical for large K. One solution is to call for an a priori bivariate distribution model; another is to work on linear transforms of the indicator variables which are less cross correlated, such as indicator principal components; see [176, 178] and Section IV.1.11 in the original edition (1992) of this book. The discussion on indicator principal component kriging (IPCK) and related codes have been removed from this second edition because IPCK has not caught on in practice notwithstanding its theoretical elegance.

Adopting an a priori bivariate distribution model amounts to forfeiting actual data-based inference of some or all of the indicator (cross-) covariances; see relation (II.7). The most widely used bivariate distribution model is the bivariate Gaussian model after normal scores transform of the original variable: $Z(\mathbf{u}) \rightarrow Y(\mathbf{u}) = \varphi(Z(\mathbf{u}))$; see [10] and [129]. All indicator (cross-) covariances are then determined from the normal score transform Y covariance by relations like (V.20). Then one is better off calling for the complete multivariate Gaussian model; see discussions on the MG approach in Sections IV.1.8 and V.2.

A slight generalization of the bivariate Gaussian model is offered by the (bivariate) isofactorial models used in disjunctive kriging (DK) [136]. The generalization is obtained by a nonlinear rescaling of the original variable $Z(\mathbf{u})$ by transforms $\psi(\cdot)$ different from the normal score transform $\varphi(\cdot)$. Besides the problem of which function $\psi(\cdot)$ to retain, the bivariate isofactorial distribution adopted for the new transform $Y(\mathbf{u}) = \psi(Z(\mathbf{u}))$ is not fundamentally different from the bivariate Gaussian model; see relation (V.22) and the related discussion. Then one may be better off using the most congenial and parsimonious[5] of all RF models, the multivariate Gaussian model.

General experience with cokriging (see Section IV.1.7 and [73, 74, 176, 185]) indicates that cokriging improves little from kriging if primary and secondary variable(s) are equally sampled, which is generally the case with indicator data defined at various cutoff values. In addition, when working with continuous variables, the corresponding cumulative indicator data do carry substantial information from one cutoff to the next one; in which case, the loss of information associated with using IK instead of coIK is not as large as it appears [74].

IV.1.11 Probability Kriging

Indicator kriging uses only the indicator data for the threshold z_{k_0} being considered [see relation (IV.29)], indicator cokriging additionally considers all indicator data [see relation (IV.41)], a middle alternative is to consider in addition the z data themselves or better their standardized rank transforms which are distributed in $[0, 1]$; see [24, 74, 86, 96, 115, 175].

The probability kriging (PK) estimate, actually a ccdf model for $Z(\mathbf{u})$, is written in its simple kriging form:

$$[i(\mathbf{u}; z_k)]^*_{PK} - F(z_k) = \sum_{\alpha=1}^{n} \lambda_\alpha(\mathbf{u}; z_k) \cdot [i(\mathbf{u}_\alpha; z_k) - F(z_k)]$$

$$+ \sum_{\alpha=1}^{n} \nu_\alpha(\mathbf{u}; z_k) \cdot [p(\mathbf{u}_\alpha) - 0.5] \qquad (IV.42)$$

---

[5]Another extremely parsimonious RF model is that corresponding to relations (IV.32) and median indicator kriging. The median IK model, like the multivariate Gaussian model, is fully characterized by a single covariance, that of $Z(\mathbf{u})$, and the marginal cdf of $Z(\mathbf{u})$.

where $p(\mathbf{u}_\alpha) = F(z(\mathbf{u}_\alpha)) \in [0, 1]$ is the uniform (cdf) transform of the datum value $z(\mathbf{u}_\alpha)$, the expected value of which is 0.5, $F(z) = \text{Prob}\{Z(\mathbf{u}) \leq z\}$ is the stationary cdf of $Z(\mathbf{u})$. The $\lambda_\alpha(\mathbf{u}; z_k)$'s and $\nu_\alpha(\mathbf{u}; z_k)$'s are the cokriging weights of the indicator data and the uniform data, respectively. Note that these weights are both location ($\mathbf{u}$) and threshold (z_k) dependent.

The corresponding simple kriging system (PK system) calls for inference and modeling of $(2K + 1)$ (cross-) covariances, K indicator covariances, K indicator-uniform cross covariances, and the single uniform transform covariance. That modeling is still demanding and represents a drawback in practice.

No explicit PK program is offered in GLSIB, but all the components for such a program are present: essentially, program histplt, declus, and/or histsmth to generate the cdf values, then the program cokb3d to perform the cokriging. Alternatively, one could get the PK code given in [86].

IV.1.12   Soft Kriging: The Markov-Bayes Model

The major advantage of the indicator kriging approach to generating posterior conditional distributions (ccdf's) is its ability to account for soft data. As long as the soft or fuzzy data can be coded into prior local probability values, indicator kriging can be used to integrate that information into a posterior probability value [6, 74, 98, 115, 198].

The prior information can take one of the following forms:

- local hard indicator data $i(\mathbf{u}_\alpha; z)$ originating from local hard data $z(\mathbf{u}_\alpha)$:

$$i(\mathbf{u}_\alpha; z) = 1, \text{ if } z(\mathbf{u}_\alpha) \leq z, = 0 \text{ if not} \qquad (\text{IV}.43)$$

or $i(\mathbf{u}_\alpha; s_k) = 1$, if $\mathbf{u}_\alpha \in$ category s_k, $= 0$ if not

- local hard indicator data $j(\mathbf{u}_\alpha; z)$ originating from ancillary information that provides hard inequality constraints on the local value $z(\mathbf{u}_\alpha)$. If $z(\mathbf{u}_\alpha) \in (a_\alpha, b_\alpha]$, then:

$$j(\mathbf{u}_\alpha; z) = \begin{cases} 0, \text{ if } z \leq a_\alpha \\ \text{undefined (missing), if } z \in (a_\alpha, b_\alpha] \\ 1, \text{ if } z > b_\alpha \end{cases} \qquad (\text{IV}.44)$$

- local soft indicator data $y(\mathbf{u}_\alpha; z)$ originating from ancillary information providing prior (preposterior) probabilities about the value $z(\mathbf{u}_\alpha)$:

$$y(\mathbf{u}_\alpha; z) = \text{Prob}\{Z(\mathbf{u}_\alpha) \leq z | \text{ local information}\} \qquad (\text{IV}.45)$$

$\in [0, 1]$, and $\neq F(z)$: global prior as defined hereafter

- *global* prior information common to all locations $\mathbf{u}$ within the stationary area A:

$$F(z) = \text{Prob}\{Z(\mathbf{u}) \leq z\}, \forall\, \mathbf{u} \in A \qquad (\text{IV}.46)$$

At any location $\mathbf{u} \in A$, prior information about the value $z(\mathbf{u})$ is characterized by any one of the four previous types of prior information. The IK process consists of a Bayesian updating of the local prior cdf into a posterior cdf using information supplied by neighboring local prior cdf's [15, 74, 103, 196, 198]:

$$[\text{Prob}\{Z(\mathbf{u}) \leq z | (n + n')\}]^*_{IK} = \qquad (IV.47)$$

$$\lambda_0(\mathbf{u})F(z) + \sum_{\alpha=1}^{n} \lambda_\alpha(\mathbf{u}; z)i(\mathbf{u}_\alpha; z) + \sum_{\alpha'=1}^{n'} \nu_{\alpha'}(\mathbf{u}; z)y(\mathbf{u}'_{\alpha}; z)$$

The $\lambda_\alpha(\mathbf{u}; z)$'s are the weights attached to the n neighboring hard indicator data of type (IV.43) *or* (IV.44), the $\nu_{\alpha'}(\mathbf{u}; z)$'s are the weights attached to the n' neighboring soft indicator data of type (IV.45), and λ_0 is the weight attributed to the global prior cdf. To ensure unbiasedness, λ_0 is usually set to:

$$\lambda_0(\mathbf{u}) = 1 - \sum_{\alpha=1}^{n} \lambda_\alpha(\mathbf{u}; z) - \sum_{\alpha'=1}^{n'} \nu_{\alpha'}(\mathbf{u}; z)$$

If $E\{Y(\mathbf{u}; z)\}$ or $E\{J(\mathbf{u}; z)\}$ are different from $F(z)$, other specific unbiasedness conditions should be considered.

The ccdf model (IV.47) can be seen as an indicator cokriging that pools information of different types: the hard i and j indicator data and the soft y prior probabilities. When the soft information is not present or is ignored $(n' = 0)$, expression (IV.47) reverts to the simple IK expression (IV.29).

Note that if different local sources of information are available at any one single location $\mathbf{u}_\alpha$, different prior probabilities of type (IV.45), $y_1(\mathbf{u}_\alpha; z) \neq y_2(\mathbf{u}_\alpha; z)$, may be derived and all used in an extended expression of type (IV.47). Care should be taken that the corresponding cokriging covariance matrices are positive definite to ensure a unique solution.

Updating

If the spatial distribution of the soft y data is modeled by the covariance $C_I(\mathbf{h}; z)$ of the hard indicator data, i.e., if:

$$
\begin{aligned}
C_Y(\mathbf{h}; z) &= \text{Cov}\{Y(\mathbf{u}; z), Y(\mathbf{u} + \mathbf{h}; z)\} \qquad (IV.48)\\
= C_{IY}(\mathbf{h}; z) &= \text{Cov}\{I(\mathbf{u}; z), Y(\mathbf{u} + \mathbf{h}; z)\}\\
= C_I(\mathbf{h}; z) &= \text{Cov}\{I(\mathbf{u}; z), I(\mathbf{u} + \mathbf{h}; z)\}
\end{aligned}
$$

then there is no updating of the prior probability values $y(\mathbf{u}'_{\alpha}; z)$ at their locations $\mathbf{u}'_{\alpha}$, i.e.,

$$[\text{Prob}\{Z(\mathbf{u}'_{\alpha}) \leq z | (n + n')\}]^*_{IK} \equiv y(\mathbf{u}'_{\alpha}; z), \ \forall \ z$$

Most often, the soft z data originate from information related to, but different from, the hard data $z(\mathbf{u}_\alpha)$. Thus the soft y indicator spatial distribution is likely different from that of the hard i indicator data; therefore:

$$C_Y(\mathbf{h}; z) \neq C_{IY}(\mathbf{h}; z) \neq C_I(\mathbf{h}; z)$$

Then the indicator *cokriging* (IV.47) amounts to a full updating of all prior cdf's that are not already hard.

At the location of a constraint interval $j(\mathbf{u}_\alpha; z)$ of type (IV.44), indicator kriging *or* cokriging amounts to in-filling the interval $(a_\alpha, b_\alpha]$ with spatially interpolated ccdf values. Thus if simulation is performed at that location, a z attribute value would be drawn necessarily from within the interval. Were quadratic programming used to impose the same interval, the solution in the case of an active constraint would be limited to either bound a_α or b_α [55, 127].

Covariance Inference

With enough data one could infer directly and model the matrix of covariance functions (one for each cutoff z): $[C_Y(\mathbf{h}; z) \neq C_{IY}(\mathbf{h}; z) \neq C_I(\mathbf{h}; z)]$, see [74].

An alternative to this tedious exercise is provided by the Markov-Bayes[6] model [196, 198], whereby:

$$\begin{aligned}
C_{IY}(\mathbf{h}; z) &= B(z)C_I(\mathbf{h}; z), \ \forall\, \mathbf{h} & \text{(IV.49)} \\
C_Y(\mathbf{h}; z) &= B^2(z)C_I(\mathbf{h}; z), \ \forall\, \mathbf{h} > 0 \\
&= |B(z)|C_I(\mathbf{h}; z), \ \mathbf{h} = 0.
\end{aligned}$$

The coefficients $B(z)$ are obtained from calibration of the soft y data to the hard z data; more precisely:

$$B(z) = m^{(1)}(z) - m^{(0)}(z) \in [-1, +1] \qquad \text{(IV.50)}$$

with:

$$\begin{aligned}
m^{(1)}(z) &= E\{Y(\mathbf{u}; z)|I(\mathbf{u}; z) = 1\} \\
m^{(0)}(z) &= E\{Y(\mathbf{u}; z)|I(\mathbf{u}; z) = 0\}
\end{aligned}$$

Consider a calibration data set $\{y(\mathbf{u}_\alpha; z), i(\mathbf{u}_\alpha; z), \alpha = 1, \ldots, n\}$ where the soft probabilities $y(\mathbf{u}_\alpha; z)$ valued in $[0, 1]$ are compared to the actual hard values $i(\mathbf{u}_\alpha; z)$ valued 0 or 1. $m^{(1)}(z)$ is the mean of the y values corresponding to $i=1$; the best situation is when $m^{(1)}(z) = 1$, that is, when all y values exactly predict the outcome $i=1$. Similarly, $m^{(0)}(z)$ is the mean of the y values corresponding to $i=0$, best being when $m^{(0)}(z) = 0$.

The parameter $B(z)$ measures how well the soft y data separate the two actual cases $i = 1$ and $i = 0$. The best case is when $B(z) = \pm 1$, and the worst case is when $B(z) = 0$; that is, $m^{(1)}(z) = m^{(0)}(z)$.

The case $B(z) = -1$ corresponds to soft data predictably wrong and is best handled by correcting the wrong probabilities $y(\mathbf{u}_\alpha; z)$ into $1 - y(\mathbf{u}_\alpha; z)$.

---

[6]The name "Markov-Bayes" model (for the ccdf) relates to the constitutive Markov approximation (IV.51) and to the Bayesian process of updating prior cdf-type data, [15, 103]. Note that (IV.49) is not an intrinsic model of coregionalization in that $B^2(z) \neq B(z)$ in the nontrivial case $B(z) \neq 1$.

When $B(z) = 1$, the soft prior probability data $y(\mathbf{u}'_\alpha; z)$ in expression (IV.47) are treated as hard indicator data; in particular, they are not updated. Conversely, when $B(z) = 0$, the soft data $y(\mathbf{u}'_\alpha; z)$ are ignored; i.e., their weights in expression (IV.47) become zero.

The Y covariance model as derived from the last two relations (IV.49) presents a strong nugget effect; thus the Markov model implies that the y data have little redundancy with one another. This may lead to giving too much weight to clustered and mutually redundant y data. One solution is to retain only the closest y datum; this limitation presents the advantage of not calling for the soft autocovariance $C_Y(\mathbf{h}; z)$ except for $\mathbf{h} = 0$; see the collocated cokriging option in Section IV.1.7 and [9, 194].

The model (IV.49) is analytically derived [74, 198] from the Markov-type hypothesis (rather, an approximation) stating that: *"hard information $I(\mathbf{u}; z)$ screens any soft collocated information $Y(\mathbf{u}; z)$,"* in the sense that:

$$\text{Prob}\{Z(\mathbf{u}') \le z | i(\mathbf{u}; z), y(\mathbf{u}; z)\} \qquad \text{(IV.51)}$$
$$= \text{Prob}\{Z(\mathbf{u}') \le z | i(\mathbf{u}; z)\}, \; \forall \, y(\mathbf{u}; z), \; \forall \, \mathbf{u}, \mathbf{u}', z$$

The previous Markov statement represents a nontrivial hypothesis. Indeed, if the soft information $y(\mathbf{u}; z)$ relates to a support (volume of definition) larger than that of the collocated hard datum $i(\mathbf{u}; z)$, it carries information beyond that hard datum. Hence the Markov models (IV.49) should be checked from sample covariances [74]. Experience has shown that the cross-covariance model (IV.49) holds better than the autocovariance model $C_Y(\mathbf{h}; z)$. Also that autocovariance model is not needed if only one secondary data $y(\mathbf{u}; z)$ is retained, usually the value collocated with the location $\mathbf{u}$ being considered.

Indicator cokriging using the Markov-Bayes model (IV.49) is implemented within the simulation program `sisim`. Indicator kriging, using the simpler model (IV.48) corresponding to $B(z){=}1$, can be performed with the program `ik3d`. A full indicator cokriging using any model for the coregionalization of hard and soft indicator data [73, 74] can be implemented through the general cokriging program `cokb3d`.

A very efficient cpu-wise Markov-Bayes model is obtained by considering that all indicator covariances $C_I(\mathbf{h}; z)$ are proportional to each other; see relation (IV.32).

Soft Data Inference

Consider the case of a primary continuous variable $z(\mathbf{u})$ informed by a related secondary variable $v(\mathbf{u})$. The series of hard indicator data valued 0 or 1, $i(\mathbf{u}_\alpha; z_k), k = 1, \ldots, K$, are derived from each hard datum value $z(\mathbf{u}_\alpha)$.

The soft indicator data series, $y(\mathbf{u}'_\alpha, z_k) \in [0, 1], k = 1, \ldots, K$, corresponding to the secondary variable value $v(\mathbf{u}'_\alpha)$, can be obtained from a calibration scattergram of z values versus collocated v values; see Figure IV.4. The range of v values is discretized into L classes $(v_{l-1}, v_l], l = 1, \ldots, L$. For class $(v_{l-1}, v_l]$, the y prior probability cdf can be modeled from the cumulative

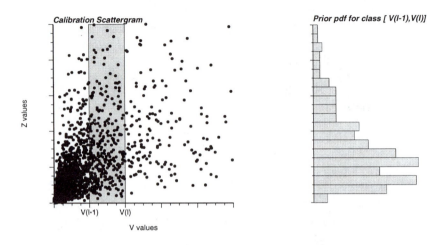

Figure IV.4: Inference of the soft prior probabilities from a calibration scattergram. The prior z probability pdf at a location $\mathbf{u}'_\alpha$ where the secondary variable is $v(\mathbf{u}'_\alpha) \in (v_{l-1}, v_l]$ is identified to the calibration conditional pdf, shown in the right of the figure.

histogram of primary data values $z(\mathbf{u}'_\alpha)$ such that the collocated secondary data values $v(\mathbf{u}'_\alpha)$ fall into class $(v_{l-1}, v_l]$:

$$y(\mathbf{u}'_\alpha; z) = \text{Prob}\{Z(\mathbf{u}'_\alpha) \leq z | v(\mathbf{u}'_\alpha) \in (v_{l-1}, v_l]\} \qquad (\text{IV.52})$$

Note that the secondary variable $v(\mathbf{u})$ need not be continuous. The classes $(v_{l-1}, v_l]$ can be replaced by categories of v values, e.g., if the information v relates to different lithofacies or mineralizations.

The calibration scattergram that provides the prior y probability values may be borrowed from a different and better sampled field. That calibration scattergram may be based on data other than those used to calibrate the covariance parameters $B(z)$ involved in relation (IV.49). Particular attention should be given to correct any bias associated to borrowing calibration data.

In the presence of sparse z-v samples, the fluctuations of the sample calibration scattergram may be smoothed using bivariate kernel functions [165] or, better, the program scatsmth presented in Section VI.2.3 and [43].

IV.1.13 Block Kriging

The linearity of the kriging algorithm allows direct estimation of *linear* averages of the attribute $z(\mathbf{u})$. For example, consider the estimation of the block average defined as:

$$z_V(\mathbf{u}) = \frac{1}{|V|} \int_{V(\mathbf{u})} z(\mathbf{u}') \, d\mathbf{u}' \simeq \frac{1}{N} \sum_{j=1}^{N} z(\mathbf{u}'_j) \qquad (\text{IV.53})$$

where $V(\mathbf{u})$ is a block of measure $|V|$ centered at $\mathbf{u}$, and the $\mathbf{u}'_j$'s are N points discretizing the volume $V(\mathbf{u})$.

One could estimate the N point values $z(\mathbf{u}'_j)$ and average the point estimates into an estimate for the block value $z_V(\mathbf{u})$. If kriging is performed with the *same* data for all N point estimates, the N point kriging systems of type (IV.4) or (IV.6) can be averaged into a single "block kriging" system (see [109], p. 306). The right-hand-side point-to-point covariance values $C(\mathbf{u} - \mathbf{u}_\alpha)$ are replaced by point-to-block covariance values defined as:

$$\overline{C}(V(\mathbf{u}), \mathbf{u}_\alpha) \simeq \frac{1}{N} \sum_{j=1}^{N} C(\mathbf{u}'_j - \mathbf{u}_\alpha) \qquad \text{(IV.54)}$$

The programs kb2d and kt3d allow both point and block kriging. The user can adjust the level of discretization N.

Warning

Block kriging is a source of common misinterpretation when applied to non-linear transforms of the original variable $z(\mathbf{u})$.

For example, the average of the log transforms $y(\mathbf{u}') = \ln z(\mathbf{u}')$ is not the log transform of the average of the $z(\mathbf{u}')$'s:

$$y_V(\mathbf{u}) = \frac{1}{|V|} \int_{V(\mathbf{u})} \ln z(\mathbf{u}') \, d\mathbf{u}' \neq \ln z_V(\mathbf{u})$$

Therefore, the antilog of a block estimate $y_V^*(\mathbf{u})$ is *not* a kriging and is a biased estimate of $z_V(\mathbf{u})$; see the related discussion in Section IV.1.8.

Similarly, when working with the indicator transform:

$$\frac{1}{|V|} \int_{V(\mathbf{u})} i(\mathbf{u}'; z) \, d\mathbf{u}' \neq i_V(\mathbf{u}; z) = \left\{ \begin{array}{ll} 1, & \text{if } z_V(\mathbf{u}) \leq z \\ 0, & \text{if not} \end{array} \right. \qquad \text{(IV.55)}$$

Therefore, the average $\frac{1}{N} \sum_{j=1}^{N} \left[i(\mathbf{u}'_j; z) \right]^*_{IK}$ as obtained from block IK should *not* be used to model the ccdf of the average $z_V(\mathbf{u})$ of the original z variable. Rather, $\frac{1}{N} \sum_{j=1}^{N} \left[i(\mathbf{u}'_j; z) \right]^*_{IK}$ is an estimate of the proportion of point values $z(\mathbf{u}'_j)$ within $V(\mathbf{u})$ that are less than the cutoff z.

Inference of the Block ccdf

Approximation of the conditional distribution of the block average $Z_V(\mathbf{u})$ could be done by correcting the variance of the point ccdf at the central location $\mathbf{u}$ of the block; several such correction algorithms are available; see [89], p. 468, [129], and [151].

Instead, we strongly recommend that the block ccdf be approached through small-scale stochastic simulations; see [67, 70, 87] and Section V.2.4. In this approach, many realizations of the small-scale distributions of point values

$z(\mathbf{u}'_j)$ within the block $V(\mathbf{u})$ are generated and the point simulated values are averaged into simulated block values. For any particular block $V(\mathbf{u})$, the proportion of simulated block values no greater than the cutoff z provides an estimate of the block indicator $I_V(\mathbf{u}; z)$ conditional expectation, i.e., an estimate of the block ccdf. The advantage of this approach is that it allows sensitivity analysis of the impact of the variogram model(s) at short scale.

IV.1.14 Cross Validation

There are so many interdependent subjective decisions in a geostatistical study that it is a good practice to validate the entire geostatistical model and kriging plan prior to any production run. The exercise of cross-validation is analogous to a dress rehearsal: It is intended to detect what could go wrong, but it does not ensure that the show will be successful.

The geostatistical model is validated by re-estimating known values under implementation conditions as close as possible to those of the forthcoming production runs. These implementation conditions include the variogram model(s), the type of kriging, and the search strategy. These re-estimation techniques are discussed in most practical statistics and geostatistics books [36, 56, 74, 89, 100, 104, 141, 181].

In cross-validation actual data are dropped one at a time and re-estimated from some of the remaining neighboring data.[7] Each datum is replaced in the data set once it has been re-estimated.

The term jackknife applies to resampling without replacement, i.e., when alternative sets of data values are re-estimated from other nonoverlapping data sets [56].

The kt3d program allows both cross-validation and the jackknife.

Analysing the Re-Estimation Scores

Consider a total of K estimation procedures or implementation variants, with the goal being to select the "best" procedure. Let $z(\mathbf{u}_j), j = 1, \ldots, N$, be the original data marked for re-estimation and let $z^{*(k)}(\mathbf{u}_j), j = 1, \ldots, N$, be their re-estimated values by procedure number k, with $k = 1, \ldots, K$.

The following criteria are suggested for analysing the re-estimation scores:

- The distribution of the N errors $\{z^{*(k)}(\mathbf{u}_j) - z(\mathbf{u}_j), j = 1, \ldots, N\}$ should be symmetric, centered on a zero mean, with a minimum spread.

- The plot of the error $z^{*(k)}(\mathbf{u}_j) - z(\mathbf{u}_j)$ versus the estimated value $z^{*(k)}(\mathbf{u}_j)$ should be centered around the zero error line, a property called "conditional unbiasedness" ([89], p. 264; [109], p. 458). Moreover, the plot should have an equal spread; i.e., the error variance should not

---

[7]The cross-validation should be as difficult as the actual estimation of unsampled values. For example, if data are aligned along drillholes or lines, then all data originating from the same hole or line should be ignored when re-estimating a datum value in order to approach the sampling density available in actual estimation.

```
                    Parameters for KB2D
                    *******************

START OF PARAMETERS:
../data/cluster.dat         \file with data
1   2   3                   \  columns for X, Y, and variable
-1.0e21   1.0e21            \  trimming limits
3                           \debugging level: 0,1,2,3
kb2d.dbg                    \file for debugging output
kb2d.out                    \file for kriged output
5    5.0   10.0             \nx,xmn,xsiz
5    5.0   10.0             \ny,ymn,ysiz
1    1                      \x and y block discretization
4    8                      \min and max data for kriging
20.0                        \maximum search radius
1    2.302                  \0=SK, 1=OK,  (mean if SK)
1    2.0                    \nst, nugget effect
1    8.0  0.0  10.0  10.0   \it, c, azm, a_max, a_min
```

Figure IV.5: An example parameter file for kb2d.

depend on whether the true value is low or high, a property known as homoscedasticity of the error variance.

- The N errors for a given procedure k should be independent one from another. This could be checked by contouring the error values: The contour map should not show any trend. Alternatively, the variogram of the error values could be calculated: It should show a pure nugget effect.

Warning

It is worth recalling that the kriging variance is but a ranking index of data configurations. Since the kriging variance does not depend on the data values, it should not be used to select a variogram model or a kriging implementation; nor should it be used as the sole criterion to determine the location of additional samples [36, 74, 99, 112].

IV.2   A Straightforward 2D Kriging Program kb2d

This is a straightforward 2D simple and ordinary kriging subroutine that can be used as is or as a basis for custom kriging programs. The code is not cluttered with extra features such as super block searching. This makes it easy to understand, but inefficient when dealing with large data sets.

The GSLIB library and kb2d program should be compiled following the instructions in Appendix B. The parameters required for the main program kb2d are listed below and shown in Figure IV.5:

- **datafl:** the input data in a simplified Geo-EAS formatted file.

- **icolx, icoly,** and **icolvr:** the columns for the x and y coordinates, and the variable to be kriged.

- **tmin** and **tmax:** all values strictly less than **tmin** and greater than or equal to **tmax** are ignored.

- **idbg:** an integer debugging level between 0 and 3. The higher the debugging level, the more output. Normally level 0 or 1 should be chosen. If there are suspected problems, or if you would like to see the actual kriging matrices, level 2 or 3 can be chosen. It is advisable to restrict the actual number of points being estimated when the debugging level is high (the debugging file can become extremely large).

- **dbgfl:** the debugging output is written to this file.

- **outfl:** the output grid is written to this file. The output file will contain both the kriging estimate and the kriging variance for all points/blocks. The output grid cycles fastest on x and then y.

- **nx, xmn, xsiz:** definition of the grid system (x axis).

- **ny, ymn, ysiz:** definition of the grid system (y axis).

- **nxdis** and **nydis:** the number of discretization points for a block. If both **nxdis** and **nydis** are set to 1, then point kriging is performed.

- **ndmin** and **ndmax:** the minimum and maximum number of data points to use for kriging a block.

- **radius:** the maximum isotropic search radius.

- **isk** and **skmean:** if **isk**=0, then simple kriging will be performed with a mean of **skmean**.

- **nst** and **c0:** the number of variogram structures and the isotropic nugget constant. The nugget constant does not count as a structure.

- For each of the **nst** nested structures one must define **it**, the type of structure; **cc**, the c parameter; **ang1**, the azimuth of the maximum range; aa_{hmax}, the maximum range; and aa_{hmin}, the minimum range. A detailed description of these parameters is given in Section II.3.

IV.3   A Flexible 3D Kriging Program kt3d

The program kt3d provides a fairly advanced 3D kriging program for points or blocks by simple kriging (SK), ordinary kriging (OK), or kriging with a polynomial trend model (KT) with up to nine monomial terms. The program works in 2D and is faster than kb2d if there are many data. One of the features that makes this program fairly fast is the super block search; see Section II.4.

```
                    Parameters for KT3D
                    ********************

START OF PARAMETERS:
../data/cluster.dat           \file with data
1    2    0    3    4          \   columns for X, Y, Z, var, sec var
-1.0e21   1.0e21              \   trimming limits
1                             \option: 0=grid, 1=cross, 2=jackknife
xvk.dat                       \file with jackknife data
1    2    0    3    0          \   columns for X,Y,Z,vr and sec var
3                             \debugging level: 0,1,2,3
kt3d.dbg                      \file for debugging output
kt3d.out                      \file for kriged output
50   0.5   1.0               \nx,xmn,xsiz
50   0.5   1.0               \ny,ymn,ysiz
1    0.5   1.0               \nz,zmn,zsiz
1    1    1                   \x,y and z block discretization
4    8                        \min, max data for kriging
0                             \max per octant (0-> not used)
20.0  20.0  20.0             \maximum search radii
 0.0   0.0   0.0             \angles for search ellipsoid
0     2.302                   \0=SK,1=OK,2=non-st SK,3=exdrift
0 0 0 0 0 0 0 0 0             \drift: x,y,z,xx,yy,zz,xy,xz,zy
0                             \0, variable; 1, estimate trend
extdrift.dat                  \gridded file with drift/mean
4                             \   column number in gridded file
1    0.2                      \nst, nugget effect
1    0.8   0.0    0.0    0.0  \it,cc,ang1,ang2,ang3
           10.0   10.0   10.0 \a_hmax, a_hmin, a_vert
```

Figure IV.6: An example parameter file for **kt3d**.

The program is set up so that a novice programmer can make changes to the form of the polynomial drift. The external drift concept has been incorporated, adding an additional unbiasedness constraint to the ordinary kriging system. When using an external drift, it is necessary to know the value of the drift variable at all data locations and all the locations that will be estimated (i.e., all grid nodes); see Section IV.1.5. The program also allows simple kriging with nonstationary (prior) means read from an input file; see expression (IV.1). The nonstationary means must be known at all data locations and all locations to be estimated.

The GSLIB library and **kt3d** program should be compiled following the instructions in Appendix B. The parameters required for the main program **kt3d** are listed below and shown in Figure IV.6:

- **datafl:** the input data in a simplified Geo-EAS formatted file.

- **icolx, icoly, icolz, icolvr,** and **icolsec:** the columns for the $x, y,$ and z coordinates, the variable to be estimated, and the external drift variable (or nonstationary mean).

- **tmin** and **tmax:** all values strictly less than **tmin** and greater than or equal to **tmax** are ignored.

- **option:** set to 0 for kriging a grid of points or blocks, to 1 for cross validation with the data in **datafl**, and to 2 for jackknifing with data in following file.

- **jackfl:** file with locations to perform estimation (jackknife option).

- **icoljx, icoljy, icoljz, icoljvr**, and **icoljsec:** the columns for the $x, y,$ and z coordinates, the variable, and the secondary variable in **jackfl**.

- **idbg:** an integer debugging level between 0 and 3. The higher the debugging level, the more output. The normal levels are 0 and 1, which summarize the results. Levels 2 and 3 provide all the kriging matrices and data used for the estimation of every point/block. It is recommended that a high debugging level not be used with a large grid.

- **dbgfl:** the debugging output is written to this file.

- **outfl:** the output grid is written to this file. The output contains the estimate and the kriging variance for every point/block on the grid, cycling fastest on x then y, and finally z. Unestimated points are flagged with a large negative number (-999.). The parameter UNEST, in the source code, can be changed if a different number is preferred.

- **nx, xmn, xsiz:** definition of the grid system (x axis).

- **ny, ymn, ysiz:** definition of the grid system (y axis).

- **nz, zmn, zsiz:** definition of the grid system (z axis).

- **nxdis, nydis**, and **nzdis:** the number of discretization points for a block. If **nxdis, nydis**, and **nzdis** are all set to 1, then point kriging is performed.

- **ndmin** and **ndmax:** the minimum and maximum number of data points to use for kriging a block.

- **noct:** the maximum number to retain from an octant (an octant search is not used if **noct=0**).

- **radius**$_{hmax}$, **radius**$_{hmin}$, and **radius**$_{vert}$: the search radii in the maximum horizontal direction, minimum horizontal direction, and vertical direction (see angles below).

- **sang1, sang2**, and **sang3:** the angle parameters that describe the orientation of the search ellipsoid. See the discussion on anisotropy specification associated with Figure II.4.

- **ikrige** and **skmean:** if **ikrige** is set to 0, then stationary simple kriging with (**skmean**) will be performed; if **ikrige** is set to 1, then ordinary kriging will be performed, if **ikrige** is set to 2, then nonstationary simple kriging with means taken from **secfile** will be performed; and if **ikrige** is set to 3, then kriging with an external drift will be performed. Note that power law variogram models (**it**=4) are not allowed with simple kriging.

- **idrif(i),i=1, ..., 9:** indicators for those drift terms to be included in the trend model. **idrif(i)** is set to 1 if the drift term number i should be included, and is set to zero if not. The nine drift terms correspond to the following:

  i = 1 linear drift in x

  i = 2 linear drift in y

  i = 3 linear drift in z

  i = 4 quadratic drift in x

  i = 5 quadratic drift in y

  i = 6 quadratic drift in z

  i = 7 cross quadratic drift in xy

  i = 8 cross quadratic drift in xz

  i = 9 cross quadratic drift in yz

- **itrend:** indicator of whether to estimate the trend (**itrend** =1) or the variable (**itrend** =0). The trend may be kriged with ordinary kriging [all **idrif(i)** values set to 0] or with any combination of trend kriging [some **idrif(i)** terms set to 1].

- **secfl:** a file for the gridded external drift variable. The external drift variable is needed at all grid locations to be estimated. The origin of the grid network, the number of nodes, and the spacing of the grid nodes should be exactly the same as the grid being kriged in kt3d. This variable is used only if **ikrige**=2 or 3.

- **iseccol:** the column number in **secfl** for the gridded secondary variable. This variable is used if **ikrige**=2 or 3.

- **nst** and **c0:** the number of variogram structures and the nugget constant. The nugget constant does not count as a structure.

- For each of the **nst** nested structures one must define **it**, the type of structure; **cc**, the c parameter; **ang1,ang2,ang3**, the angles defining the geometric anisotropy; $\mathbf{aa}_{hmax}$, the maximum horizontal range; $\mathbf{aa}_{hmin}$, the minimum horizontal range; and $\mathbf{aa}_{vert}$, the vertical range. A detailed description of these parameters is given in Section II.3.

The maximum dimensioning parameters are specified in the file $\boxed{\text{kt3d.inc}}$. All memory is allocated in common blocks declared in this file. This file is included in all functions and subroutines except for the low-level routines which are independent of the application. The advantage of this practice is that modifications can be made easily in one place without altering multiple source code files and long argument lists. One disadvantage is that a programmer must avoid all variable names declared in the include file (a global parameter could accidentally be changed). Another potential disadvantage is that the program logic can become obscure because no indication has to be given about the variables that are changed in each subroutine. Our intention is to provide adequate in-line documentation.

IV.4   Cokriging Program `cokb3d`

The construction of the cokriging matrix requires the linear model of coregionalization ([74], [89], p. 390, 175). The input variogram parameters are checked for positive definiteness. The power model is not allowed. A specific search is done for secondary data (same for all secondary) allowing the option of collocated cokriging; see Section IV.1.7.

A cokriging program for scattered points and cross validation is not provided; programs `cokb3d` and `kt3d` could be combined for this purpose.

The GSLIB library and `cokb3d` program should be compiled following the instructions in Appendix B. The parameters required for the main program `cokb3d` are listed below and shown in Figure IV.7:

- **datafl:** the input data in a simplified Geo-EAS formatted file.

- **nvar:** the number of variables (primary plus all secondary). For example, **nvar=2** if there is only one secondary variable.

- **icolx, icoly, icolz**, and **icolvr()**: the columns for the x, y, and z coordinates, the primary variable to be kriged, and all secondary variables.

- **tmin** and **tmax:** all values (for all variables) strictly less than **tmin** and greater than or equal to **tmax** are ignored.

- **coloc:** set to 1 if performing co-located cokriging with a gridded secondary variable, otherwise set to 0 $\longrightarrow$ **not implemented!**

- **secfl:** if co-located cokriging, the file with the gridded secondary variable $\longrightarrow$ **not implemented!**

- **icolsec:** if co-located cokriging, the column number for the secondary variable in **secfl** $\longrightarrow$ **not implemented!**

- **idbg:** an integer debugging level between 0 and 3. The higher the debugging level, the more output. Normally level 0 or 1 should be chosen.

- **dbgfl:** the debugging output is written to this file.

```
                    Parameters for COKB3D
                    **********************

START OF PARAMETERS:
somedata.dat                          \file with data
3                                     \   number of variables primary+other
1    2    0    3    4    5             \   columns for X,Y,Z and variables
-0.01      1.0e21                      \   trimming limits
0                                     \co-located cokriging? (0=no, 1=yes)
somedata.dat                          \   file with gridded covariate
4                                     \   column for covariate
3                                     \debugging level: 0,1,2,3
cokb3d.dbg                            \file for debugging output
cokb3d.out                           \file for output
50   0.5   1.0                        \nx,xmn,xsiz
50   0.5   1.0                        \ny,ymn,ysiz
10   0.5   1.0                        \nz,zmn,zsiz
1    1    1                           \x, y, and z block discretization
1    12     8                         \min primary,max primary,max all sec
25.0  25.0  25.0                      \maximum search radii: primary
10.0  10.0  10.0                      \maximum search radii: all secondary
 0.0   0.0   0.0                      \angles for search ellipsoid
2                                     \kriging type (0=SK, 1=OK, 2=OK-trad)
3.38  2.32   0.00   0.00              \mean(i),i=1,nvar
1    1                                \semivariogram for "i" and "j"
1    11.0                             \   nst, nugget effect
1    39.0   0.0    0.0    0.0         \   it,cc,ang1,ang2,ang3
            60.0   60.0   60.0        \   a_hmax, a_hmin, a_vert
1     2                               \semivariogram for "i" and "j"
1    0.0                              \   nst, nugget effect
1    14.5   0.0    0.0    0.0         \   it,cc,ang1,ang2,ang3
            60.0   60.0   60.0        \   a_hmax, a_hmin, a_vert
1     3                               \semivariogram for "i" and "j"
1    0.0                              \   nst, nugget effect
1    5.0    0.0    0.0    0.0         \   it,cc,ang1,ang2,ang3
            60.0   60.0   60.0        \   a_hmax, a_hmin, a_vert
2     2                               \semivariogram for "i" and "j"
1    9.0                              \   nst, nugget effect
1    15.0   0.0    0.0    0.0         \   it,cc,ang1,ang2,ang3
            60.0   60.0   60.0        \   a_hmax, a_hmin, a_vert
2     3                               \semivariogram for "i" and "j"
1    0.0                              \   nst, nugget effect
1    3.8    0.0    0.0    0.0         \   it,cc,ang1,ang2,ang3
            60.0   60.0   60.0        \   a_hmax, a_hmin, a_vert
3     3                               \semivariogram for "i" and "j"
1    1.1                              \   nst, nugget effect
1    1.8    0.0    0.0    0.0         \   it,cc,ang1,ang2,ang3
            60.0   60.0   60.0        \   a_hmax, a_hmin, a_vert
```

Figure IV.7: An example parameter file for cokb3d.

- **outfl:** the output grid is written to this file. The output file will contain both the kriging estimate and the kriging variance for all points/blocks. The output grid cycles fastest on x then y then z.

- **nx, xmn, xsiz:** definition of the grid system (x axis).

- **ny, ymn, ysiz:** definition of the grid system (y axis).

- **nz, zmn, zsiz:** definition of the grid system (z axis).

- **nxdis, nydis,** and **nzdis:** the number of discretization points for a block. If **nxdis, nydis,** and **nzdis** are set to 1 then point cokriging is performed.

- **ndmin, ndmaxp,** and **ndmaxs:** the minimum and maximum number of primary data, and the maximum number of secondary data (regardless of which secondary variable) to use for kriging a block.

- **radius$_{hmax}$, radius$_{hmin}$,** and **radius$_{vert}$** : the search radii in the maximum horizontal direction, minimum horizontal direction, and vertical direction for primary data (see angles below).

- **radiuss$_{hmax}$, radiuss$_{hmin}$,** and **radiuss$_{vert}$** : the search radii in the maximum horizontal direction, minimum horizontal direction, and vertical direction for all secondary data types (see angles below).

- **sang1, sang2,** and **sang3:** the angle parameters that describe the orientation of the search ellipsoid. See the discussion on anisotropy specification associated with Figure II.4.

- the angles defining the common orientation of the search ellipsoids for primary and secondary data.

- **ktype:** the kriging type must be specified: $0 =$ simple cokriging, $1 =$ standardized ordinary cokriging with recentered variables and a single unbiasedness constraint; see relation (IV.24), $2 =$ traditional ordinary cokriging.

- **mean():** the mean of the primary and all secondary variables are required if either simple cokriging or standardized ordinary cokriging are used. The program calculates the data residuals from these means.

The direct and cross variograms may be specified in any order; they are specified according to the variable **number**. Variable "1" is the primary (regardless of its column ordering in the input data files) and the secondary variables are numbered from "2" depending on their ordered specification in **icolvr()**. It is unnecessary to specify the j to i cross variogram if the i to j cross variogram has been specified; the cross variogram is expected to be symmetric (as from theory). For each i to j variogram the following are required:

- **nst** and **c0:** the number of variogram structures and the isotropic nugget constant. The nugget constant does not count as a structure.

- For each of the **nst** nested structures one must define **it**, the type of structure (the power model is not allowed); **cc**, the c parameter; **ang1,ang2,ang3**, the angles defining the geometric anisotropy; aa_{hmax}, the maximum horizontal range; aa_{hmin}, the minimum horizontal range; and aa_{vert}, the vertical range. A detailed description of these parameters is given in Section II.3.

The semivariogram matrix corresponding to the input of Figure IV.7 is (see [74] and [109], p. 259):

$$
\begin{bmatrix}
\gamma_Z(\mathbf{h}) & \gamma_{ZU}(\mathbf{h}) & \gamma_{ZV}(\mathbf{h}) \\
\gamma_{UZ}(\mathbf{h}) & \gamma_U(\mathbf{h}) & \gamma_{UV}(\mathbf{h}) \\
\gamma_{VZ}(\mathbf{h}) & \gamma_{VU}(\mathbf{h}) & \gamma_V(\mathbf{h})
\end{bmatrix}
=
\begin{bmatrix}
11.0 & 0.0 & 0.0 \\
0.0 & 9.0 & 0.0 \\
0.0 & 0.0 & 1.1
\end{bmatrix} \gamma_0(\mathbf{h})
$$

$$
+
\begin{bmatrix}
39.0 & 14.5 & 5.0 \\
14.5 & 15.0 & 3.8 \\
5.0 & 3.8 & 1.8
\end{bmatrix} \gamma_1(\mathbf{h})
$$

where $\gamma_0(\mathbf{h})$ is a nugget effect model and $\gamma_1(\mathbf{h})$ is a spherical structure of sill 1 and range 60. The program checks for a positive definite covariance model. The user must tell the program, by keying in a *yes*, to continue if the model is found not to be positive definite.

IV.5   Indicator Kriging Program ik3d

The program ik3d considers ordinary or simple indicator kriging of either categorical and cdf-type indicators; it performs all order relation corrections, and, optionally, allows the direct input of already transformed indicator data, including possibly soft data and constraint intervals. As opposed to program sisim with the Markov-Bayes option, soft data of type (IV.43) are not updated.

The GSLIB library and ik3d program should be compiled following the instructions in Appendix B. The parameters required for the main program ik3d are listed below and shown in Figure IV.8:

- **vartype:** the variable type (1=continuous, 0=categorical).

- **ncat:** the number of thresholds or categories.

- **cat:** the thresholds or categories (there should be **ncat** values on this line of input).

- **pdf:** the global cdf or pdf values (there should be **ncat** values on this line of input).

```
                        Parameters for IK3D
                        *******************

START OF PARAMETERS:
1                                       \1=continuous(cdf), 0=categorical(pdf)
1                                       \option: 0=grid, 1=cross, 2=jackknife
jack.dat                                \file with jackknife data
1    2    0    3                        \    columns for X,Y,Z,vr
5                                       \number thresholds/categories
0.5   1.0   2.5   5.0   10.0            \    thresholds / categories
0.12  0.29  0.50  0.74  0.88            \    global cdf / pdf
../data/cluster.dat                     \file with data
1    2    0    3                        \    columns for X,Y,Z and variable
direct.ik                               \file with soft indicator input
1    2    0    3   4   5   6            \    columns for X,Y,Z and indicators
-1.0e21    1.0e21                       \trimming limits
2                                       \debugging level: 0,1,2,3
ik3d.dbg                                \file for debugging output
ik3d.out                                \file for kriging output
10    2.5      5.0                       \nx,xmn,xsiz
10    2.5      5.0                       \ny,ymn,ysiz
1     0.0      5.0                       \nz,zmn,zsiz
1     8                                  \min, max data for kriging
20.0  20.0  20.0                         \maximum search radii
 0.0   0.0   0.0                         \angles for search ellipsoid
0                                        \max per octant (0-> not used)
1    2.5                                 \0=full IK, 1=Median IK(threshold num)
1                                        \0=SK, 1=OK
1    0.15                                \One   nst, nugget effect
1    0.85 0.0    0.0    0.0              \      it,cc,ang1,ang2,ang3
          10.0   10.0   10.0             \      a_hmax, a_hmin, a_vert
1    0.1                                 \Two   nst, nugget effect
1    0.9  0.0    0.0    0.0              \      it,cc,ang1,ang2,ang3
          10.0   10.0   10.0             \      a_hmax, a_hmin, a_vert
1    0.1                                 \Three nst, nugget effect
1    0.9  0.0    0.0    0.0              \      it,cc,ang1,ang2,ang3
          10.0   10.0   10.0             \      a_hmax, a_hmin, a_vert
1    0.1                                 \Four  nst, nugget effect
1    0.9  0.0    0.0    0.0              \      it,cc,ang1,ang2,ang3
          10.0   10.0   10.0             \      a_hmax, a_hmin, a_vert
1    0.15                                \Five  nst, nugget effect
1    0.85 0.0    0.0    0.0              \      it,cc,ang1,ang2,ang3
          10.0   10.0   10.0             \      a_hmax, a_hmin, a_vert
```

Figure IV.8: An example parameter file for ik3dm.

- **datafl:** the input continuous data in a simplified Geo-EAS formatted file.

- **icolx, icoly, icolz,** and **icolvr:** the columns for the x, y, and z coordinates, and the continuous variable to be estimated.

- **directik:** already transformed indicator values are read from this file. Missing values are identified as less than **tmin**, which would correspond to a constraint interval. Otherwise, the cdf data should steadily increase from 0 to 1 and soft categorical probabilities must be between 0 to 1 and sum to 1.0.

- **icolsx, icolsy, icolsz,** and **icoli:** the columns for the x, y, and z coordinates, and the indicator variables.

- **tmin** and **tmax:** all values strictly less than **tmin** and greater than or equal to **tmax** are ignored.

- **idbg:** an integer debugging level between 0 and 3. The higher the debugging level, the more output.

- **dbgfl:** the debugging output is written to this file.

- **outfl:** the output grid is written to this file. The output file will contain the order relation corrected conditional distributions for all nodes in the grid. The conditional probabilities or ccdf values will be written to the same line or "record"; there are **ncat** variables in the output file. Unestimated nodes are identified by all ccdf values equal to -9.9999.

- **nx, xmn, xsiz:** definition of the grid system (x axis).

- **ny, ymn, ysiz:** definition of the grid system (y axis).

- **nz, zmn, zsiz:** definition of the grid system (z axis).

- **ndmin** and **ndmax:** the minimum and maximum number of data points to use for kriging a block.

- **radius**$_{hmax}$, **radius**$_{hmin}$, and **radius**$_{vert}$: the search radii in the maximum horizontal direction, minimum horizontal direction, and vertical direction (see angles below).

- **sang1, sang2,** and **sang3:** the angle parameters that describe the orientation of the search ellipsoid. See the discussion on anisotropy specification associated with Figure II.4.

- **noct:** the maximum number of samples per octant (octant search is not used if **noct**=0).

- **mik** and **mikcut:** if **mik**=0, then a full IK is performed. If **mik**=1, then a median IK approximation is performed where the kriging weights from the category or cutoff closest to **mikcut** are used for all cutoffs.

- **ktype:** simple kriging is performed if **ktype**=0; ordinary kriging is performed if **ktype**=1.

A variogram is required for each of the **ncat** thresholds/categories:

- **nst** and **c0:** the number of variogram structures and the nugget constant, for the corresponding indicator variogram.

- for each of the **nst** nested structures one must define **it**, the type of structure (the power model is not allowed); **cc**, the c parameter; **ang1,ang2,ang3**, the angles defining the geometric anisotropy; $\mathbf{aa}_{hmax}$, the maximum horizontal range; $\mathbf{aa}_{hmin}$, the minimum horizontal range; and $\mathbf{aa}_{vert}$, the vertical range. A detailed description of these parameters is given in Section II.3.

IV.6   Application Notes

- In general, there are two different objectives when using kriging as a mapping algorithm: (1) to show a smooth picture of global trends, and (2) to provide local estimates to be further processed (e.g., selection of high grade areas, flow simulation, etc.). In the first case a large number of samples, hence, a large search neighborhood, should be used for each kriging in order to provide a smooth continuous surface. In the second case one may have to limit the number of samples within the local search neighborhood. This minimizes smoothing if the local data mean is known to fluctuate considerably and also speeds up the matrix inversion.

- It is not good practice to limit systematically the search radius to the largest variogram range. Although a datum location $\mathbf{u}_\alpha$ may be beyond the range, i.e., if $C(\mathbf{u} - \mathbf{u}_\alpha) = 0$, the datum value $z(\mathbf{u}_\alpha)$ still provides information about the unknown mean value $m(\mathbf{u})$ at the location $\mathbf{u}$ being estimated.

- When using quadrant and octant data searches, the maximum number of samples should be set larger than normal. The reason is that quadrant and octant searches cause artifacts when the data are spaced regularly and aligned with the centers of the blocks being estimated; i.e., from one point/block to the next there can be dramatic changes in the data configuration. The artifacts take the form of artificial short-scale fluctuations in the estimated trend or mean surface. This problem is less significant when more samples are used. The user can check for this problem by mapping the trend with program kt3d.

- When using KT it is good practice to contour or post the estimated trend with color or gray scale on some representative 2D sections. This may help to obtain a better understanding of the data spatial distribution, and it will also indicate potential problems with erratic estimation and extrapolation. Problems are more likely to occur with higher-order drift terms. The trend is estimated instead of the actual variable by setting the parameter **itrend**=1 in program kt3d. Recall the warning given in Section IV.1.4 against overzealous modeling of the trend. In most interpolation situations, OK with a moving data neighborhood will suffice [74, 112].

- The KT estimate (IV.10) or (IV.12) does not depend on a factor multiplying the residual covariance $C_R(\mathbf{h})$, but the resulting KT (kriging) variances do. Thus if KT is used only for its estimated values, the scaling of the residual covariance is unimportant. But if the kriging variance is to be used, e.g., as in lognormal or multi-Gaussian simulation (see Section V.2.3), then the scaling of the residual covariance is important. In such cases, even if the KT or OK estimate is retained, it is the smaller SK (kriging) variance that should be used; see [94, 152] and the discussion "SK or OK?" in Section V.2.3.

- When using OK, KT, or any kriging system with unbiasedness-type conditions on the weights, to avoid matrix instability the units of the unbiasedness constraints should be scaled to that of the covariance. This rescaling is done automatically in all GSLIB kriging programs.

  Whenever the KT matrix is singular, reverting to OK often allows a diagnostic of the source of the problem.

- There are two ways to handle an exhaustively sampled secondary variable: kriging with an external drift and cokriging. Kriging with an external drift only uses the secondary variable to inform the shape of the primary variable mean surface, whereas cokriging uses the secondary variable to its full potential (subject to the limitations of the unbiasedness constraint and the linear model of coregionalization). Checking hypothesis (IV.15) should always be attempted before applying the external drift formalism.

- If one can calibrate the non-stationary mean of the primary variable to some known function of the secondary variable, simple kriging of the corresponding (primary variable) residual could be performed. The kriged residual is then added to that mean to provide an estimate of the primary variable.

- If the data are on a regular grid, then it is strongly recommended to customize the search procedure to save CPU time; e.g., a spiral search may be most efficient.

- The kriging algorithm, particularly in its SK version, is a nonconvex interpolation algorithm. It allows for negative weights and for the estimate to be valued outside the data interval. This feature has merits whenever the variable being interpolated is very continuous, or in extrapolation cases. It may create unacceptable estimated values such as negative grades or percentages greater than 100%. One solution is to reset the estimated value to the nearest bound of the physical interval [127], another is to constrain the kriging weights to be nonnegative [13, 44], for example, by adding the modulus of the most negative weight to all weights and resetting the sum to one [155]. A more demanding solution is to consider the physical interval as a constraint interval applicable everywhere [see relation (IV.35)], and to retain the E-type estimate (IV.39) possibly using median IK to save CPU time.

- The IK cutoffs should be chosen judiciously to separate any significant mode in the sample histogram, for example, the mode originating from zero or below-detection limit values.

- Fixing the IK cutoffs for all locations $\mathbf{u}$ may be inappropriate depending on which z data values are present in the neighborhood of $\mathbf{u}$. For example, if there are no z data values in the interval $(z_{k-1}, z_k]$, separating indicator krigings at both cutoff values z_{k-1} and z_k is artificial. It would be better to take for cutoff values the z data values actually falling in the neighborhood of $\mathbf{u}$: This solution could be easily implemented [28, 42]. Note that the variogram parameters (nugget constant, sills, and ranges) would have to be modeled with smoothly changing parameters so that the variogram at any specified cutoff could be derived; see relation (IV.40).

- Using too many constraint intervals results in an underestimation of the proportions of extreme low or high z values. This bias can be reduced by using SK with prior mean values taken from the unbiased global cdf $F(z) = E\{I(\mathbf{u}); z\}$ for the means of the j values defined in (IV.44).

IV.7   Problem Set Three: Kriging

The goals of this problem set are to experiment with kriging in order to judge the relative importance of variogram parameters, data configuration, screening, support effect, and cross validation. The GSLIB programs kb2d and kt3d are called for. This problem set is quite extensive and important since, as mentioned in Section II.1.4, kriging is used extensively in stochastic simulation techniques in addition to mapping.

The data for this problem set are based on the now-familiar 2D irregularly spaced data in cluster.dat. A subset of 29 samples in data.dat is used for some of the problems, and three other small data sets, $\boxed{\text{parta.dat}}$, $\boxed{\text{partb.dat}}$, and $\boxed{\text{partc.dat}}$, are used to illuminate various aspects of kriging.

Part A: Variogram Parameters

When performing a kriging study, it is good practice to select one or two representative locations and have a close look at the kriging solution including the kriging matrix, the kriging weights, and how the estimate compares to nearby data.

The original clustered data are shown in the upper half of Figure IV.9. An enlargement of the area around the point $x = 41$ and $y = 29$ is shown in the lower half. This part of the problem set consists of estimating point $x = 41, y = 29$ using OK with different variogram parameters. The goal is to gain an intuitive understanding of the relative importance of the various variogram parameters. Note in most cases a standardized semivariogram is used such that the nugget constant C_0 plus the variance contributions of all nested structures sum to 1.0. Further, if you are unsure of your chosen variogram parameters, then use the program vmodel in conjunction with vargplt to plot the variogram in a number of directions.

A data file containing the local data needed, $\boxed{\text{parta.dat}}$, is located in the data subdirectory on the diskettes. Start with these data to answer the following questions:

1. Consider an isotropic spherical variogram with a range of 10 plus a nugget effect. Vary the relative nugget constant from 0.0 (i.e., $C_0 = 0.0$ and $C = 1.0$) to 1.0 (i.e., $C_0 = 1.0$ and $C = 0.0$) in increments of 0.2. Prepare a table showing the kriging weight applied to each datum, the kriging estimate, and the kriging variance. Comment on the results.

2. Consider an isotropic spherical variogram with a variance contribution $C = 0.8$ and nugget constant $C_0 = 0.2$. Vary the isotropic range from 1.0 to 21.0 in increments of 2.0. Prepare a table showing the kriging weight applied to each datum, the kriging estimate, and the kriging variance. Comment on the results.

3. Consider an isotropic spherical variogram with a range of 12.0. Vary the nugget constant and variance contribution values without changing their relative proportions [i.e., keep $C_0/(C_0 + C) = 0.2$]. Try $C_0 = 0.2$ ($C = 0.8$), $C_0 = 0.8$ ($C = 3.2$), and $C_0 = 20.0$ ($C = 80.0$). Comment on the results.

4. Consider alternative isotropic variogram models with a variance contribution $C = 1.0$ and nugget constant $C_0 = 0.0$. Keep the range constant at 15. Consider the spherical, exponential, and Gaussian variogram models. Prepare a table showing the kriging weight applied to each datum, the kriging estimate, and the kriging variance. Use the programs

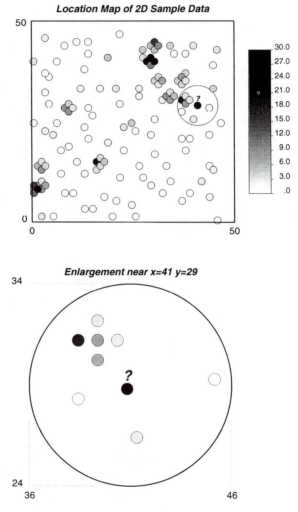

Figure IV.9: Map of the sample data shown on the original data posting and enlarged around the point being estimated.

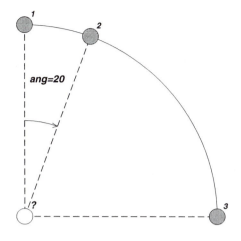

Figure IV.10: Three data points aligned on an arc of radius 1.0 from the point (?) to be estimated.

vmodel and vargplt to plot the variogram models if you are unfamiliar with their shapes. Comment on the results.

5. Consider a spherical variogram with a long range of 10.0, a variance contribution $C = 0.8$, and nugget constant $C_0 = 0.2$. Consider a geometric anisotropy with the direction of major continuity at 75 degrees clockwise from the y axis and anisotropy ratios of 0.33, 1.0, and 3.0. Prepare a table showing the kriging weight applied to each datum, the kriging estimate, and the kriging variance. Comment on the results.

Part B: Data Configuration

This part aims at developing intuition about the influence of the data configuration. Consider the situation shown in Figure IV.10 with the angle **ang** varying between $0°$ and $90°$.

A data file containing these data, $\boxed{\text{partb.dat}}$, is located in the **data** subdirectory. Start with these data, an isotropic spherical variogram with a range of 2.0, a variance contribution of $C = 0.8$, and nugget constant $C_0 = 0.2$, and use kb2d to answer the following:

1. How does the weight applied to data point 2 vary with **ang**? Vary **ang** in 5 or 10 degree increments (you may want to customize the kriging program to recompute automatically the coordinates of the second data point). Plot appropriate graphs and comment on the results.

2. Repeat question 1 with $C_0 = 0.9$ and $C = 0.1$.

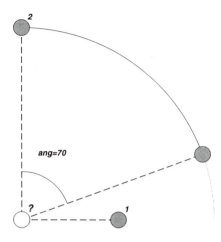

Figure IV.11: The first data point is fixed at one unit away, and the second describes a circle two units away from the point being estimated.

Part C: Screen Effect

This part will reaffirm the intuition developed in Part A about the influence of data screening. Consider the situation shown in Figure IV.11 with the angle **ang** varying between $0°$ and $90°$. The first data point is fixed at 1.0 unit away, while the second is at 2.0 units away from the point being estimated.

A data file containing these two data, $\boxed{\text{partc.dat}}$, is located in the **data** subdirectory. Start with this file, an isotropic spherical variogram with a range of 5.0, a variance contribution of $C = 1.0$, and nugget constant $C_0 = 0.0$, and use **kb2d** to answer the following:

1. How does the weight applied to point 2 vary with **ang**? Vary **ang** in 5 or 10 degree increments (you may want to customize the kriging program to recompute automatically the coordinates of the second data point). Plot appropriate graphs and comment on the results.

2. Repeat question 1 with $C_0 = 0.9$ and $C = 0.1$.

Part D: Support Effect

Consider the 29 samples of example in Part A, except this time estimate a block 2.0 miles by 2.0 miles centered at $x = 41$ and $y = 29$. See Figure IV.12 for an illustration of the area being estimated.

Start with the same files as part A for **kb2d**. Consider a spherical variogram with a range of 10.0, a variance contribution of $C = 0.9$, and nugget constant $C_0 = 0.1$.

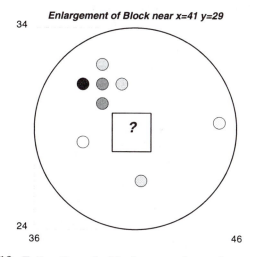

Figure IV.12: Estimation of a block centered at point $x = 41$, $y = 29$.

1. Change the number of discretization points from 1 by 1 to 4 by 4 and 10 by 10. Note the effect on the kriging weights, the estimate, and the kriging variance when estimating a block rather than a point. Comment on the results.

2. Repeat question 1 with $C_0 = 0.9$ and $C = 0.1$.

Part E: Cross Validation

Consider the data set data.dat and the crossvalidation option within program kt3d.

1. Cross validate the 29 data with kt3d. Consider the exhaustive variogram model built in problem Set 2. Perform the following:

   - Plot scatterplots of the true values versus the estimates and comment on conditional unbiasedness and homoscedasticity of error variance.

   - Create a data file that contains the error (estimate minus true), the true value, the estimate, and the estimation variance.

   - Plot a histogram of the errors and comment.

   - Plot a map of the 29 errors (contour or gray scale) and comment.

   - Plot a scatterplot of the error absolute value versus the corresponding kriging variance and comment.

2. Justify your choice of search radius, minimum and maximum number of data, and other kriging parameters. Back up your justification with additional crossvalidation runs if necessary.

3. Try simple kriging and the declustered mean you established in problem Set 1. Compare the results to ordinary kriging and comment.

4. Try linear drift terms in both the x and y directions. Compare the results to ordinary kriging and comment.

Part F: Kriging on a Grid

Consider the complete 50 by 50 square mile area informed by $\boxed{\text{data.dat}}$. For this part use the program kt3d and answer the following:

1. Plot a gray-scale map of the true values in $\boxed{\text{true.dat}}$ using pixelplt.

2. Perform kriging on a one mile by one mile point grid by simple kriging (with your declustered mean from problem Set 1). Note the mean and variance of the kriging estimates, plot a gray-scale map, and comment on the smoothing effect.

3. Use program trans (described in Section VI.2.7) to correct for the smoothing effect of kriging. The target distribution is the declustered histogram of the 140 sample values. Plot the histogram and the gray-scale map of the corrected estimates and compare the reference figures. Note preservation of the data exactitude.

4. Study the effect of the maximum number of data by considering ndmax at 4, 8, and 24. Note the variance of the kriging estimates and comment on the look of the gray-scale map.

5. Study the effect of the search radius by setting the isotropic radius at 5, 10, and 20. Keep ndmax constant at 24. Again, note the mean and variance of the kriging estimates and the look of the gray scale map.

6. Repeat question 2 with ordinary kriging. Comment on the results. Compare the mean of the kriging estimates to the declustered mean established in problem Set 1. How could you determine the declustering weight that should be applied to each datum from such kriging?

7. Repeat question 2 using kriging with a trend model (linear drift terms in both the x and y directions). Comment on the results.

8. Repeat question 2 with ordinary block kriging (grid cell is discretized 4 by 4). Comment on the results.

9. Estimate the mean/trend surface (by resetting the **itrend** flag to 1) with the KT parameter file used earlier. Note that you will have an estimate of the point trend at the node locations. Comment on the results. Compare these results to those obtained from block kriging.

IV.8 Problem Set Four: Cokriging

The goal of this problem set is to experiment with cokriging and kriging with an external drift for mapping purposes. The GSLIB programs gamv, vmodel, vargplt are needed to compute and model the matrix of covariances; programs kt3d and cokb3d are required.

The 2D irregularly spaced data in data.dat are considered as the sample data set for the primary variable. An exhaustive secondary variable data set is given in ydata.dat in the **data** subdirectory.

The 2D estimation exercise consists of estimating the primary variable, denoted $z(\mathbf{u})$, at all 2500 nodes using:

1. A relatively sparse sample of 29 primary data in file data.dat and duplicated in ydata.dat . These 29 data are preferentially located in high-z concentration zones.

2. In addition, a much larger sample of 2500 secondary data, denoted $y(\mathbf{u})$, in file ydata.dat .

Since the secondary information is available at all nodes to be estimated and at all nodes where primary information is available, it can be interpreted as an external drift for the primary variable $z(\mathbf{u})$; see Section IV.1.5.

All the estimates can be checked versus the actual z values in true.dat .

Part A: Cokriging

1. Calculate the 3 sample covariances and cross covariances, i.e., $C_Z(\mathbf{h})$, $C_Y(\mathbf{h})$, $C_{ZY}(\mathbf{h})$, and model them with a legitimate model of coregionalization; see [89], p. 390.

2. Use program cokb3d to perform cokriging of the primary variable at all 2500 nodes of the reference grid. Document which type of cokriging and unbiasedness conditions you have retained; you may want to consider all 3 types of condition coded in cokb3d; see Section IV.1.7.

3. Consider also using collocated cokriging with the Markov-type model (IV.25). Note that the secondary data is available at all 2500 nodes.

4. **Validation:** Compare the cokriging estimates to the reference true values given in true.dat . Plot the histogram of the errors (estimate minus true) and the scattergram of the errors versus the true values. Comment on the conditional unbiasedness and homoscedasticity of the error distribution. Plot a gray (or color) scale map of the cokriging estimates and error values. Compare to the corresponding map of true values and direct kriging estimates. Comment on any improvement brought by cokriging.

Part B: Kriging with an External Drift

1. Use program `kt3d` with the external drift option[8] to estimate the primary variable at all 2500 nodes of the reference grid.

2. **Validation:** Compare the resulting estimates to the previous cokriging estimates and to the reference true values. Use any or all of the previously proposed comparative graphs and plots. Comment on the respective performance of kriging, cokriging, and kriging with an external drift.

IV.9 Problem Set Five: Indicator Kriging

The goal of this problem set is to provide an introduction to indicator kriging as used to build probabilistic models of uncertainty (ccdf's) at unsampled locations. From these models, optimal estimates (possibly non least squares) can be derived together with various nonparametric probability intervals, ([102], Lesson 4).

The GSLIB programs `gamv`, `ik3d`, and `postik` are called for. The 140 irregularly spaced data in $\boxed{\text{cluster.dat}}$ are considered as the sample data. The reference data set is in $\boxed{\text{true.dat}}$.

The following are typical steps in an IK study. A detailed step-by-step description of the practice of IK is contained in [100].

1. **Determination of a representative histogram:** Retain the declustered (weighted) cdf based on all 140 z data; see problem Set 1. Determine the nine decile cutoffs $z_k, k = 1, \ldots, 9$ from the declustered cdf $F(z)$.

2. **Indicator variogram modeling:** Use `gamv` to compute the corresponding nine indicator variograms (once again the 140 clustered data in $\boxed{\text{cluster.dat}}$ are used). Standardize these variograms by their respective (nondeclustered) data variances and model them. Use the same nested structures at all cutoffs with the same total unit sill. The relative sills and range parameters may vary; see the example of relation (IV.40). The graph of these parameters versus the cutoff value z_k should be reasonably smooth.

3. **Indicator kriging:** Using `ik3d`, perform an indicator kriging at each of the 2500 nodes of the reference grid. The program returns the $K = 9$ ccdf values at each node with order relation problems corrected.

---

[8]This example is a favorable case for using the external drift concept in that the secondary y data were generated as spatial averages of z values (although taken at a different level from an original 3D data set). Compare Figures A.36 and A.37.

Program postik provides the E-type estimate, i.e., the mean of the calculated ccdf, any conditional p quantile, and the probability of (not) exceeding any given threshold value.

4. **E-type estimate:** Retrieve from postik the 2500 E-type estimates. Plot the corresponding gray scale map and histogram of errors ($z^* - z$). Compare to the corresponding figures obtained through ordinary kriging and cokriging (problem sets three and four).

5. **Quantile and probability maps:** Retrieve from postik the 2500 conditional 0.1 quantile and probability of exceeding the high threshold value $z_9 = 5.39$. The corresponding isopleth maps indicate zones that are most certainly high z valued. Compare the M-type estimate map (0.5 quantile) to the E-type estimate and comment.

6. **Median indicator kriging:** Assume a median IK model of type (IV.32); i.e., model all experimental indicator correlograms at all cutoffs by the same model better inferred at the median cutoff. Compute the E-type estimate and probability maps and compare.

7. **Soft structural information:** Beyond the 140 sample data, soft structural information indicates that high-concentration z values tend to be aligned NW-SE, whereas median-to-low-concentration values are more isotropic in their spatial distribution. Modify correspondingly the various indicator variogram models to account for that structural information, and repeat the runs of questions 3 to 5. Comment on the flexibility provided by multiple IK with multiple indicator variograms.

Chapter V

Simulation

The collection of stochastic simulation programs included with GSLIB is presented in this chapter. Section V.1 presents the principles of stochastic simulation including the important concept of sequential simulation.

Section V.2 presents the Gaussian-related algorithms which are the algorithms of choice for most continuous variables. Sections V.3 and V.4 present the indicator simulation algorithms that are better suited to categorical variables, cases where the Gaussian RF model is inappropriate, or where accounting for additional soft/fuzzy data is important. Sections V.5 and V.6 present object-based and simulated annealing algorithms, respectively.

Sections V.7 and V.8 give detailed descriptions of the Gaussian and indicator simulation programs in GSLIB. Different programs for many different situations are available for each of these major approaches. Elementary programs for Boolean simulation and simulations based on simulated annealing are presented in Sections V.9 and V.10.

Some useful application notes are presented in Section V.11. Finally, a problem set is proposed in Section V.12, which allows the simulation programs to be tested and understood.

V.1 Principles of Stochastic Simulation

Consider the distribution over a field A of one (or more) attribute(s) $z(\mathbf{u}), \mathbf{u} \in A$. Stochastic simulation is the process of building alternative, equally probable, high-resolution models of the spatial distribution of $z(\mathbf{u})$; each realization is denoted with the superscript l: $\{z^{(l)}(\mathbf{u}), \mathbf{u} \in A\}$. The simulation is said to be "conditional" if the resulting realizations honor data values at their locations:

$$z^{(l)}(\mathbf{u}_\alpha) = z(\mathbf{u}_\alpha), \ \forall \, l \qquad (V.1)$$

The variable $z(\mathbf{u})$ can be categorical, e.g., indicating the presence or absence of a particular rock type or vegetation cover, or it can be continuous such as porosity in a reservoir or concentrations over a contaminated site.

Simulation differs from kriging or any interpolation algorithm, in two major aspects:

1. In most interpolation algorithms, including kriging, the goal is to provide a "best," hence unique, local estimate of the variable or any of its trend components without specific regard to the resulting spatial statistics of the estimates taken together. In simulation, reproduction of global features (texture) and statistics (histogram, covariance) take precedence over local accuracy. Kriging provides a set of local representations, say $z^*(\mathbf{u}), \mathbf{u} \in A$, where local accuracy prevails. Simulation provides alternative global representations, $z^{(l)}(\mathbf{u}), \mathbf{u} \in A$, where reproduction of patterns of spatial continuity prevails; see Figure V.1.

2. Except if a Gaussian model for errors is assumed, kriging provides only an incomplete measure of local accuracy (see discussion in Section II.1.4), and no appreciation of joint accuracy when several locations are considered together. Simulations are designed specifically to provide such measures of accuracy, both local and involving several locations. These measures are given by the differences between L alternative simulated values at any location (local accuracy) or the L alternative simulated fields (global or joint accuracy).

Different simulation algorithms impart different global statistics and spatial features on each realization. For example, simulated categorical values can be made to honor specific geometrical patterns as in object-based simulations (see [48, 149] and Section V.4) or the covariance of simulated continuous values can be made to honor a prior covariance model as in Gaussian-related simulations; see Section V.2. A hybrid approach could be considered to generate simulated numerical models that reflect widely different types of features. For example, one may start with an object-based process or categorical indicator simulation to generate the geometric architecture of the various lithofacies, following with a Gaussian algorithm to simulate the distribution of continuous petrophysical properties within each separate lithofacies, then a simulated annealing process could be used to modify locally the petrophysical properties to match, say, well test data [42].

No single simulation algorithm is flexible enough to allow the reproduction of the wide variety of features and statistics encountered in practice. It is the responsibility of each user to select the set of appropriate algorithms and customize them if necessary. At present, no single paper or book provides an exhaustive list or classification of available simulation algorithms. See the following publications and their reference lists [16, 18, 27, 46, 54, 78, 106, 120, 125, 149, 157, 179].

The GSLIB software offers code for the simulation algorithms frequently used in mining, hydrogeology, and petroleum applications, and for which the authors have a working experience. Notable omisions are:

- Frequency-domain simulations [18, 29, 76, 128]. These algorithms can

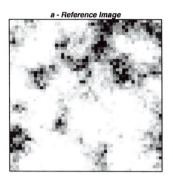
a - Reference Image

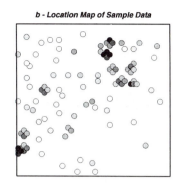
b - Location Map of Sample Data

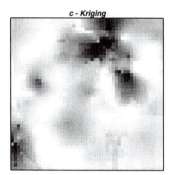
c - Kriging

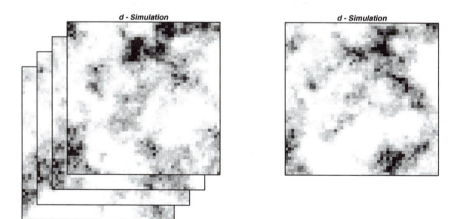
d - Simulation   d - Simulation

Figure V.1: Kriging versus stochastic simulation: The reference image (a) has been sparsely sampled (b). The kriging estimates based on the sparse sample are given in a gray-scale map in (c); note the smoothing effect more pronounced in areas of sparser sampling. The same sample was used to generate multiple stochastic images [shown in (d)]; note the difference between the various stochastic images and the better reproduction of the reference spatial variability.

be seen as duals of Gaussian algorithms, where the covariance is reproduced through sampling of its discretized Fourier transform (spectrum).

- Random fractal simulation [29, 57, 80, 81, 184], which can be seen as particular implementations of frequency-domain (spectral) simulations.

- Simulation of marked point processes [20, 150, 157, 174]. These processes can be seen as generalizations of the Boolean process discussed in Section V.5.

- Deterministic simulation of physical processes, e.g., of sedimentation, erosion, transport, redeposition, etc. [21, 119, 179]. Although such simulations can result in alternative images, their emphasis is on reproducing the genetic process not on the reproduction of local data. Conversely, random-function-based stochastic simulations tend to bypass the actual genetic process and rely on global statistics that are deemed representative of the *end* phenomenon actually observed.

The following five subsections provide a brief discussion of paradigms and implementation procedures that are common to many stochastic simulation algorithms.

V.1.1   Reproduction of Major Heterogeneities

Major large-scale heterogeneities are usually the most consequential features of the numerical models generated. For example, facies boundaries are often critical for flow performance in a reservoir or for slope stability in an open pit mine. Similarly, boundaries between different types of vegetation may condition the spatial distribution of a particular insect. Those heterogeneities should be given priority in the simulation process. A continuous RF model $Z(\mathbf{u})$ cannot reproduce severe heterogeneities or discontinuities, as found when crossing a physical boundary such as that of a lithotype.

If the phenomenon under study is a mixture of different physical and/or statistical populations, the geometry of the mixture should be modeled and simulated first; then the attribute(s) within each homogeneous population can be simulated.[1] For example, the spatial architecture of major lithofacies in an oil reservoir, or mineralizations in a mining deposit, should be represented first, e.g., through prior rock type/facies zoning or the simulation of categorical-type variables. Then the distribution of petrophysical properties or mineral grades within each facies or mineralization can be simulated [7, 14, 33, 48, 110, 177, 197]. A two-step approach to simulation is not only more consistent with the underlying physical phenomenon (e.g., geology); it

---

[1]There are cases when that sequence should be reversed, e.g., when one population is of quasizero measure (volume), as in the case of fracture lines in 2D, or planes in 3D. Such fractures and the corresponding displacements can be simulated afterwards and superimposed on the prior continuous simulation.

also avoids stretching the stationarity/homogeneity requirement[2] underlying most continuous RF models.

Similarly, reproduction of patterns of spatial continuity of extreme values should be given priority, especially in situations where connected patterns of extreme values condition the results obtained from processing the simulated numerical models [105, 106]. For example, the reproduction of the geometry of fractures intersecting multiple facies may be more important than reproduction of the facies geometry.

V.1.2 Joint Simulation of Several Variables

In many applications reproduction of the spatial dependence of several variables is critical. For example, a realistic model for a porphyry copper-gold deposit must account for the spatial cross-correlation of the copper and gold grades. An oil reservoir model typically calls for joint knowledge of porosity, permeability, and fluid saturations. As another example, the concentration of competing insect species should be simulated jointly.

Most random-function-based simulation algorithms can be generalized, at least in theory, to joint simulation of several variables by considering a vectorial random function $\mathbf{Z}(\mathbf{u}) = \{Z_1(\mathbf{u}), Z_2(\mathbf{u}), \ldots, Z_K(\mathbf{u})\}$; see [25, 70, 71, 74, 125, 144, 183]. The problem resides in the inference and modeling of the cross-covariance matrix

$$\Sigma = [C_{k,k'}(\mathbf{h}) = Cov\{Z_k(\mathbf{u}), Z_{k'}(\mathbf{u}+\mathbf{h})\}, k, k' = 1, \ldots, K]$$

Because of implementation problems, joint simulation of several variables is rarely implemented to the full extent allowed by theory [73]. Rather, various approximations are considered; the two most common are:

1. Replace the simulation of the K dependent variables $Z_k(\mathbf{u})$ by that of K *independent* factors $Y_k(\mathbf{u})$ from which the original Z variables can be reconstituted:

$$\text{if } \mathbf{Y} = \varphi(\mathbf{Z}), \text{ then } \mathbf{z}^{(l)} = \varphi^{-1}(\mathbf{y}^{(l)}) \qquad (\text{V.2})$$

with $\mathbf{y}^{(l)} = [y_1^{(l)}(\mathbf{u}), \ldots, y_K^{(l)}(\mathbf{u}), \mathbf{u} \in A]$ being the set of simulated independent factors $y_k^{(l)}(\mathbf{u})$.

$\mathbf{z}^{(l)} = [z_1^{(l)}(\mathbf{u}), \ldots, z_K^{(l)}(\mathbf{u}), \mathbf{u} \in A]$ is the set of resulting simulated values $z_k^{(l)}(\mathbf{u})$ whose interdependence is guaranteed by the common inverse transform φ^{-1}.

Except for specific cases where physics-based non-linear factors $\mathbf{y}$ are evident and particularly cases involving constraints on the sums of variables (as in chemical analyses), the vast majority of applications consider only linear factors based on an orthogonal decomposition of some

---

[2]The degree of spatial correlation of the attribute $z(\mathbf{u})$ is usually quite different within a facies than across facies, unless there has been important diagenetic remobilization.

covariance matrix Σ of the original K variables $Z_k(\mathbf{u})$; see [74, 178]. The decomposition of the covariance matrix,

$$\Sigma(\mathbf{h}_o) = [Cov\{Z_k(\mathbf{u}), Z_{k'}(\mathbf{u} + \mathbf{h}_o)\}, k, k' = 1, \ldots, K]$$

corresponds to a single separation vector $\mathbf{h}_o$. The Z cross-correlations are reproduced only for that specific separation vector. The zero distance, $\mathbf{h}_o = 0$, is often chosen for convenience, the argument being that cross-correlation is maximum at the zero distance $\mathbf{h}_o = 0$.

2. The second approximation is much simpler. The most important or better autocorrelated variable, hereafter called the primary variable $Z_1(\mathbf{u})$, is simulated first; then all other covariates $Z_k(\mathbf{u}), k \neq 1$ are simulated by drawing from specific conditional distributions. For example, typically in reservoir modeling [8]:

   - The porosity $\Phi(\mathbf{u}) = Z_1(\mathbf{u})$ within a given lithofacies is simulated first, since its spatial variability is reasonably smooth and easy to infer.

   - The horizontal permeability $K_h(\mathbf{u}) = Z_2(\mathbf{u})$ is then simulated from the conditional distribution

   $$\text{Prob}\{K_h(\mathbf{u}) \leq z_2 | \Phi(\mathbf{u}) = z_1^{(l)}\} \qquad (\text{V}.3)$$

   of $K_h(\mathbf{u})$ given the simulated porosity value $z_1^{(l)}$ at the same location. Such conditional distributions are directly inferred from the sample scattergram of K_h versus Φ corresponding to the facies prevailing at $\mathbf{u}$; see Figure IV.4.

   - Next, given the simulated value for horizontal permeability $K_h(\mathbf{u}) = z_2^{(l)}(\mathbf{u})$, the simulated value for vertical permeability is drawn from the sample scattergram of K_h versus K_v, again specific to the facies prevailing at $\mathbf{u}$.

The cross-correlation between K_v and Φ is thus indirectly approximated. Similarly, the spatial autocorrelations of the secondary variables $K_h(\mathbf{u})$ and $K_v(\mathbf{u})$ are indirectly reproduced through that of the primary variable $\Phi(\mathbf{u})$. Therefore, this approximation to joint simulation is not recommended when it is important to reproduce accurately cross-covariance(s) beyond lag $\mathbf{h} = 0$ and covariance(s) of secondary variable(s) that differ markedly from that of the primary variable.

A variant of this approach builds on the concept of collocated cokriging with a Markov-type model; see relation (IV.26) in Section IV.1.7. The cosimulation is done in the space of normal scores transforms (for all variables), then the results are backtransformed; see Sections V.2.1 to V.2.3 and [9, 74, 194].

V.1.3 The Sequential Simulation Approach

Approximation (V.3) allows drawing the value of a variable $Z(\mathbf{u})$ from its conditional distribution given the value of the most related covariate at the same location $\mathbf{u}$. The sequential simulation principle is a generalization of that idea: the conditioning is extended to include all data available within a neighborhood of $\mathbf{u}$, including the original data and all previously simulated values [87, 90, 102, 157, 159].

Consider the joint distribution of N random variables Z_i with N possibly very large. The N RVs Z_i may represent the same attribute at the N nodes of a dense grid discretizing the field A, or they can represent N different attributes measured at the same location, or they could represent a combination of K different attributes defined at the N' nodes of a grid with $N = K \cdot N'$.

Next consider the conditioning of these N RV's by a set of n data of *any* type symbolized by the notation $|(n)$. The corresponding N variate ccdf is denoted:

$$F_{(N)}(z_1, \ldots, z_N|(n)) = \text{Prob}\{Z_i \leq z_i, i = 1, \ldots, N|(n)\} \qquad (\text{V.4})$$

Expression (V.4) is completely general with no limitation; some or all of the variables Z_i could be categorical.

Successive application of the conditional probability relation shows that drawing an N variate sample from the ccdf (V.4) can be done in N successive steps, each involving a univariate ccdf with increasing level of conditioning:

- Draw a value $z_1^{(l)}$ from the univariate ccdf of Z_1 given the original data (n). The value $z_1^{(l)}$ is now considered as a conditioning datum for all subsequent drawings; thus, the information set (n) is updated to $(n+1) = (n) \cup \{Z_1 = z_1^{(l)}\}$.

- Draw a value $z_2^{(l)}$ from the univariate ccdf of Z_2 given the updated data set $(n+1)$, then update the information set to $(n+2) = (n+1) \cup \{Z_2 = z_2^{(l)}\}$.

- Sequentially consider all N RVs Z_i.

The set $\{z_i^{(l)}, i = 1, \ldots, N\}$ represents a simulated joint realization of the N dependent RVs Z_i. If another realization is needed, $\{z_i^{(l')}, i = 1, \ldots, N\}$, the entire sequential drawing process is repeated.

This sequential simulation procedure requires the determination of N univariate ccdf's, more precisely:

$$\text{Prob}\{Z_1 \leq z_1|(n)\} \qquad (\text{V.5})$$
$$\text{Prob}\{Z_2 \leq z_2|(n+1)\}$$
$$\text{Prob}\{Z_3 \leq z_3|(n+2)\}$$

$$\vdots$$

$$\mathrm{Prob}\{Z_N \leq z_N|(n+N-1)\}$$

The sequential simulation principle is independent of the algorithm or model used to establish the sequence (V.5) of univariate ccdf's. In the program sgsim, all ccdf's (V.5) are assumed Gaussian and their means and variances are given by a series of N simple kriging systems; see Sections IV.1.1 and V.2. In the program sisim, the ccdf's are obtained directly by indicator kriging; see Sections IV.1.9 and V.3.

An Important Theoretical Note

In most applications the N RV's whose N-variate ccdf (V.4) is to be reproduced correspond to N locations $\mathbf{u}_i$ of a RF $Z(\mathbf{u})$ related to a single attribute z. The sequential simulation algorithm requires knowledge of the N univariate ccdf's (V.5). Except if $Z(\mathbf{u})$ is modeled as a Gaussian RF, these univariate ccdf's are not Gaussian nor are they fully determined by their sole means and variances. Fortunately, if one seeks only reproduction of the z-covariance (or variogram) by the sets of simulated values $\{z^{(l)}(\mathbf{u}_i), i = 1, \ldots, N\}, \forall l$, it suffices that each of the N univariate ccdf's in the sequence (V.5) identify the simple kriging mean and variance of the corresponding RV $Z(\mathbf{u}_i)$ given the $(n+i-1)$ data considered, see [74, 103, 115]. The ccdf type of any of these N univariate ccdf's can be anything, for example it can be a 3-parameter distribution, the third parameter being used to impart to the simulated values $z^{(l)}(\mathbf{u}_i)$ properties beyond the Z-covariance model.

Implementation Considerations

- Strict application of the sequential simulation principle calls for the determination of more and more complex ccdf's, in the sense that the size of the conditioning data set increases from (n) to $(n+N-1)$. In practice, the argument is that the closer[3] data screen the influence of more remote data; therefore, only the closest data are retained to condition any of the N ccdf's (V.5). Since the number of previously simulated values may become overwhelming as i progresses from 1 to $N \gg n$, one may want to give special attention to the original data (n) even if they are more remote.

- The neighborhood limitation of the conditioning data entails that statistical properties of the $(N+n)$ set of RVs will be reproduced only up to the maximum sample distance found in the neighborhood. For

---

[3] "Closer" is not necessarily taken in terms of Euclidean distance, particularly if the original data set (n) and the N RVs include different attribute values. The data "closest" to each particular RV Z_i being simulated are those that have the most influence on its ccdf. This limitation of the conditioning data to the closest is also the basis of the Gibbs sampler as used in [64].

example, the search must extend at least as far as the distance to which the variogram is to be reproduced; this requires extensive conditioning as the sequence progresses from 1 to N. One solution is provided by the multiple-grid concept which is to simulate the N nodal values in two or more steps [70, 71, 180]:

- First, a coarse grid, say each tenth node is simulated using a large data neighborhood. The large neighborhood allows the reproduction of large-scale variogram structures.

- Second, the remaining nodes are simulated with a smaller neighborhood.

This multiple grid concept is implemented in programs sgsim and sisim.

- Theory does not specify the sequence in which the N nodes should be simulated. Practice has shown, however, that it is better to consider a random sequence [87]. Indeed, if the N nodes are visited, e.g., row-wise, any departure from rigorous theory[4] may entail a corresponding spread of artifacts along rows.

V.1.4 Error Simulation

The smoothing effect of kriging, and more generally, of any low-pass-type interpolation, is due to a missing error component. Consider the RF $Z(\mathbf{u})$ as the sum of the estimator and the corresponding error:

$$Z(\mathbf{u}) = Z^*(\mathbf{u}) + R(\mathbf{u}) \tag{V.6}$$

Kriging, for example, would provide the smoothly varying estimator $Z^*(\mathbf{u})$. To restore the full variance of the RF model, one may think of simulating a realization of the random function error with zero mean and the correct variance and covariance. The simulated z value would be the sum of the unique estimated value and a simulated error value:

$$z_c^{(l)}(\mathbf{u}) = z^*(\mathbf{u}) + r^{(l)}(\mathbf{u}) \tag{V.7}$$

The idea is simple, but requires two rather stringent conditions:

1. The error component $R(\mathbf{u})$ must be independent or, at least, orthogonal to the estimator $Z^*(\mathbf{u})$. Indeed, the error value $r^{(l)}(\mathbf{u})$ is simulated independently, and is then added to the estimated value $z^*(\mathbf{u})$.

2. The RF $R(\mathbf{u})$ modeling the error must have the same spatial distribution or, at least, the same covariance as the actual error. Then the simulated value in (V.7) will have the same covariance, hence the same variance, as the true values $z(\mathbf{u})$.

---

[4]The limited data neighborhood concept coupled with ordinary kriging (as opposed to simple kriging) privileges the influence of close data.

Condition (1) is met if the estimator $Z^*(\mathbf{u})$ is obtained by orthogonal projection of $Z(\mathbf{u})$ onto some (Hilbert) space of the data, in which case the error vector $Z^*(\mathbf{u}) - Z(\mathbf{u})$ is orthogonal to the estimator $Z^*(\mathbf{u})$ [109, 124]. Condition (1) is met if $Z^*(\mathbf{u})$ is a *simple* kriging estimator of $Z(\mathbf{u})$.

Condition (2) is difficult to satisfy since the error covariance is unknown; worse, it is not stationary even if the RF model $Z(\mathbf{u})$ is stationary since the data configuration used from one location $\mathbf{u}$ to another is usually different. The solution is to pick (simulate) the error values from a simulated estimation exercise performed with exactly the same data configuration ([109], p. 495; [125], p. 86).

More precisely, consider a nonconditional simulated realization $z^{(l)}(\mathbf{u})$ available at all nodal locations and actual data locations. This reference simulation shares the covariance of the RF model $Z(\mathbf{u})$. The simulated values $z^{(l)}(\mathbf{u}_\alpha), \alpha = 1, \ldots, n$, at the n data locations are retained and the estimation algorithm performed on the actual data is repeated. This provides a simulation of the estimated values $z^{*(l)}(\mathbf{u})$, hence also of the simulated errors $r^{(l)}\mathbf{u}) = z^{(l)}(\mathbf{u}) - z^{*(l)}(\mathbf{u})$. These simulated errors are then simply added to the actual estimated value $z^*(\mathbf{u})$ using the actual data values $z(\mathbf{u}_\alpha)$:

$$z_c^{(l)}(\mathbf{u}) = z^*(\mathbf{u}) + [z^{(l)}(\mathbf{u}) - z^{*(l)}(\mathbf{u})] \tag{V.8}$$

The notation $z_c^{(l)}$ with subscript c differentiates the conditional simulation from the nonconditional simulation $z^{(l)}(\mathbf{u})$.

If the estimation algorithm used is exact, as in kriging, the data values are honored at the data locations $\mathbf{u}_\alpha$, thus; $z^{(l)}(\mathbf{u}_\alpha) = z^{*(l)}(\mathbf{u}_\alpha)$ entailing: $z_c^{(l)}(\mathbf{u}_\alpha) = z^*(\mathbf{u}_\alpha) = z(\mathbf{u}_\alpha), \forall \alpha = 1, \ldots, n$. Last, it can be shown that the variogram of $Z_c^{(l)}(\mathbf{u})$ is identical to that of the original RF model $Z(\mathbf{u})$; see [109], p. 497.

The conditioning algorithm characterized by relation (V.8) requires a prior nonconditional simulation to generate the $z^{(l)}(\mathbf{u})$'s and two krigings to generate the $z^*(\mathbf{u})$ and $z^{*(l)}(\mathbf{u})$'s. Actually, only one set of kriging weights is needed since both the covariance function and the data configuration are common to both estimates $z^*(\mathbf{u})$ and $z^{*(l)}(\mathbf{u})$. A worthwhile addition to GSLIB would be a utility program to perform such kriging. Meanwhile the kriging programs kb2d and kt3d can be adapted.

Relation (V.8) is used as a postprocessor with nonconditional simulation algorithms such as the turning band algorithm (program tb3d given in the 1992 edition of this book) and random fractals simulation [28, 29, 80].

V.1.5   Questions of Ergodicity

A simulation algorithm aims at drawing realizations that reflect the statistics modeled from the data. The question is how well these statistics should be reproduced. Given that the model statistics are inferred from sample statistics that are uncertain because of limited sample size, exact reproduction of the model statistics by each simulated realization may not be desirable.

Consider the example of drawing from a stationary Gaussian-related RF model $Z(\mathbf{u})$, fully characterized by its cdf $F(z)$ and the covariance $C_Y(\mathbf{h})$ of its normal score transform $Y(\mathbf{u}) = G^{-1}(F(Z(\mathbf{u})))$, where $G(y)$ is the standard normal cdf; see Section V.2.1. Let $\{z^{(l)}(\mathbf{u}), \mathbf{u} \in A\}$ be a particular realization (number l) of the RF model $Z(\mathbf{u})$ over a finite field A, and $F^{(l)}(z)$, $C_Y^{(l)}(\mathbf{h})$ be the corresponding cdf and normal score covariance, the latter being the covariance of the Gaussian simulated values $y^{(l)}(\mathbf{u})$. "Ergodic fluctuations" refer to the discrepancy between the realization statistics $F^{(l)}(z)$, $C_Y^{(l)}(\mathbf{h})$ and the corresponding model parameters $F(z)$, $C_Y(\mathbf{h})$.

The stationary RF $Z(\mathbf{u})$ is said to be "ergodic" in the parameter μ if the corresponding realization statistics $\mu^{(l)}, \forall\, l$, tends toward μ as the size of the field A increases. The parameter μ is usually taken as the stationary mean of the RF, i.e., the mean of its cdf $F(z)$ [50]. The notion of ergodicity can be extended to any other model parameter, e.g., the cdf value $F(z)$ itself for a given cutoff z or the entire covariance model $C_Y(\mathbf{h})$.

Thus, provided that $Z(\mathbf{u})$ is stationary and ergodic and the simulation field is large enough,[5] one should expect the statistics of any realization l to reproduce the model parameters exactly.

The size of a simulation field A large enough to allow for ergodicity depends on the type of RF chosen and the range of its heterogeneities, as characterized, e.g., by the type and range[6] of the covariance model. The more continuous the covariance at $\mathbf{h} = 0$ and the larger the covariance range, the larger the field A has to be before ergodicity is achieved.

Figure V.2 shows the 20 histograms and semivariograms of 20 independent realizations of a standard Gaussian RF model with a zero mean, unit variance, and spherical variogram with a unit range and 20% nugget effect. The simulated field is a square with each side 4 times the range. The discretization is 1/25th of the range; thus each simulation comprises 10,000 nodes. Although the field is 4 times the range, a very favorable case in practice, the fluctuations of the realization statistics are seen to be large.

Figure V.3 illustrates the ergodic fluctuations that result from a different RF model with the same variogram. Figure V.3 is that implicit to the median IK model, whereby all standardized indicator variograms and cross variograms are equal to each other and equal to the attribute variogram; see relation (IV.32).

The fluctuations of the realization variograms of Figure V.3 for the median IK model are seen to be larger than those shown in Figure V.2 for the Gaussian RF model, although both models share the same histogram and variogram. The median IK model, and in general all RF models implicit to the indicator simulation approach, have poorer ergodic properties than

---

[5]What matters is the relative dimensions of A with respect to the range(s) of the covariance of $Z(\mathbf{u})$, *not* the density of discretization of A.

[6]Some phenomena, for which the spatial variance keeps increasing with the field size, are modeled with power variogram models $\gamma(\mathbf{h}) = |\mathbf{h}|^{\omega}$, i.e., with an implicit infinite range.

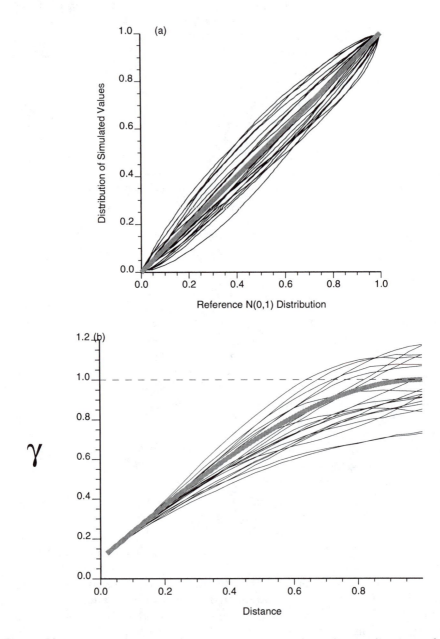

Figure V.2: Fluctuations of the variograms of 20 independent realizations of a Gaussian RF model: (a) shows the P-P plots of each realization showing the fluctuation in the histogram from the reference standard normal distribution; (b) shows the fluctuation in the variogram for 20 realizations.

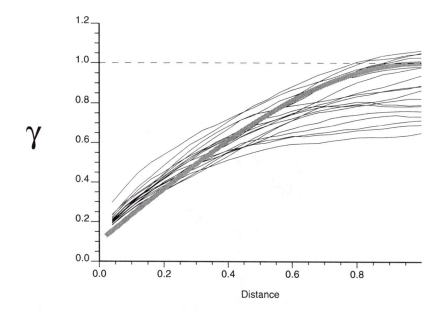

Figure V.3: Fluctuations of the variograms of 20 independent realizations of a median IK-type RF with a standard normal cdf and the same variogram as used for Figure V.2.

Gaussian-related RF models.[7] The larger fluctuations next to the origin seen in Figure V.3 are due to the discretization entailed by the indicator approach: within-class simulated values are drawn independently one from another.[8]

Observation of the fluctuations shown in Figures V.2 and V.3 may lead to choosing naively the RF model with better ergodic properties, or even looking for the simulation algorithm that would reproduce exactly the imposed statistics. This inclination would be correct only if the original sample statistics are considered exact and unquestionable.

In most applications, the model statistics are inferred from sparse samples and cannot be deemed exactly representative of the population statistics. One must allow for departure from the model statistics. The statistical fluctuations shown by different realizations of the RF model are useful. One could even argue that, everything else being equal as between Figures V.2 and V.3, one should prefer the RF model with poorer ergodic properties since it would provide the largest, hence most conservative, assessment of uncertainty

---

[7]Gaussian-related RF models are about the only practical models whose multivariate cdf is fully, and analytically, explicit. Most other RF models, including those implicit to indicator simulations, Boolean processes, and marked point processes, are not fully determined. Only some of their statistics are explicitly known (other statistics can be sampled from simulations). Be aware that an explicit analytical representation does not necessarily make a RF model more appropriate for a specific application.

[8]The nugget effect could be reduced by artificially lowering the nugget constant of the model.

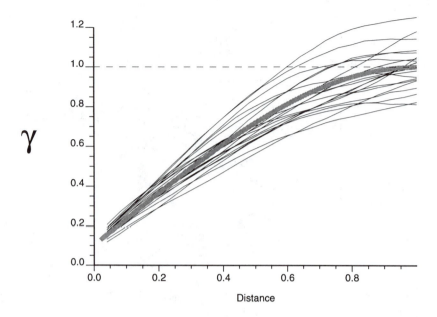

Figure V.4: The variograms of 20 realizations that have been transformed (with **trans**) to honor a $N(0, 1)$ distribution exactly. The variograms prior to transformation are shown in Figure V.2(b). Note that the fluctuations in the variogram and the reproduction of the input model are similar before and after transformation.

[42, 114].

It is worth noting that even a little data-conditioning significantly reduces ergodic fluctuations [125].

Identifying a Target Histogram

In some applications it may be imperative to reproduce closely the sample histogram or some target histogram. For example, if only one single simulated realization is to be processed, then it makes sense that this realization matches all sample or model statistics including the model or target histogram. Program **trans** introduced in Section VI.2.7 allows the simulated realization $\{z^{(l)}(\mathbf{u}), \mathbf{u} \in A\}$ to be postprocessed to match any target histogram while honoring the original data values at their locations. If the target histogram is not too different from the histogram of the simulated values $z^{(l)}(\mathbf{u})$, the variogram of these values would not be changed significantly [113].

Figures V.2 (a) and (b) give the 20 histograms and semivariograms of 20 independent realizations obtained by the Gaussian-based simulation program. The model histogram is normal with a unit mean and variance. The model semivariogram is spherical with a unit range and 20% nugget effect. The simulated field is square, with each side 4 times the unit range. The 20 realizations were postprocessed with **trans** to identify the model histogram: The fluctuations seen in Figure V.2.a are completely erased (figure not shown). Figure V.4 gives the semivariograms of the 20 realizations after postprocessing [compare with Figure V.2(b)]: The histogram identification did not significantly affect the variogram fluctuations and model reproduction.

However, we warn against systematic identification of sample histograms, for it gives a false sense of certainty in sample statistics.

Selecting Realizations

Issues of ergodicity often arise because practitioners tend to retain only one realization, i.e., use stochastic simulation as an improved interpolation algorithm. In the worst case, the first realization drawn is retained without any postprocessing, then the user complains that the model statistics are not matched! The only match that simulation theory guarantees is an average (expected value) over a large number of realizations. The less ergodic the RF model chosen, the more realizations are needed to approach this expected value.

In other cases, several realizations are drawn and one (or very few) is retained for further processing, e.g., through a flow simulator. Various criteria are used to select the realization, from closeness of the realization statistics to the model parameters, to subjective aesthetic appreciation of overall appearance or, better, selecting the realizations that match a datum or property that was not initially input to the model, e.g., travel times in a flow simulator using a simulated permeability field. Selecting realizations amounts to further conditioning by as yet unused information, be it subjective appreciation or hard production data for matching purposes. The realizations selected are better conditioned to actual data, whether hard or soft: In this sense they are better numerical models of the phenomenon under study. On the other hand, the selection process generally wipes out the measure of uncertainty[9] that the simulation approach was originally designed to provide.

How many realizations should one draw? The answer is definitely more than one to get some sense of the uncertainty. If two images, although both a priori acceptable, yield widely different results, then more images should be drawn. The number of realizations needed depends on how many are deemed sufficient to model (bracket) the uncertainty being addressed. Note that an evaluation of uncertainty need not require that each realization cover the entire field or site: simulation of a typical or critical subarea or section may suffice.

In indicator-type simulations, a systematic poor match of the model statistics may be due to too many order relation problems, themselves possibly due to inconsistent indicator variograms from one cutoff to the next one, or to variogram models inconsistent with the (hard or soft) conditioning data used.

---

[9]An alternative to selection is ranking of the realizations drawn according to some property of each realization. From such ranking, a few quantile-type realizations can be selected that allow preserving a measure of uncertainty [12, 190].

Interpolation or Simulation?

Although stochastic simulation was developed to provide measures of spatial uncertainty, simulation algorithms are increasingly used in practice to provide a single improved "estimated" map. Indeed, stochastic simulation algorithms have proven to be much more versatile than traditional interpolation algorithms in reproducing the full spectrum of data spatial variability *and* in accounting for data of different types and sources, whether hard or soft.

Inasmuch as a simulated realization honors the data[10] deemed important, it can be used as an interpolated map for those applications where reproduction of spatial features is more important than local accuracy. The warning that there can be several such realizations applies equally to the interpolated map: there can be alternative estimated images depending on the interpolation/estimation algorithm used and its specific implementation.

V.1.6   Going Beyond a Discrete CDF

Continuous cdf's and ccdf's are always informed at a discrete number of cutoff values. Regardless of whether a Gaussian-based method or an indicator-based method is used, the maximum level of discretization is provided by the following sample cumulative histogram with at most n step functions if all n data values $z(\mathbf{u}_\alpha)$ are different:

$$F^*(z) = \sum_{\alpha=1}^{n} a_\alpha i(\mathbf{u}_\alpha; z) \qquad (V.9)$$

$$\text{with: } a_\alpha \in [0,1], \quad \sum_{\alpha=1}^{n} a_\alpha = 1$$

where $i(\mathbf{u}_\alpha; z)$ is the indicator datum, set to one if $z(\mathbf{u}_\alpha) \leq z$, to zero otherwise. a_α is the declustering weight attached to datum location $\mathbf{u}_\alpha$; see [41, 74] and Section VI.2.2. An equally weighted histogram would correspond to $a_\alpha = 1/n$, $\forall \alpha$.

If the sample size n is small, assumptions must be made for extrapolation beyond the smallest z datum value (lower tail), the largest z datum value (upper tail), and for interpolation between two consecutively ranked z data. Being a data expansion technique, stochastic simulation will generate many more values than there are original data. Therefore, the maximum resolution provided by the sample cdf (V.9) is likely insufficient, particularly in the two tails.

This lack of resolution is particularly severe when using indicator-related algorithms with only a few cutoff values such as the nine deciles of the sample

---

[10] Any realization can be postprocessed to reproduce the sample histogram; hence the sample mean and variance; see previous discussion related to Figure V.4.

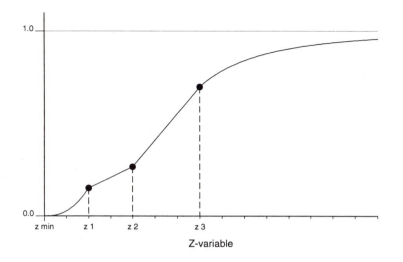

Figure V.5: Interpolation and extrapolation beyond the ccdf initially known at three cutoff values z_1, z_2, and z_3.

cdf (V.9). In this case, the procedures for interpolation between the IK-derived ccdf values and, *most important*, for extrapolating beyond the two extreme ccdf values are critical.

Figure V.5 shows an example of a ccdf $F(\mathbf{u}; z)$ known at 3 cutoff values (z_1, z_2, z_3), which could be the three quartiles of the marginal cdf. Linear cdf interpolation within the two central classes $(z_1, z_2], (z_2, z_3]$ amounts to assuming a uniform within-class distribution. The lower tail $(z_{min}, z_1]$ has been extrapolated toward a fixed minimum z_{min} using a power model (V.10) corresponding to a negatively skewed distribution. The upper tail $(z_3, +\infty]$ has been extrapolated up to a potentially infinite upper bound using a positively skewed hyperbolic model (V.11). The alternative of smoothing the corresponding histogram is discussed later in this section.

Interpolation Models

The within-class cdf interpolation models considered in GSLIB are:

Power model: For a finite class interval $(z_{k-1}, z_k]$ and a parameter $\omega > 0$ (the power), this cdf model is written:

$$F^{\omega}_{z_{k-1}, z_k}(z) = \begin{cases} 0, & \forall z \leq z_{k-1} \\ \left[\frac{z - z_{k-1}}{z_k - z_{k-1}}\right]^{\omega}, & \forall z \in (z_{k-1}, z_k] \\ 1, & \forall z \geq z_k \end{cases} \tag{V.10}$$

In practice, this cdf model is scaled between the calculated cdf values at z_{k-1} and z_k rather than between 0 and 1. Distributions with $\omega < 1$ are positively

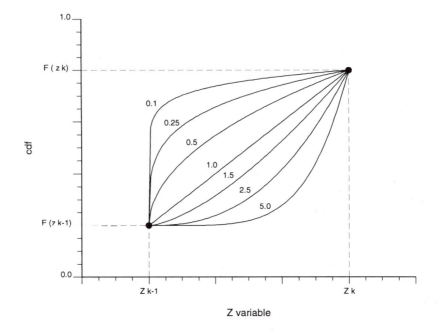

Figure V.6: Some power models for ω=0.1, 0.25, 0.5, 1.0, 1.5, 2.5 and 5.0. ω less than 1.0 leads to positive skewness, ω=1.0 gives the uniform distribution, and ω greater than 1.0 leads to negative skewness.

skewed, $\omega = 1$ corresponds to the linear cdf model (uniform distribution), and distributions with $\omega > 1$ are negatively skewed; see Figure V.6. If ω is set very small, the interpolated value will be set to z_{k-1}; if ω is set very large, the interpolated value will be set to z_k.

Linear interpolation between tabulated bound values: This option considers a fixed number of subclasses with given bound values within each class $(z_{k-1}, z_k]$. For example, the three bound values a_{k1}, a_{k2}, a_{k3} can be tabulated defining four subclasses $(z_{k-1}, a_{k1}], (a_{k1}, a_{k2}], (a_{k2}, a_{k3}], (a_{k3}, z_k]$ that share the probability p_k calculated for class $(z_{k-1}, z_k]$; p_k is shared equally unless specified otherwise. Then linear cdf interpolation is performed separately within each subclass.

This option allows the user to add detail to the distribution within the classes defined by the cutoffs z_k. That detail, i.e., the subclasses bound values, can be identified to some or all of the original data values falling within each class $(z_{k-1}, z_k]$ of the marginal (sample) distribution. Thus some of the resolution lost through discretization by the z_k values can be recovered.[11]

More generally, the subclass bound values a_k can be taken from any parametric model, e.g., beta or gamma distribution.

---

[11]However, original data values are marginal quantile values, whereas conditional quantile values, specific to each ccdf, would be needed.

Hyperbolic model: This last option is to be used only for the upper tail of a positively skewed distribution. Decisions regarding the upper tail of ccdf's are often the most consequential[12]; therefore, a great deal of flexibility is needed, including the possibility of a very long tail.

The hyperbolic cdf upper tail model for a strictly positive variable is a two-parameter distribution:

$$F_{\omega,\lambda}(z) = 1 - \frac{\lambda}{z^\omega}, \quad \omega \geq 1, \quad z^\omega > \lambda > 0 \qquad (V.11)$$

The scaling parameter λ allows identification of any precalculated quantile value, for example, the p quantile z_p such that $F_{\omega,\lambda}(z_p) = p$; then:

$$\lambda = z_p^\omega(1 - p)$$

The parameter $\omega > 1$ controls how fast the cdf reaches its upper limit value 1; the smaller ω, the longer the tail of the distribution.

The mean z value above the p quantile value z_p is: $m_p = \frac{\omega}{\omega-1}z_p > z_p$. Hence the smaller ω, the larger the mean above z_p. At its minimum value, $\omega = 1$ identifies the Pareto distribution, which has an infinite mean $m_p, \forall p$, corresponding to a very long tail.

The parameter ω should be chosen to be realistic and conservative for the application being considered. When considering an environmental issue with z representing concentration of a pollutant, $\omega = 1$ is a safe choice because a long tail is usually a conservative (pessimistic) assessment. If a metal distribution is being considered, a value between $\omega = 2$ and 3 will cause the tail to drop off reasonably fast without being too optimistic about the occurrence of very large values. Figure V.7 shows hyperbolic distributions for ω between 1 and 5. Practice has shown that a distribution with $\omega = 1.5$ is a general-purpose model that yields acceptable results in a wide variety of applications.

The GSLIB programs allow different sets of options depending on whether interpolation is needed within the middle classes or extrapolation for the lower and upper tails. The available options are:

Lower Tail: below the first calculated cdf value:

1. Linear model (uniform distribution)
2. Power model
3. Tabulated bound values

Middle: between any two calculated cdf values:

---

[12]Even when using a multivariate Gaussian RF model (after normal score transform), a decision about the shape of the cdf in the upper tail is required for the back transform; see program sgsim. It is naive to believe that a fully defined parametric RF model allows escaping critical decisions about the distribution of extreme values; that decision is built into the RF model chosen. In the case of indicator-based methods that decision is made fully explicit.

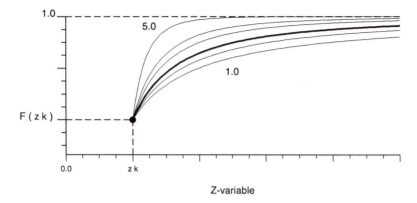

Figure V.7: The family of hyperbolic models; the distribution models for $\omega=1.0$, 1.25, 1.5, 2.0, 2.5, and 5.0 are shown. The models for $\omega=1.0$ and 5.0 are labeled and the model for $\omega=1.5$ is bolder than the rest. All models have a positive skewness.

1. Linear model (uniform distribution)
2. Power model
3. Tabulated bound values

Upper Tail: above the last calculated cdf value:

1. Linear model (uniform distribution)
2. Power model
3. Tabulated bound values
4. Hyperbolic model

The user is asked for a specific model for each of these regions (the integer number identifying each model in the list above is used).

The Smoothing Alternative

An alternative to interpolating between sample cdf values is to smooth the sample histogram. The `histsmth` program proposed in Section VI.2.3 not only smooths out sample fluctuations of the input histogram, it can also add resolution by increasing the number of classes and extrapolates toward specified minimum and maximum values beyond the sample minimum and maximum. Unlike kernel smoothing, program `histsmth` allows the target mean and variance, usually informed by declustered sample statistics, to be identified.

Program `histsmth` uses a simulated annealing procedure [43] to obtain a smoothed distribution that includes a smoothness criterion, closeness to input quantiles, and closeness to target mean and variance values. The number of classes can be very large, possibly larger than the number of original data, and these classes can span a range larger than that of the data (extrapolation).

V.2 Gaussian-Related Algorithms

The Gaussian RF model is unique in statistics for its extreme analytical simplicity and for being the limit distribution of many analytical theorems globally known as "central limit theorems" [10, 91].

In a nutshell, if the continuous[13] spatial phenomenon $\{z(\mathbf{u}), \mathbf{u} \in A\}$ is generated by the sum of a (not too large) number of *independent* sources $\{y_k(\mathbf{u}), \mathbf{u} \in A\}, k = 1, \ldots, K$, with similar spatial distributions, then its spatial distribution can be modeled by a multivariate Gaussian RF model:

$$Z(\mathbf{u}) = \sum_{k=1}^{K} Y_k(\mathbf{u}) \approx \text{Gaussian}$$

The limiting constraint is not the number K, or the fact that the components $Y_k(\mathbf{u})$ are equally distributed, but the hypothesis of *independence* of the $Y_k(\mathbf{u})$'s.

If human and measurement errors can sometimes be considered as independent events or processes, in the earth sciences the diverse geological/biological processes that have generated the observed phenomenon are rarely independent one from another, nor are they additive. This notwithstanding, multivariate Gaussian models are extremely congenial, they are well understood, and they have an established record of successful applications. These heuristic considerations are enough to make the Gaussian model the privileged choice for modeling continuous variables, unless proven inappropriate.

The traditional definition of a multivariate Gaussian (or normal) distribution as expressed through the matrix expression of its multivariate pdf can be found in any multivariate statistics book; see, e.g., [10]. That expression is of little practical use. It is better to define the multivariate Gaussian RF model through its *characteristic* properties.

The RF $Y(\mathbf{u}) = \{Y(\mathbf{u}), \mathbf{u} \in A\}$ is multivariate normal if and only if[14]:

- all subsets of that RF, for example $\{Y(\mathbf{u}), \mathbf{u} \in B \subset A\}$ are also multivariate normal.

- all linear combinations of the RV components of $Y(\mathbf{u})$ are (univariate) normally distributed, i.e.,

$$X = \sum_{\alpha=1}^{n} \omega_\alpha Y(\mathbf{u}_\alpha) \text{ is normally distributed,} \qquad (V.12)$$

$\forall n$, $\forall$ the weights ω_α, as long as $\mathbf{u}_\alpha \in A$

---

[13] "Continuous" as opposed to a phenomenon $z(\mathbf{u})$ characterized by discrete categorical values or by the juxtaposition of different populations.

[14] These characteristic properties are not all independent. For example, the second property of normality of all linear combinations suffices to characterize a multivariate normal distribution.

- zero covariance (or correlation) entails full independence:

  If $Cov\{Y(\mathbf{u}), Y(\mathbf{u}')\} = 0$, the two RVs $Y(\mathbf{u})$ and $Y(\mathbf{u}')$ (V.13)
  are not only uncorrelated, they are also independent.

- all conditional distributions of any subset of the RF $Y(\mathbf{u})$, given realizations of any other subset, are (multivariate) normal. For example, the conditional distribution of the K RVs $\{Y_k(\mathbf{u}'_k), k = 1, \ldots, K, \mathbf{u}'_k \in A\}$, given the realizations $y(\mathbf{u}_\alpha) = y_\alpha, \alpha = 1, \ldots, n$, is K variate normal, $\forall K, \forall \mathbf{u}'_k, \forall n, \forall \mathbf{u}_\alpha, \forall y_\alpha$.

The case of $K = 1, \mathbf{u}'_1 = \mathbf{u}_o$, where the RV $Y(\mathbf{u}_o)$ models the uncertainty about a specific unsampled value $y(\mathbf{u}_o)$ is of particular interest: The ccdf of $Y(\mathbf{u}_o)$, given the n data y_α, is normal and fully characterized by:

- its mean, or conditional expectation, identified to the SK (linear regression) estimate of $y(\mathbf{u}_o)$

$$E\{Y(\mathbf{u}_o)|y(\mathbf{u}_\alpha) = y_\alpha, \alpha = 1, \ldots, n\} \equiv [y(\mathbf{u}_o)]^*_{SK} \qquad (V.14)$$

$$= m(\mathbf{u}_o) + \sum_{\alpha=1}^{n} \lambda_\alpha [y_\alpha - m(\mathbf{u}_\alpha)]$$

where $m(\mathbf{u}) = E\{Y(\mathbf{u})\}$ is the expected value of the not necessarily stationary RV $Y(\mathbf{u})$. The n weights λ_α are given by the SK system (IV.2):

$$\sum_{\beta=1}^{n} \lambda_\beta C(\mathbf{u}_\beta, \mathbf{u}_\alpha) = C(\mathbf{u}_o, \mathbf{u}_\alpha), \alpha = 1, \ldots, n \qquad (V.15)$$

where $C(\mathbf{u}, \mathbf{u}') = Cov\{Y(\mathbf{u}), Y(\mathbf{u}')\}$ is the covariance, not necessarily stationary, of the RF $Y(\mathbf{u})$.

- its variance, or conditional variance, is the SK variance:

$$Var\{Y(\mathbf{u}_o)|y(\mathbf{u}_\alpha) = y_\alpha, \alpha = 1, \ldots, n\} = \qquad (V.16)$$

$$C(\mathbf{u}_o, \mathbf{u}_o) - \sum_{\alpha=1}^{n} \lambda_\alpha C(\mathbf{u}_o, \mathbf{u}_\alpha)$$

Note the homoscedasticity of the conditional variance: It does not depend on the data values y_α, but only on the data configuration (geometry) and the covariance model.

For stochastic sequential simulation the normality of all ccdf's is a true blessing: indeed, the determination of the sequence (V.5) of successive ccdf's reduces to solving a corresponding sequence of SK systems.

V.2.1 Normal Score Transform

The first necessary condition for the stationary RF $Y(\mathbf{u})$ to be multivariate normal is that its univariate cdf be normal, i.e.,

$$\text{Prob}\{Y(\mathbf{u}) \leq y\} = G(y), \ \forall y \qquad (\text{V.17})$$

where $G(\cdot)$ is the standard Gaussian cdf; $Y(\mathbf{u})$ is assumed to be standardized, i.e., with a zero mean and unit variance.

Unfortunately, most earth sciences data do not present symmetric Gaussian histograms. This is not a major problem since a nonlinear transform can transform any continuous cdf into any other cdf; see [49, 90] and program trans in Section VI.2.7.

Let Z and Y be two RVs with cdf's $F_Z(z)$ and $F_Y(y)$ respectively. The transform $Y = \varphi(Z)$ identifies the cumulative probabilities corresponding to the Z and Y p quantiles:

$$F_Y(y_p) = F_Z(z_p) = p, \ \forall p \in [0, 1], \text{ hence:}$$

$$y = F_Y^{-1}(F_Z(z)) \qquad (\text{V.18})$$

with $F_Y^{-1}(\cdot)$ being the inverse cdf, or quantile function, of the RV Y:

$$y_p = F_Y^{-1}(p), \ \forall p \in [0, 1]$$

If Y is standard normal with cdf $F_Y(y) = G(y)$, the transform $G^{-1}(F_Z(\cdot))$ is the normal score transform.

In practice, the n z sample data are ordered in increasing value:

$$z^{(1)} \leq z^{(2)} \leq \cdots \leq z^{(n)}$$

The cumulative frequency corresponding to the kth largest z datum is $F_Z(z^{(k)}) = k/n$, or $F_Z(z^{(k)}) = \sum_{j=1}^{k} \omega_j \in [0, 1]$ if a set of declustering weights ω_j have been applied to the n data.

Then the normal score transform of $z^{(k)}$ is the k/n quantile of the standard normal cdf, i.e.,

$$y^{(k)} = G^{-1}\left(\frac{k}{n}\right) \qquad (\text{V.19})$$

Implementation details of the normal score transformation, including the treatment of the last value $G^{-1}\left(\frac{n}{n}\right) = +\infty$, and back transform, $Z \rightleftharpoons Y$, are given in Sections VI.2.5 and VI.2.6; see programs nscore and backtr.

If the marginal sample cdf $F_Z(z)$ is too discontinuous due, for example, to too few data, one might consider smoothing it prior to normal score transform. Program histsmth, introduced in Section VI.2.3, could be used for this purpose. The same smoothed cdf should be considered for the back transform.

If weights are used to correct (decluster) the sample cdf $F_Z(z)$, it is the correspondingly weighted histogram of the normal score data that is standard normal, not the unweighted normal scores histogram.

V.2.2   Checking for Bivariate Normality

The normal score transform (V.19) defines a new variable Y, which is, by construction, (univariate) normally distributed. This is a necessary but not a sufficient condition for the spatially distributed values $Y(\mathbf{u}), \mathbf{u} \in A$, to be multivariate normal. The next necessary condition is that the bivariate cdf of any pair of values $Y(\mathbf{u}), Y(\mathbf{u} + \mathbf{h})$, $\forall \mathbf{u}, \forall \mathbf{h}$, be normal.

There are various ways to check[15] for bivariate normality of a data set $\{y(\mathbf{u}_\alpha), \alpha = 1, \ldots, n\}$ whose histogram is already standard normal [10, 17, 91, 111].

The most relevant check directly verifies that the experimental bivariate cdf of any set of data pairs $\{y(\mathbf{u}_\alpha), y(\mathbf{u}_\alpha + \mathbf{h}), \alpha = 1, \ldots, N(\mathbf{h})\}$ is indeed standard bivariate normal with covariance function $C_Y(\mathbf{h})$. There exist analytical and tabulated relations linking the covariance $C_Y(\mathbf{h})$ with any standard normal bivariate cdf value [2, 74, 102, 191]:

$$\text{Prob}\{Y(\mathbf{u}) \le y_p, Y(\mathbf{u} + \mathbf{h}) \le y_p\} = \qquad (\text{V.20})$$

$$p^2 + \frac{1}{2\pi} \int_0^{\arcsin C_Y(\mathbf{h})} \exp\left(-\frac{y_p^2}{1 + \sin\theta}\right) d\theta$$

where $y_p = G^{-1}(p)$ is the standard normal p quantile and $C_Y(\mathbf{h})$ is the correlogram of the standard normal RF $Y(\mathbf{u})$.

Now, the bivariate probability (V.20) is the noncentered indicator covariance for the threshold y_p:

$$\text{Prob}\{Y(\mathbf{u}) \le y_p, Y(\mathbf{u} + \mathbf{h}) \le y_p\} \qquad (\text{V.21})$$

$$= E\{I(\mathbf{u}; p) \cdot I(\mathbf{u} + \mathbf{h}; p)\} = p - \gamma_I(\mathbf{h}; p)$$

where $I(\mathbf{u}; p) = 1$, if $Y(\mathbf{u}) \le y_p$; zero, if not; and $\gamma_I(\mathbf{h}; p)$ is the indicator semivariogram for the p quantile threshold y_p.

The check consists of comparing the sample indicator semivariogram $\gamma_I(\mathbf{h}; p)$ to the theoretical bivariate normal expression (V.20). Program `bigaus` calculates the integral expression (V.20) for various p quantile values. Simple parametric approximations to (V.20) are given in [111].

Remarks

- The indicator variable $I(\mathbf{u}; p)$ is the same whether defined on the original z data or on the y normal score data, as long as the cutoff values z_p and y_p are both p quantiles of their respective cdf's. In other words, indicator variograms, i.e., bivariate cdf's parametrized in cdf values p, are invariant by any linear or nonlinear monotonic increasing transform.

---

[15] A check differs from a formal statistical test in that it does not provide any measure of accepting wrongly the hypothesis. Unfortunately, most formal tests typically require, in practice, independent data or data whose multivariate distribution is known a priori.

The normal score transform (V.18) and, more generally, any monotonic transform of the original variable z does *not* change the essence of its bivariate cdf:

$$\text{Prob}\{Y(\mathbf{u}) \leq y_p, Y(\mathbf{u} + \mathbf{h}) \leq y_p\} \qquad (V.22)$$

$$= \text{Prob}\{Z(\mathbf{u} \leq z_p, Z(\mathbf{u} + \mathbf{h}) \leq z_p\}, \, \forall p \in [0, 1]$$

for any monotonic increasing[16] transform $Y = \varphi(Z)$. $y_p = \varphi(z_p)$ and z_p are the marginal p quantiles of the Y and Z distributions.

Relation (V.22) is methodologically important since it indicates that it is naive to expect a univariate monotonic transform $\varphi(\cdot)$ to impart new bivariate or multivariate properties to the transformed y data. Either those properties already belong to the original z data, or they do not and the transform $\varphi(\cdot)$ will not help. Bivariate and multivariate normality are either properties of the z data or they are not; the nonlinear rescaling of the z units provided by the normal score transform does not help.

- Symmetric normal quantiles are such that $y_p = -y_{p'}$, i.e., $y_p^2 = y_{p'}^2$, $\forall p' = 1 - p$. Simple arithmetic between relations (V.20) and (V.21) yields

$$\gamma_I(\mathbf{h}; p) = \gamma_I(\mathbf{h}; p'), \forall p' = 1 - p \qquad (V.23)$$

Moreover, as p tends toward its bound values, 0 or 1, $y_p^2 \to +\infty$, and expression (V.20) tends toward the product p^2 of the two marginal probabilities (independence). Consequently, the semivariogram $\gamma_I(\mathbf{h}; p)$ tends toward its sill value $p(1 - p)$ no matter how small the separation vector $\mathbf{h}$. Occurrence of Gaussian extreme values, whether high or low, is purely random: Extreme Gaussian values do not cluster in space.

A simple way to check for binormality is to check for the symmetric "destructuration" of the occurrence of extreme z values: The practical ranges of the indicator variograms $\gamma_I(\mathbf{h}; p)$ should decrease symmetrically and continuously as p tends toward its bound values of 0 and 1; see [111].

Bivariate normality is necessary but not sufficient for multivariate normality. Beyond bivariate normality, one should check that all trivariate, quadrivariate, ..., K variate experimental frequencies match theoretical Gaussian expressions of type (V.20) given the covariance model $C_Y(\mathbf{h})$. The problem does not reside in computing these theoretical expressions, but in the inference of the corresponding experimental multivariate frequencies: Unless the

---

[16]If the transform φ is monotonic decreasing, then an equivalent relation holds:

$$\text{Prob}\{Y(\mathbf{u}) > y_p, Y(\mathbf{u} + \mathbf{h}) > y_p\}$$

$$= 1 - 2p + \text{Prob}\{Y(\mathbf{u} \leq y_p, Y(\mathbf{u} + \mathbf{h}) \leq y_p\}$$
$$= \text{Prob}\{Z(\mathbf{u} \leq z_p, Z(\mathbf{u} + \mathbf{h}) \leq z_p\}$$

data are numerous and gridded, there are rarely enough triplets, quadruplets, ..., K tuples sharing the same geometric configuration to allow such inference. Therefore, in practice, if one cannot from sample statistics show that bivariate Gaussian properties are violated, the multivariate Gaussian RF model should be the prime choice for continuous variable simulation.

V.2.3   Sequential Gaussian Simulation

The most straightforward algorithm for generating realizations of a multivariate Gaussian field is provided by the sequential principle described in Section V.1.3. Each variable is simulated sequentially according to its normal ccdf fully characterized through an SK system of type (V.15). The conditioning data consist of all original data and all previously simulated values found within a neighborhood of the location being simulated.

The conditional simulation of a *continuous* variable $z(\mathbf{u})$ modeled by a Gaussian-related stationary RF $Z(\mathbf{u})$ proceeds as follows:

1. Determine the univariate cdf $F_Z(z)$ representative of the entire study area and not only of the z sample data available. Declustering may be needed if the z data are preferentially located; smoothing with extrapolation may also be needed.

2. Using the cdf $F_Z(z)$, perform the normal score transform of z data into y data with a standard normal cdf; see Section V.2.1 and program `nscore` in Section VI.2.5.

3. Check for bivariate normality of the normal score y data; see Section V.2.2. If the multivariate Gaussian model cannot be retained, then consider alternative models such as a mixture of Gaussian populations [197] or an indicator-based algorithm for the stochastic simulation.

4. If a multivariate Gaussian RF model can be adopted for the y variable, proceed with program `sgsim` and sequential simulation, i.e.,

    - Define a random path that visits each node of the grid (not necessarily regular) once. At each node $\mathbf{u}$, retain a specified number of neighboring conditioning data including both original y data and previously simulated grid node y values.

    - Use SK with the normal score variogram model to determine the parameters (mean and variance) of the ccdf of the RF $Y(\mathbf{u})$ at location $\mathbf{u}$.

    - Draw a simulated value $y^{(l)}(\mathbf{u})$ from that ccdf.

    - Add the simulated value $y^{(l)}(\mathbf{u})$ to the data set.

    - Proceed to the next node, and loop until all nodes are simulated.

5. Backtransform the simulated normal values $\{y^{(l)}(\mathbf{u}), \mathbf{u} \in A\}$ into simulated values for the original variable $\{z^{(l)}(\mathbf{u}) = \varphi^{-1}(y^{(l)}(\mathbf{u})), \mathbf{u} \in A\}$. Within-class interpolations and tail extrapolations are usually called for; see Section V.1.6.

Multiple Realizations

If multiple realizations are desired $\{z^{(l)}(\mathbf{u}), \mathbf{u} \in A\}, l = 1, \dots, L$, the previous algorithm is repeated L times with either of these options:

1. The same random path visiting the nodes. In this case the data configuration, hence the SK systems, are the same from one realization to another: They need be solved only once. CPU time is reduced considerably; see footnote 19.

2. A different random path for each realization. The sequence of data configurations is different; thus different SK systems must be set up and solved. CPU time is then proportional to the number of realizations. This last option has been implemented in program sgsim.

SK or OK?

The prior decision of stationarity requires that simple kriging (SK) with zero mean be used in step 4 of the sGs algorithm. If data are abundant enough to consider inference of a nonstationary RF model, one may

1. Consider an array of local SK prior means (different from zero) to be used in the expression of the local SK estimates (V.14). This amounts to bypass the OK process by providing local nonstationary means obtained from some other information source, or

2. Split the area into distinct subzones and consider for each subzone a different stationary RF model, which implies inference of a different histogram and a different normal score variogram for each subzone, or

3. Consider a stationary normal score variogram, inferred from the entire pool of data, and a nonstationary mean for $Y(\mathbf{u})$. The nonstationary mean, $E\{Y(\mathbf{u})\}$, at each location $\mathbf{u}$, is implicitly re-estimated from the neighborhood data through ordinary kriging (OK); see Section IV.1.2. The price of such local rescaling of the model mean is, usually, a poorer reproduction of the stationary Y histogram and variogram model.

Similarly, sequential Gaussian simulation with an external drift can be implemented by replacing the simple kriging estimate and system (V.14, V.15) by a kriging estimate of type (IV.16) where the local normal score trend value is linearly calibrated to the collocated value of a smoothly varying secondary variable (different from z). Whether OK or KT with an external drift is used to replace the SK estimate, in all rigor it is the SK variance (V.16) that should be used for the variance of the Gaussian ccdf; see [94]. This was not implemented in program `sgsim`. An approximation would consist in multiplying each locally derived OK or KT (kriging) variance by a constant factor obtained from prior calibration.

Simple kriging should be the preferred algorithm unless proven inappropriate.

Declustered Transform

If the original z-sample is preferentially clustered, the declustered sample cdf $F_Z(z)$ should be used for both the normal score transform and backtransform. As a consequence the unweighted mean of the normal score data is not zero, nor the variance 1: the normal score covariance model should be fitted as best as possible to these data then restandardized to unit variance: $C_Y(0) = 1$.

V.2.4   LU Decomposition Algorithm

When the total number of conditioning data *plus* the number of nodes to be simulated is small (fewer than a few hundred) and a large number of realizations is requested, simulation through LU decomposition of the covariance matrix provides the fastest solution [5, 38, 61, 67].

Let $Y(\mathbf{u})$ be the stationary Gaussian RF model with covariance $C_Y(\mathbf{u})$. Let $\mathbf{u}_\alpha, \alpha = 1, \ldots, n$, be the locations of the conditioning data and $\mathbf{u}'_i, i = 1, \ldots, N$, be the N nodes to be simulated. The large covariance matrix $(n + N) \cdot (n + N)$ is partitioned into the data-to-data covariance matrix, the node-to-node covariance matrix, and the two node-to-data covariance matrices:

$$\mathbf{C}_{(n+N)(n+N)} = \left[\begin{array}{cc} [C_Y(\mathbf{u}_\alpha - \mathbf{u}_\beta)]_{n \cdot n} & [C_Y(\mathbf{u}_\alpha - \mathbf{u}'_j)]_{n \cdot N} \\ [C_Y(\mathbf{u}'_i - \mathbf{u}_\beta)]_{N \cdot n} & [C_Y(\mathbf{u}'_i - \mathbf{u}'_j)]_{N \cdot N} \end{array} \right] = \mathbf{L} \cdot \mathbf{U} \quad (V.24)$$

The large matrix $\mathbf{C}$ is decomposed into the product of a lower and an upper triangular matrix, $\mathbf{C} = \mathbf{L} \cdot \mathbf{U}$. A conditional realization $\{y^{(l)}(\mathbf{u}'_i), i = 1, \ldots, N\}$ is obtained by multiplication of $\mathbf{L}$ by a column matrix $\omega^{(l)}_{(N+n) \cdot 1}$ of normal deviates:

$$\mathbf{y}^{(l)} = \left[\begin{array}{c} [y(\mathbf{u}_\alpha)]_{n \cdot 1} \\ [y^{(l)}(\mathbf{u}'_i)]_{N \cdot 1} \end{array} \right] = \mathbf{L} \cdot \omega^{(l)} = \left[\begin{array}{cc} \mathbf{L}_{11} & \mathbf{0} \\ \mathbf{L}_{21} & \mathbf{L}_{22} \end{array} \right] \cdot \left[\begin{array}{c} \omega_1 \\ \omega^{(l)}_2 \end{array} \right] \quad (V.25)$$

where $[y(\mathbf{u}_\alpha)]_{n \cdot 1}$ is the column matrix of the n normal score conditioning data and $[y^{(l)}(\mathbf{u}'_i)]_{N \cdot 1}$ is the column matrix of the N conditionally simulated y values.

Identification of the conditioning data is written as $\mathbf{L}_{11}\omega_1 = [y(\mathbf{u}_\alpha)]$; thus matrix ω_1 is set at:

$$\omega_1 = [\omega_1]_{n \cdot 1} = \mathbf{L}_{11}^{-1} \cdot [y(\mathbf{u}_\alpha)] \quad (V.26)$$

The column vector $\omega^{(l)}_2 = \left[\omega^{(l)}_2 \right]_{N \cdot 1}$ is a vector of N *independent* standard normal deviates.

Additional realizations, $l = 1, \ldots, L$, are obtained at very little additional cost by drawing a new set of normal deviates $\omega_2^{(l)}$, then by applying the matrix multiplication (V.25). The major cost and memory requirement is in the upfront LU decomposition of the large matrix $\mathbf{C}$ and in the identification of the weight matrix ω_1.

The LU decomposition algorithm requires that *all* nodes and data locations be considered simultaneously in a single covariance matrix $\mathbf{C}$. The current practical limit of the number $(n + N)$ is no greater than a few hundred. The code `lusim` provided in GSLIB is a full 3D implementation of the LU decomposition algorithm.

Implementation variants have been considered, relaxing the previous size limitation by considering overlapping neighborhoods of data locations [5]. Unfortunately, artifact discontinuities appear if the correlation between all simulated nodes is not fully accounted for.

The LU decomposition algorithm is particularly appropriate when a large number (L) of realizations is needed over a small area or block ($n + N$ is then small). A typical application is the evaluation of block ccdf's, a problem known in geostatistical jargon as "change of support"; see [67] and discussion in Section IV.1.13. A block or any small subset of the study area, $B \subset C$, can be discretized into N points. The normal score values at these N points can be simulated repetitively ($l = 1, \ldots, L$) through the LU decomposition algorithm and backtransformed into simulated point z values: $\{z^{(l)}(\mathbf{u}_i'), i = 1, \ldots, N; \mathbf{u}_i' \in B \subset A\}, l = 1, \ldots, L$. Each set of N simulated point values can then be averaged to yield a simulated block value:

$$z_B^{(l)} = \frac{1}{N} \sum_{i=1}^{N} z^{(l)}(\mathbf{u}_i')$$

This is a linear average, but any other averaging expression could be considered, e.g., the geometric average of the N simulated values. The distribution of the L simulated block values $z_B^{(l)}, l = 1, \ldots, L$, provides a numerical approximation of the probability distribution (ccdf) of the block average, conditional to the data retained.

V.2.5 The Turning Band Algorithm

The turning band algorithm was the first large-scale 3D Gaussian simulation algorithm actually implemented [27, 93, 125, 128, 135]. It provides only nonconditional realizations of a standard Gaussian field $Y(\mathbf{u})$ with a given covariance $C_Y(\mathbf{h})$. The conditioning to local normal score y data calls for a postprocessing of the nonconditional simulation by kriging [see relation (V.8) in Section V.1.4].

The turning band algorithm is very fast, as fast as fast-Fourier spectral domain techniques [18, 76] or random fractal algorithms [29, 80, 184]. The additional kriging step needed to condition to local data make all these techniques less appealing. For comparison, the sequential Gaussian algorithm

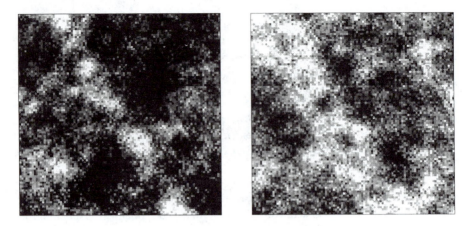

Figure V.8: Two 2D slices through a 3D realization generated by turning band simulation. Note the artifact banding.

(sGs) also requires the solution of one kriging system per grid node, whether the simulation is conditional or not. In addition, the application of sGs is much more straightforward than that of the turning band algorithm.

The turning band algorithm is fast because it achieves a 3D simulation through a series of 1D simulations along the 15 lines that constitute a regular partitioning of the 3D space; see [109], p. 499. Each node in the 3D space is projected onto a particular point on each of the 15 lines, and the simulated value at that node is the sum of the simulated values at the 15 projected points. The covariance of the 1D process to be simulated independently on each line is deduced by a deconvolution process from the imposed 3D isotropic covariance. This deconvolution process is reasonably straightforward in 3D but awkward in 2D [23, 128].

Aside from being slow and cumbersome when used to generate *conditional* simulations, the turning band algorithm presents two important drawbacks:

1. The realizations show artifact banding due to the limitation to the maximum of 15 lines that provide a regular but coarse partitioning of the 3D space [76]; see Figure V.8. There is no such maximum in 2D.

2. Anisotropy directions other than the coordinate axes of the simulation grid cannot be handled easily.

The *non*conditional simulation program tb3d given in the original (1992) edition of GSLIB has been removed from this new edition.

V.2.6 Multiple Truncations of a Gaussian Field

If there are only two categories ($K=2$), one being the complement of the other, one can obtain nonconditional realizations of the unique indicator variable

$i(\mathbf{u})$ by truncating the continuous realizations $\{y^{(l)}(\mathbf{u}), \mathbf{u} \in A\}$ of a standard Gaussian RF $Y(\mathbf{u})$:

$$i^{(l)}(\mathbf{u}) = 1, \text{if } y^{(l)}(\mathbf{u}) \le y_p \qquad (\text{V.27})$$
$$= 0, \text{ if not}$$

with $y_p = G^{-1}(p)$ being the standard normal p quantile and p being the desired proportion of indicators equal to one:

$$E\{I(\mathbf{u})\} = p$$

Since the Gaussian RF model is fully determined by its covariance $C_Y(\mathbf{h})$, there exists a one-to-one relation between that covariance $C_Y(\mathbf{h})$ and the indicator covariance after truncation at the p quantile. By inverting relation (V.20), one can determine the covariance $C_Y(\mathbf{h})$ of the Gaussian RF that will yield the required indicator variogram [110].

Since the covariance $C_Y(\mathbf{h})$ is the unique parameter (degree of freedom) of the Gaussian RF model $Y(\mathbf{u})$, it cannot be used to reproduce more than one indicator covariance. Multiple truncations of the same realization $\{y^{(l)}(\mathbf{u}), \mathbf{u} \in A\}$ at different threshold values would yield multiple categorical indicators with the correct marginal proportions [138, 192], but with indicator covariances and cross correlations arbitrarily controlled by the Gaussian model. Uncontrolled indicator covariances imply realizations with uncontrolled spatial variability.

The series of threshold values that define the various categories can be made variable in space. This allows varying the proportions of each category, for example, within different depositional environments or within a stratigraphic sequence.

Another consequence of multiple truncations of the same continuous realization is that the spatial sequence of simulated categories is fixed. For example, along *any* direction the sequences of categories $1, 2, 3, 4$ or $4, 3, 2, 1$ will always be found. The algorithm does not allow generating discontinuous sequences, such as $1, 4, 3, 2$. This constraint is useful in some cases, limiting in other cases. The common continuous distribution origin of all simulated categories results in nicely continuous and nested spatial distributions for these categories.

Last, since the Gaussian RF $Y(\mathbf{u})$ being simulated is continuous, conditioning to categorical data requires further approximations [192].

GSLIB gives a utility program, gtsim, for simulation of categorical variables by multiple truncations of Gaussian realizations, e.g., as generated by program sgsim. Program gtsim is documented in Section V.7.3; see also [192].

V.3 Indicator-Based Algorithms

Indicator random function models, being binary, are ideally suited for simulating categorical variables controlled by two-point statistics such as probabilities of the type:

$$\text{Prob}\{I(\mathbf{u}) = 1, I(\mathbf{u} + \mathbf{h}) = 1\} \quad = \quad E\{I(\mathbf{u}) \cdot I(\mathbf{u} + \mathbf{h})\} \qquad (V.28)$$
$$= \quad \text{Prob}\{I(\mathbf{u}) = 1 | I(\mathbf{u} + \mathbf{h}) = 1\} \cdot p$$

$I(\mathbf{u})$ is the RF modeling the binary indicator variable $i(\mathbf{u})$ set to 1 if a certain category or event prevails at location $\mathbf{u}$, zero if not, and $E\{I(\mathbf{u})\} = p$ is the stationary proportion.

Note that the probability of transition (V.28) is the noncentered indicator covariance. Thus reproduction of an indicator covariance model through stochastic indicator simulation allows identifying a series of transition probabilities of type (V.28) for different separation vectors $\mathbf{h}$. The question is whether two-point statistics of the type (V.28) suffice to characterize the geometry of the category $\{\mathbf{u} \text{ such that } i(\mathbf{u}) = 1\}$.

If the answer to the previous question is no, then one needs to condition the indicator RF model $I(\mathbf{u})$ to higher-order statistics such as three-point statistics:

$$\text{Prob}\{I(\mathbf{u}) = 1, I(\mathbf{u} + \mathbf{h}) = 1, I(\mathbf{u} + \mathbf{h}') = 1\} = E\{I(\mathbf{u}) \cdot I(\mathbf{u} + \mathbf{h}) \cdot I(\mathbf{u} + \mathbf{h}')\}$$
$$(V.29)$$

or implicit multivariate statistics such as the shape and size of units s such that $i(\mathbf{u}) = 1, \forall \mathbf{u} \in s$.

Conditioning to explicit $3,4,\ldots,K$ variate statistics of the type (V.29) can be obtained through an annealing procedure (Section V.6 and [42]), or generalized indicator kriging and normal equations, the latter being beyond the scope of this manual [75, 105]. Conditioning to implicit multipoint statistics, such as distributions of shape and size parameters, is better done by object-based processes and other marked point processes; see below and Section V.5.

The most important contribution of the indicator formalism is the direct evaluation of conditional probabilities as required by the sequential simulation principle. Indeed,

- If the variable to be simulated is already a binary indicator $i(\mathbf{u})$, set to 1 if the location $\mathbf{u}$ belongs to category $s \subset A$, to zero otherwise, then:

$$\text{Prob}\{I(\mathbf{u}) = 1 | (n)\} = E\{I(\mathbf{u}) | (n)\} \qquad (V.30)$$

- If the variable $z(\mathbf{u})$ to be simulated is continuous, its ccdf can also be written as an indicator conditional expectation; see also relation (IV.27):

$$\text{Prob}\{Z(\mathbf{u}) \leq z | (n)\} = E\{I(\mathbf{u}; z) | (n)\} \qquad (V.31)$$

with $I(\mathbf{u}; z) = 1$ if $Z(\mathbf{u}) \leq z, = 0$ otherwise.

In both cases, the problem of evaluating the conditional probability is mapped onto that of evaluating the conditional expectation of a specific indicator RV. The evaluation of any conditional expectation calls for well-established regression theory, i.e., kriging. Kriging or cokriging applied to the proper indicator RV provides the "best"[17] estimates of the indicator conditional expectation, hence of the conditional probabilities (V.30) and (V.31).

V.3.1 Simulation of Categorical Variables

Consider the spatial distribution of K mutually exclusive categories $s_k, k = 1, \ldots, K$. This list is also exhaustive; i.e., any location $\mathbf{u}$ belongs to one and only one of these K categories.

Let $i(\mathbf{u}; s_k)$ be the indicator of class s_k, set to 1 if $\mathbf{u} \in s_k$, zero otherwise. Mutual exclusion and exhaustivity entail the following relations:

$$i(\mathbf{u}; s_k) \cdot i(\mathbf{u}; s_{k'}) = 0, \ \forall \, k \neq k' \qquad (V.32)$$

$$\sum_{k=1}^{K} i(\mathbf{u}; s_k) = 1$$

Direct kriging of the indicator variable $i(\mathbf{u}; s_k)$ provides an estimate (actually, a model) for the probability that s_k prevails at location $\mathbf{u}$. For example, using simple indicator kriging:

$$\text{Prob}^*\{I(\mathbf{u}; s_k) = 1|(n)\} = p_k + \sum_{\alpha=1}^{n} \lambda_\alpha [I(\mathbf{u}_\alpha; s_k) - p_k] \qquad (V.33)$$

where $p_k = E\{I(\mathbf{u}; s_k)\} \in [0, 1]$ is the marginal frequency of category s_k inferred, e.g., from the declustered proportion of data of type s_k.

The weights λ_α are given by an SK system of type (IV.4) using the indicator covariance of category s_k.

In cases where the average proportions p_k vary locally, one can explicitly provide the simple indicator kriging systems with smoothly varying local proportions $p_k(\mathbf{u})$'s, see [138, 192] and relation (V.29), or (implicitly) re-estimate these proportions from the indicator data available in the neighborhood of location $\mathbf{u}$. Such local re-estimation of p_k amounts to using ordinary kriging; see Sections IV.1.2 and IV.1.9.

If one category, say s_{k_0}, is predominant, its probability of occurrence may be deduced as the complement to 1 of the $(K - 1)$ other probabilities:

$$\text{Prob}^*\{I(\mathbf{u}; s_{k_0}) = 1|(n)\} = 1 - \sum_{k \neq k_0} \text{Prob}^*\{I(\mathbf{u}; s_k)|(n)\}$$

---

[17]Whenever the least-squares error criterion is used, i.e., kriging, the qualifier "best" is justified only when estimating an expected value or conditional expectation; then the least-squares criterion *must* be used in preference to any other. Specific adaptations of this LS criterion for robustness and resistance of the estimator may be in order [84].

Sequential Simulation

See Section V.1.3 and [7, 74, 102, 108]. At each node $\mathbf{u}$ along the random path, indicator kriging followed by order relation correction provides K estimated probabilities $p_k^*(\mathbf{u}|(\cdot)), k = 1, \ldots, K$. The conditioning information $(\cdot)$ consists of both the original data and the previously simulated indicator values for category s_k.

Next define *any* ordering of the K categories, say $1,2,\ldots,K$. This ordering defines a cdf-type scaling of the probability interval $[0, 1]$ with K intervals, say:

$$[0, p_1^*(\cdot)], (p_1^*(\cdot), p_2^*(\cdot) + p_1^*(\cdot)], \ldots, \left(1 - \sum_{k=1}^{K-1} p_k^*(\cdot), 1\right]$$

Draw a random number p uniformly distributed in $[0, 1]$. The interval in which p falls determines the simulated category at location $\mathbf{u}$. Update *all* K indicator data sets with this new simulated information, and proceed to the next location $\mathbf{u}'$ along the random path. The arbitrary ordering of the K probabilities $p_k^*(\cdot)$ does not affect which category is drawn nor the spatial distribution of categories [7], because of the uniform distribution of p. Sequential indicator simulation for categorical variables is implemented in program sisim presented in Section V.8.1.

V.3.2 Simulation of Continuous Variables

The spatial distribution of a continuous variable $z(\mathbf{u})$ discretized into K mutually exclusive classes $s_k : (z_{k-1}, z_k], k = 1, \ldots, K$ can be interpreted and simulated as the spatial distribution of K class indicators. The within-class resolution lost can be recovered in part by using some a priori within-class distribution, such as a power model; see Section V.1.6.

One advantage of considering the continuous variable $z(\mathbf{u})$ as a paving (mosaic) of K classes is the flexibility to model the spatial distribution of each class by a different indicator variogram. For example, the class of highest gold grades corresponding to a complex network of veinlets in fractured rocks, may be modeled by a zonal anisotropic indicator variogram with the maximum direction of continuity in the fracture direction; while the geometry of the classes of low-to-median gold grades would be modeled by more isotropic indicator variograms. The indicator formalism allows for modeling mixtures of populations loosely[18] defined as classes of values of a continuous attribute $z(\mathbf{u})$ [35, 97].

---

[18]Recall the discussion of Section V.1.1. Major heterogeneities characterized by actual categorical variables, such as lithofacies types, should be dealt with (simulated) first, e.g., through categorical indicator simulation or object-based algorithms. There are cases and/or scales where the only property recorded is a continuous variable such as a mineral grade or acoustic log; yet experience tells us that this continuous variable is measured across heterogeneous populations. In this case the indicator formalism allows a "loose" separation of populations through discretization of the range of the continuous attribute measured [97].

Another advantage of the indicator formalism over most other approaches to estimation and simulation is the possibility of accounting for soft information; see Section IV.1.12 and [6, 99, 198].

A single Gaussian RF model does not have the flexibility, i.e., enough free parameters, to handle a mixture of populations or account for soft information. Mixtures of Gaussian models with different covariances, e.g., one per lithotype, requires prior categorical data to separate the facies types: such categorical data may not be available at the scale required. Quadratic programming used to include soft information, such as inequality constraints [55], is cumbersome, CPU intensive, and provides only a partial solution. For example, quadratic programming used to enforce the active constraint $z(\mathbf{u}_\alpha) \in (a_\alpha, b_\alpha]$ limits the solutions to the bounds of the interval, a_α or b_α, rather than allowing a solution within the interval itself [99, 127].

Cumulative Class Indicators

As opposed to truly categorical populations, which need not have any order, the classes $(z_{k-1}, z_k], k = 1, \ldots, K$, of a continuous variable $z(\mathbf{u})$ are ordered sequentially. For this reason, it is better to characterize the classes with cumulative class indicators:

$$i(\mathbf{u}; z_k) = 1, \text{ if } z(\mathbf{u}) \le z_k \qquad (V.34)$$
$$= 0, \text{ otherwise}$$

The class $(z_{k-1}, z_k]$ is defined by the product $i(\mathbf{u}; z_k)[1 - i(\mathbf{u}; z_{k-1})] = 1$.

Except for the first and last classes, inference of cumulative indicator variograms is easier than inference of class indicator variograms, particularly if the classes have small marginal probabilities. Also, considering cumulative (cdf-type) indicators instead of class (pdf-type) indicators allows using indicator data that carry information across cutoffs; thus there is less of a loss of information. Last, cumulative indicators are directly related to the ccdf of the continuous variable under study; see relation (V.31).

Variogram Reproduction

Simulation from an indicator-derived ccdf guarantees reproduction of the indicator variograms for the threshold z_k's considered, up to ergodic fluctuations; see Sections V.1.3 and V.1.5 and [103]. It does not guarantee reproduction of the original z variogram, unless a full indicator cokriging with a very fine discretization (large number of thresholds K) is implemented.

Rather, it is the madogram $2\gamma_M(\mathbf{h}) = E\{|Z(\mathbf{u}) - Z(\mathbf{u} + \mathbf{h})|\}$, as inferred from expression (III.8), that is reproduced since this structural function is the integral of all indicator variograms; see [6, 74]. There is no a priori reason to privilege reproduction of the variogram over that of the madogram.

If for some reason the traditional z variogram must be reproduced, one can use the median IK option [Section IV.1.9 and relation (IV.30)] or revert to a

Gaussian-based algorithm. Note that in this latter option it is the variogram of the normal score transforms of z that is reproduced. Another option is to postprocess the z simulations by annealing to impose the required variogram; see Section V.6 and [42].

Recall that the finite class discretization introduces an additional noise (nugget effect); see related discussion in Section V.1.6. Also, corrections for order relations impact the reproduction of the indicator variograms.

Sequential Simulation

See Section V.1.3 and [72, 102, 106]. At each node $\mathbf{u}$ to be simulated along the random path indicator kriging (SK or OK) provides a ccdf model through the K probability estimates:

$$F^*(\mathbf{u}; z_k|(n)) = \text{Prob}^*\{Z(\mathbf{u}) \leq z|(n)\}, k = 1, \ldots, K$$

Within-class interpolation provides the continuum for all threshold values $z \in [z_{min}, z_{max}]$; see Section V.1.6.

Monte Carlo simulation of a realization $z^{(l)}(\mathbf{u})$ is obtained by drawing a uniform random number $p^{(l)} \in [0, 1]$ and retrieving the ccdf $p^{(l)}$ quantile:

$$z^{(l)}(\mathbf{u}) = F^{*-1}(\mathbf{u}; p^{(l)}|(n)) \tag{V.35}$$

$$\text{such that } F^*(\mathbf{u}; z^{(l)}(\mathbf{u})|(n)) = p^{(l)}$$

The indicator data set (for all cutoffs z_k) is updated from the simulated value $z^{(l)}(\mathbf{u})$, and one proceeds to the next location $\mathbf{u}'$ along the random path.

Once all locations $\mathbf{u}$ have been simulated, a stochastic image $\{z^{(l)}(\mathbf{u}), \mathbf{u} \in A\}$ is obtained. The entire sequential simulation process with a new[19] random path can be repeated to obtain another independent realization $\{z^{(l')}(\mathbf{u}), \mathbf{u} \in A\}, l' \neq l$.

Sequential indicator simulation for both continuous and categorical variables is implemented in program sisim presented in Section V.8.1. A combination of simple and ordinary indicator kriging is allowed. Soft information coded as missing indicator data [constraint intervals (IV.44)] or prior probability data (IV.45) can be considered, as long as the same indicator covariance $C_I(\mathbf{h}; z)$ is used for both hard and soft indicator data; see Section IV.1.12. If a different indicator covariance is considered for the soft prior probability data, then indicator cokriging or the Markov-Bayes algorithm [74], the latter implemented in program sisim, should be considered.

---

[19]The CPU time can be reduced considerably by keeping the same random path for each new realization. Then the sequence of conditioning data configurations, hence the kriging systems, remain the same. The sequence of kriging weights can be stored and used identically for all realizations. This savings comes at the risk of drawing realizations that are too similar. Another major saving in CPU time is possible with median indicator kriging; see Section IV.1.9 and system (IV.34). Provided there are only hard indicator data, median IK calls for solving only one kriging system (instead of K) per location $\mathbf{u}$. The price is the loss of modeling flexibility allowed by multiple indicator variograms.

Markov-Bayes Simulation

The Markov-Bayes implementation to account for soft indicator data amounts to an indicator cokriging, where the soft indicator data covariances and cross-covariances are calibrated from the hard indicator covariance models; see relations (IV.49). In all other aspects the Markov-Bayes algorithm is similar to sequential indicator simulation. The soft indicator data, i.e., the prior probability cdf's of type (IV.45), are derived from calibration scattergrams using program `bicalib`.

Warning: Artificially setting the accuracy measures $B(z)$ to high values close to one does not necessarily pull the simulated z realizations closer to the secondary data map if, simultaneously, the variances of the pre-posterior distributions $y(\mathbf{u}_\alpha; z)$ derived from the secondary data are not lowered; see definition (IV.47). Indeed with $B(z)=1$, the cdf values $y(\mathbf{u}_\alpha; z)$ are not updated and the simulation at location $\mathbf{u}_\alpha$ amounts to drawing from that cdf independently of neighboring hard or soft data: a *noisy* z simulated realization may result.

V.4 p-Field Simulation

Sequential simulation succeeds in reproducing the covariance models by treating each simulated value as a conditioning datum for simulation at all subsequent nodes. The ccdf's at each node to be simulated must then be reconstructed for each new realization. Much speed would be gained if these ccdf's could remain the same from one realization to another, e.g., by being conditioned on the sole original data. The correlation between simulated values is obtained by ensuring that the probability values used to draw from the ccdf's are themselves correlated instead of being independent one from another as in the sequential approach; see [74, 169, 170].

Let $F(\mathbf{u}; z|(n))$ and $F(\mathbf{u}'; z|(n))$ be the ccdf's at locations $\mathbf{u}$ and $\mathbf{u}'$ conditioned on *only* the original n data values. These original data values can be continuous or indicators or categories. These ccdf's can be obtained through multi-Gaussian kriging performed on normal scores transforms of the z continuous data or through indicator kriging performed on indicator data. The z simulated values are then drawn from these ccdf's using spatially correlated probability values $p^{(l)}(\mathbf{u})$ and $p^{(l)}(\mathbf{u}')$, the superscript (l) denoting the lth realization:

$$z^{(l)}(\mathbf{u}) = F^{*-1}(\mathbf{u}; p^{(l)}(\mathbf{u})|(n)) \tag{V.36}$$

such that $F^*(\mathbf{u}; z^{(l)}(\mathbf{u})|(n)) = p^{(l)}(\mathbf{u})$

The probability values $p^{(l)}(\mathbf{u})$, $p^{(l)}(\mathbf{u}')$ are spatially correlated in that they come from the same realization (l) of a "p-field," or RF $P(\mathbf{u})$, with stationary uniform distribution in $[0, 1]$ and covariance modeled from the sample covariance of the uniform transforms of the data.

Note that the p-field realizations $\{p^{(l)}(\mathbf{u}), \mathbf{u} \in A\}$ need not be conditional since the ccdf's are already conditioned to the original data. At a datum location $\mathbf{u}_\alpha$, the ccdf $F(\mathbf{u}_\alpha; z|(n))$ has zero variance and is centered on the datum value $z(\mathbf{u}_\alpha)$; hence whatever the probability value $p^{(l)}(\mathbf{u}_\alpha)$ that ccdf will always return the datum value: $z^{(l)}(\mathbf{u}_\alpha) = z(\mathbf{u}_\alpha)$.

As opposed to the sequential approach [see relation (V.35)], the p-field approach dissociates the task of conditioning, done through the ccdf's $F(\mathbf{u}; z|(n))$, and the task of covariance reproduction, done through the probability values $p^{(l)}(\mathbf{u})$.

The main advantage of the p-field approach is speed:

- the ccdf's $F(\mathbf{u}; z|(n))$ are conditioned only to the n original data; they are calculated only once and stored,

- because the p-field realizations $\{p^{(l)}(\mathbf{u}), \mathbf{u} \in A, \ l = 1, \ldots, L\}$ are non-conditional, any fast simulation algorithm can be used, such as spectral techniques [18, 76], nonconditional fractals [29, 80] or simply a stochastic moving average [109], p. 504 and [125, 170].

- each p realization is then used to draw the conditional z realization from the previsously stored ccdf's; see expression (V.36).

p-Field simulation of a variable, either continuous or categorical, is implemented in program pfsim in Section V.8.2. Both the local ccdf's and the p-field probability values are read in the program pfsim; hence they should be generated prior to using the program.

V.5   Boolean Algorithms

Boolean algorithms cover a vast category of categorical simulation algorithms and would require a whole book to discuss completely. Boolean algorithms and their extension, marked point processes [30, 157, 174], are beyond the scope of this guidebook. GSLIB is oriented more toward models characterized by two-point statistics. The practical importance of Boolean algorithms and more generally the study of spatial distributions of objects deserves mention [77, 150].

Boolean processes are generated by the distribution of geometric objects in space according to some probability laws. The spatial distribution of the object centroids constitutes a point process. A marked point process is a point process attached to (marked with) random processes defining the type, shape, and size of the random objects. For example, the point process generated by the impacts of German V2 bombs during WW II was studied by Allied scientists for an indication of remote guidance. The distribution of impervious shales in a clastic reservoir can be modeled as a Boolean or marked point process of Poisson-distributed ellipsoids or parallelepipeds with a given distribution of volumes and elongation ratios [77].

Let $\mathbf{U}$ be a vector of coordinate RVs. Further, let $\mathbf{X}_k$ be a vector of parameter RVs characterizing the geometry (shape, size, orientation) of category k. The geometry could be defined by a parametric analytical expression or by a digitized template of points. A wide variety of shapes could be generated by coordinate transformations of an original template.

The point process $\mathbf{U}$ is "marked" by the *joint* distribution of the shape random process $\mathbf{X}_k$ and the indicator random process for occurrence of category s_k. In geostatistical notation, one would consider the *joint* distribution of the 2 x K random functions (RFs) $\mathbf{X}_k(\mathbf{u}), I(\mathbf{u}; s_k), k = 1, \ldots, K, \mathbf{u} \in$ study area A, with $i(\mathbf{u}; s_k)$ set to 1 if $\mathbf{u}$ is the center of an object of category k, to zero if not.

Correlation between shape characteristics, say aspect ratio and size of an object of type k_0, could be modeled by the covariance between the two corresponding RV's of $\mathbf{X}_{k_0}$. Similarly, attraction between two objects of types k, k' could be modeled by a positive correlation between the two indicator RFs $I(\mathbf{u}; s_k)$ and $I(\mathbf{u}; s_{k'})$.

In the earth sciences, the major problem with Boolean and other marked point processes is inference. Geological lithofacies or mineral bodies are rarely of a simple parametric shape, nor are they distributed uniformly within the reservoir or deposit. The available data, even as obtained from extensive 3D outcrop sampling, rarely allows determination of the complex joint distribution of $\{\mathbf{X}_k(\mathbf{u}), I(\mathbf{u}; s_k), k = 1, \ldots, K\}$. Consequently the determination of a Boolean model is very much a trial-and-error process where various combinations of parameter distributions and interactions are tried until the final stochastic images are deemed visually satisfactory. Calibration of a Boolean model is more a matter of art (in the best sense of the term) than statistical inference. This allows for the extreme flexibility of Boolean models in reproducing very complex geometric shapes that are beyond the reach of better understood models based on only two-point or multiple-point statistics.

Other problems with Boolean processes are:

- They are difficult to condition to local data, such as the lithofacies series intersected by a well, particularly with closely spaced conditioning data. Specific adhoc solutions are available.

- The probability law (multivariate distribution) of most marked point processes is usually too complex to be analytically defined and understood. The same could be said about most RF models implicit to stochastic simulation algorithms, except the multivariate Gaussian model.

Boolean methods are typically custom-built[20] to reproduce a particular

---

[20]Most marked point processes rely on congenial rather than realistic factorable exponential-type probability distributions such as:

$$\text{Prob}\{\mathbf{X}_k = \mathbf{x}_k, k = 1, \ldots, K\} = C\exp \sum_{k} \sum_{k'} \varphi_{kk'}(\mathbf{x}_k, \mathbf{x}_{k'})$$

type of image [20, 48, 149]. There cannot be a single general Boolean conditional simulation program. The program `ellipsim` provided in GSLIB is a very simple program for generating 2D randomly distributed ellipses with specified distributions for the radius and the direction of ellipse elongation. This program could be easily modified to handle more difficult parametric shapes.

V.6   Simulated Annealing

Generating alternate conditional stochastic images of either continuous or categorical variables with the aid of the numerical technique known as "simulated annealing" is a relatively new approach [1, 42, 60, 64]. The technique has the potential of combining the reproduction of two-point statistics with complex multiple-point spatial statistics implicit to, say, geometrical shapes.

The basic idea of simulated annealing is to perturb continuously an original image until it matches some prespecified characteristics written into an objective function. Each perturbation is accepted or not depending on whether it carries the image toward the objective. To avoid local optima some unfavorable perturbations are accepted.

Although initially developed as an optimization algorithm [1], simulated annealing has been used to generate stochastic images, i.e., alternative realizations that match exactly or approximately the objective. Indeed, the number of degrees of freedom (e.g., the number of pixels that can be perturbed) usually vastly exceeds the constraints of the objective function; also there is usually tolerance on how well that objective is matched; this defines a family of "acceptable" realizations that is sampled by the stochastic process of simulated annealing [46].

The technique can be extremely CPU-intensive if used in a brute force way. The key to success of a simulated annealing application is the ability to judge very quickly the quality of the image between two perturbations and decide whether to keep the perturbation or not [45, 46]. The technique has been used to modify (postprocess) prior rough stochastic images generated by faster simulation algorithms, such as sequential Gaussian or indicator algorithms [142]; it has also been used to modify prior locations and/or sizes of objects dropped with a Boolean process until local data are properly honored [48].

Two straightforward implementations are provided in GSLIB. In the first program (`sasim`) realizations of continuous variables are achieved by allowing reproduction of a variogram model, indicator variograms, linear correlation with a secondary variable, and the reproduction of a conditional distributions

---

which allows the conditional probability of X_{k_0} to be expressed as:

$$\text{Prob}\{X_{k_0} = x_{k_0} | X_k = x_k, \forall k \neq k_0\} = \exp \sum_k \varphi_{k_0 k}(x_{k_0}, x_k)$$

C is a standardization factor and the functions φ are chosen to provide specific properties to the conditional probability [64, 174].

derived from a secondary variable. The second program (`anneal`) postprocesses or finishes prior realizations by imposing reproduction of a series of transition probabilities as obtained from a training image.

V.6.1   Simulation by Simulated Annealing

An initial image is created by relocating the conditioning data to the nearest grid nodes and then drawing all remaining node values at random from the user-specified histogram. This initial image is sequentially modified by perturbing or redrawing the value at a randomly selected grid node which is not a conditioning datum location. A perturbation is accepted if the objective function (average squared difference between the experimental and the model variogram) is lowered. Not all perturbations that raise the objective function are rejected; the success of the method depends on a slow *cooling* of the realization controlled by a temperature function that decreases with time [60, 117, 139]. The higher the temperature (or control parameter), the greater the probability that an unfavorable perturbation will be accepted. The simulation is complete when the realization is frozen, i.e., when further perturbations do not lower the objective function or when a specified minimum objective function value is reached.

Data values are reproduced by relocating them to grid nodes and disallowing any perturbation.

In the context of 3D numerical modeling, the *annealing* process may be simulated through the following steps:

1. An initial 3D numerical model (analogous to the initial melt in true annealing) is created by assigning a random value at each grid node by drawing from the population distribution. Note that if the problem is cosimulation, i.e., a secondary variable is already available, then the initial values could be assigned by drawing from the appropriate conditional distribution obtained from a calibration scatterplot.

2. An objective function (analogous to the Gibbs free energy in true annealing) is defined as a measure of difference between desired spatial features and those of the realization (e.g., the difference between the variogram of the realization and a model variogram).

3. The image is perturbed by drawing a new value for a randomly selected location (this mimics the thermal vibrations in true annealing).

4. The perturbation (thermal vibration) is always accepted if the objective is decreased; it is accepted with a certain probability if the objective is increased (the Boltzmann probability distribution of true annealing).

5. The perturbation procedure is continued while reducing the probability with which unfavorable swaps are accepted (lower the temperature parameter of the Boltzmann distribution) until a low objective function state is achieved.

Low objective function states correspond to plausible 3D realizations.

Any combination of the following five objective functions are allowed in the sasim program:

A histogram or univariate distribution is one statistical measure that stochastic realizations should honor. The cumulative distribution $F^*(z)$ of the simulated realization should match the prespecified cumulative distribution $F(z)$ for some number of z values [chosen equally to discretize the reference cumulative distribution $F(z)$]:

$$O_1 = \sum_z [F^*(z) - F(z)]^2 \qquad (V.37)$$

It is unnecessary explicitly to constrain the histogram when a series of conditional distributions specifies the marginal histogram.

A semivariogram captures the two-point spatial variability in the realization. The semivariogram $\gamma^*(h)$ of the simulated realization should match the prespecified semivariogram model $\gamma(h)$. The objective function is written:

$$O_2 = \sum_h \frac{[\gamma^*(h) - \gamma(h)]^2}{\gamma(h)^2} \qquad (V.38)$$

Division by the square of the model semivariogram value at each lag standardizes the units and gives more weight to closely spaced (low variogram) values.

Indicator semivariograms allow the explicit specification of greater or lesser continuity at low/high threshold values. An objective function to include indicator variogram(s) is written:

$$O_3 = \sum_{j=1}^{n_c} \sum_h \frac{\left[\gamma_j^*(h_i) - \gamma_j(h_i)\right]^2}{\gamma_j(h_i)^2} \qquad (V.39)$$

where n_c is the number of indicator variograms.

A correlation coefficient between the primary variable being simulated and a secondary variable (available at the same resolution) captures any linear correlation. The component objective function is:

$$O_4 = [\rho^* - \rho]^2 \qquad (V.40)$$

where ρ^* is the correlation between the primary and secondary values at all grid nodes of the realization and ρ is the target correlation coefficient, e.g., obtained from calibration data.

Conditional distributions between the primary variable being simulated and a secondary variable captures much more than a linear correlation coefficient. The objective function is written:

$$O_5 = \sum_{i=0}^{n_s} \sum_{j=0}^{n_p} [f_i^*(j) - f_i(j)]^2 \qquad (V.41)$$

where n_s and n_p are the number of secondary and primary classes, respectively; the notation $f_i(j)$ is used for the conditional distribution of the primary variable ($j = 1, \ldots, n_p$) given that the collocated secondary variable is in class i. Note that $\sum_{j=1}^{n_p} f_i(j) = 1, \ \forall i$.

The correlation coefficient and the conditional distributions allow the realizations to be constrained to a secondary variable. The implicit assumption in most cokriging is that the volume supports of the primary and secondary data are comparable. In some cases, however, the secondary variable relates to a significantly larger volume than primary variable. For example, in the context of petroleum reservoir characterization, seismic attribute data measure a complex vertical average of the petrophysical properties, which is imprecisely related to the vertical average of porosity. To handle this situation, the two objective functions (V.40) and (V.41) may be established between a vertical average of the primary variable and the secondary variable rather than a point-to-point correlation.

In general, the objective function O is made up of the weighted sum of C components:

$$O = \sum_{c=1}^{C} w_c O_c \qquad (V.42)$$

where w_c and O_c are the weights and component objective functions, respectively, and C may be a maximum of 5 in `sasim`. Each component objective function O_c could be expressed in widely different units of measurement.

The weights w_c allow equalizing the contributions of each component in the global objective function (see [42, 45]). All decisions of whether to accept or reject a perturbation are based on the change to the objective function,

$$\Delta O \ = \ O_{new} - O_{old}, \ \text{with}$$
$$\Delta O \ = \ \sum_{c=1}^{C} w_c [O_{c_{new}} - O_{c_{old}}] = \sum_{c=1}^{C} w_c \Delta O_c \qquad (V.43)$$

The weights $w_c, c = 1, \ldots, C$, are established so that, on average, each component contributes equally to the change in the objective function ΔO. That is, each weight w_c is made inversely proportional to the average change in absolute value of its component objective function:

$$w_c = \frac{1}{|\Delta O_c|}, \quad c = 1, \ldots, C \qquad (V.44)$$

In practice, the average change of each component $\overline{|\Delta O_c|}$ may not be computed analytically; it can, however, be numerically approximated by evaluating the average change due to a certain number M (say 1000) of independent perturbations:

$$\overline{|\Delta O_c|} = \frac{1}{M} \sum_{m=1}^{M} |O_c^{(m)} - O_c|, \quad c = 1, \ldots, C \qquad (V.45)$$

where $\overline{|\Delta O_c|}$ is the average change for component c, $O_c^{(m)}$ is the perturbed objective value, and O_c is the initial objective value. Each of the M perturbations, $m = 1, \ldots, M$, arises from the perturbation mechanism that will be employed for the annealing simulation. These weights are calculated within the sasim program.

An efficient coding of the objective function is essential to reduce CPU time. For example, when a perturbation is considered, the variogram lags are updated rather than recalculated; i.e., if a value z is a lag distance h from the location of z_i, the previous contribution of z_i is subtracted from the variogram and the new contribution due to z_j is added:

$$\gamma_{new}^*(h) = \gamma^*(h) + \frac{1}{2N(h)} \left[(z - z_j)^2 - (z - z_i)^2 \right] \qquad (V.46)$$

All five component objective functions in sasim allow fast updating.

In addition to the definition of an objective function, a critical aspect of simulated annealing-based simulation algorithms is a prescription for when to accept or reject a given perturbation. The acceptance probability distribution is given by the Boltzmann distribution [1]:

$$P\{\text{accept}\} = \begin{cases} 1, & \text{if } O_{new} \leq O_{old} \\ e^{\frac{O_{old} - O_{new}}{t}}, & \text{otherwise} \end{cases} \qquad (V.47)$$

All favorable perturbations ($O_{new} \leq O_{old}$) are accepted and unfavorable perturbations are accepted with an exponential probability distribution. The parameter t of the exponential distribution is analogous to the "temperature" in annealing. The higher the temperature, the more likely an unfavorable perturbation will be accepted.

The temperature t must not be lowered too fast or else the image may get trapped in a suboptimal situation and never converge. However, if lowered too slowly, then convergence may be unnecessarily slow. The specification of how to lower the temperature t is known as the "annealing schedule." There are mathematically based annealing schedules that guarantee convergence [1, 64]; however, they are much too slow for a practical application. The following empirical annealing schedule is one practical alternative [60, 154].

The idea is to start with an initially high temperature t_0 and lower it by some multiplicative factor λ whenever enough perturbations have been accepted (K_{accept}) or too many have been tried (K_{max}). The algorithm is stopped when efforts to lower the objective function become sufficiently discouraging. The following parameters describe this annealing schedule:

- t_0: the initial temperature.

- λ: the reduction factor $0 < \lambda < 1$.

- K_{max}: the maximum number of attempted perturbations at any one temperature (on the order of 100 times the number of nodes). The temperature is multiplied by λ whenever K_{max} is reached.

- K_{accept}: the acceptance target. After K_{accept} perturbations are accepted, the temperature is multiplied by λ (on the order of 10 times the number of nodes).

- S: the stopping number. If K_{max} is reached S times then the algorithm is stopped (usually set at 2 or 3).

- ΔO: a low objective function indicating convergence.

To illustrate the flexibility of the sasim program, consider the data shown in Figure V.9. A sasim run was set up to honor the 50 quantiles of the porosity histogram, 50 variogram lags, and a 0.6 correlation with the vertical average of porosity and the seismic data. This correlation coefficient of 0.60 is established from calibration data. Two realizations and the reproduction of the 0.6 correlation coefficient are shown in Figure V.10. These realizations were created in 20 seconds on an SGI workstation.

The interesting aspect of the simulated annealing algorithm is the ability to incorporate additional constraints into the objective function, e.g., specific n variate statistics with $n > 2$. For example, a series of indicator connectivity functions [105] defined as:

$$\phi(n; z_c) = E\left\{\prod_{j=1}^{n} I(\mathbf{u} + (j-1)\mathbf{h}; z_c)\right\} \qquad (V.48)$$

$$= \mathrm{Prob}\{Z(\mathbf{u} + (j-1)\mathbf{h}) \leq z_c, j = 1, \ldots, n\}$$

could be considered as part of the objective function

$$O = \sum_{k} [\phi^*(k; z_c) - \phi(k; z_c)]^2 \qquad (V.49)$$

An indicator connectivity function such as (V.48) describes the multiple-point connectivity in the direction of $\mathbf{h}$: It relates to the probability of having a string of n values z jointly lesser (greater) than a given cutoff value z_c.

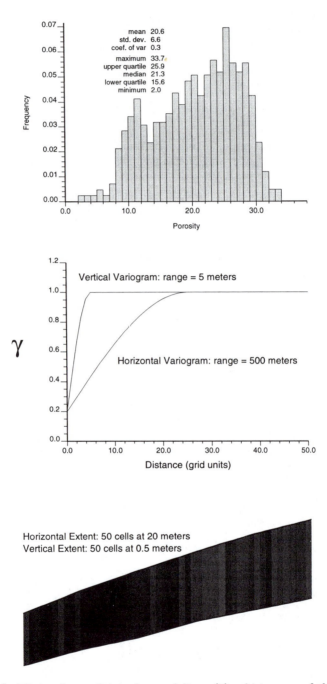

Figure V.9: Three pieces of data for modeling: (1) a histogram of the fine-scale porosity values, (2) vertical and horizontal variogram, and (3) a section of vertically averaged seismic-derived porosity.

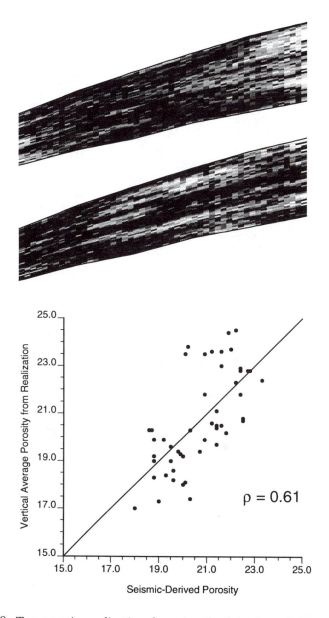

Figure V.10: Two porosity realizations honoring the data shown in Figure V.9, i.e., the histogram of porosity values, the vertical and horizontal variogram, and correlation with the vertically averaged seismic-derived porosity. Note the reproduction of the target 0.6 correlation coefficient.

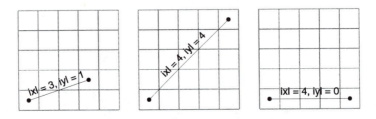

Figure V.11: Three illustrations of lag vectors defined by x and y offsets. In three dimensions there would a third z lag offset required to define each vector.

V.6.2   Postprocessing with Simulated Annealing

A useful application of simulated annealing is to postprocess images that already have desired spatial features, e.g., the result of sequential Gaussian or indicator simulation. The initial random realization of program sasim is replaced by a realization that already possesses some of the desired features, and an objective function is constructed that imposes additional spatial features.

The program anneal provided in GSLIB is meant to be a starter program for postprocessing with annealing. There are many possible control statistics that could be used in addition to the two-point histograms considered in anneal. Examples are multiple-point covariances [75], transition probabilities [142], connectivity functions [105], fidelity to secondary data (through reproduction of aspects of a calibration scatterplot), and fidelity to well test interpreted effective permeability [42]. Only two-point histograms of categorical variables are coded into anneal. Given a random variable Z that can take one of K outcomes $(k = 1, \ldots, K)$ the two-point histogram for a particular lag separation vector $\mathbf{h}$ is the set of all bivariate transition probabilities [59]:

$$p_{k,k'}(\mathbf{h}) = \text{Prob}\left\{ \begin{array}{l} Z(\mathbf{u}) \in \text{ category } k, \\ Z(\mathbf{u} + \mathbf{h}) \in \text{ category } k' \end{array} \right\} \qquad (\text{V.50})$$

independent of $\mathbf{u}$; for all $k, k' = 1, \ldots, K$. Figure V.11 gives three illustrations of lag vectors specified by x and y offsets. The objective function corresponding to this control statistic is as follows:

$$O = \sum_{h} \left(\sum_{k=1}^{K} \sum_{k'=1}^{K} \left[p_{k,k'}^{training}(\mathbf{h}) - p_{k,k'}^{realization}(\mathbf{h}) \right]^2 \right) \qquad (\text{V.51})$$

where $p_{k,k'}^{training}(\mathbf{h})$ are the target transition probabilities, for example, read directly from a training image and $p_{k,k'}^{realization}(\mathbf{h})$ are the corresponding frequencies of the realization image.

Figure V.12 illustrates an example application of the anneal program. The upper two figures are a training image and the corresponding normal score variogram used for sequential Gaussian simulation; the middle two figures are two Gaussian realizations (generated with `sgsim`) that are postprocessed into the two realizations shown on the bottom. The objective was to reproduce the two-point histogram of eight gray-level classes, for 10 lags, in two orthogonal directions aligned with the sides of the image. Note how the general character, i.e., the positioning of highs and lows, has not been changed and yet more features from the training image have been imparted to the bottom realizations.

For postprocessing applications the annealing schedule should not be started at too high a temperature; otherwise, the initial image will be randomized before it starts converging. To avoid this problem and to speed convergence the MAP (maximum a posteriori) algorithm [16, 46, 52] is used to control the acceptance mechanism. That is, the temperature is fixed at zero and the program repeatedly cycles over all grid nodes along a random path. At each grid node location all of the possible K codes are checked and the one that reduces the objective function the most is kept. True simulated annealing could be implemented by accepting unfavorable code values with a certain probability. When starting from a fairly advanced image, the problems are rarely difficult enough to warrant a full implementation of simulated annealing.

Program `anneal` does not allow for any local conditioning. The source code for conditioning to local data values may be taken from `sasim` if required.

V.6.3 Iterative Simulation Techniques

The idea behind iterative simulation techniques is similar to the basic idea underlying simulated annealing; the image is built through sets of successive perturbations of an original image. Rather than accepting or rejecting the perturbations by looking at the progress of an objective function, each set of perturbations is accepted, the hope being that the trend of modifications is always positive. Iterative techniques differ by their perturbation process and convergence criterion.

A particular implementation using the Gibbs sampler [64, 171] proceeds as follows:

1. Start from an original raster image, usually purely random but with the correct histogram. The variable involved can be continuous or categorical.

2. Precalculate the indicator weights necessary to build the ccdf at any particular node. Since all pixels are filled (occupied by a datum), except for the border nodes there is only one data configuration; hence one can afford solving a large and complex system, for example, an indicator cokriging system accounting for all indicator cross covariances.

3. Visit all nodes along a random path. At each node, discard the present value and replace it by a new simulated value drawn from that node

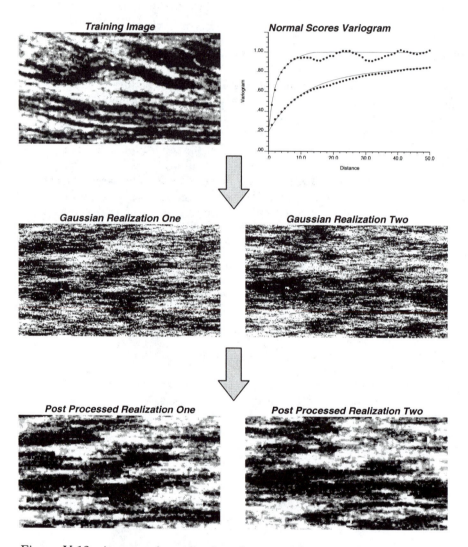

Figure V.12: An example application of `anneal`. A training image and model of the normal score variogram is shown at the top, two Gaussian realizations (generated with `sgsim`) are shown in the middle, and finally the outputs of `anneal` after postprocessing of the Gaussian realizations are shown at the bottom.

```
                    Parameters for LUSIM
                    ********************

START OF PARAMETERS:
parta.dat                          \file with data
1    2    0    3                   \   columns for X,Y,Z, normal scores
-1.0e21    1.0e21                  \   trimming limits
3                                  \debugging level: 0,1,2,3
lusim.dbg                          \file for debugging output
lusim.out                          \file for realization(s)
100                                \number of realizations
4    40.25    0.5                  \nx,xmn,xsiz
4    28.25    0.5                  \ny,ymn,ysiz
1     0.00    1.0                  \nz,zmn,zsiz
112063                             \random number seed
1    0.2                           \nst, nugget effect
1    0.8   0.0    0.0    0.0       \it,cc,ang1,ang2,ang3
           10.0   10.0   10.0      \a_hmax, a_hmin, a_vert
```

Figure V.13: An example parameter file for `lusim`.

ccdf. Building that ccdf involves only a matrix multiplication of the constant set of weights by the neighboring indicator data (these are not constant).

4. After each visit of all nodes, compute the global statistics of the image [histogram and variogram(s)] and stop the iteration if these are deemed close enough to the targets. If not, then return to step 3.

Many iterations are usually needed and convergence is not guaranteed by any theorem. The method is, however, remarkably fast because it requires only a series of matrix multiplications.

GSLIB does not presently offer a program for simulation using iterative techniques. The program ik3d can be used to establish the constant set of weights.

V.7 Gaussian Simulation Programs

V.7.1 LU Simulation `lusim`

This program requires standard normal data and writes standard normal simulated values. Normal score transforms and back transforms are to be performed outside this program. Some dimensioning parameters are entered in an include file `lusim.inc`. The GSLIB library and `lusim` program should be compiled following the instructions in Appendix B. The parameters required for the main program `lusim` are listed below and shown in Figure V.13:

- **datafl:** the input data are in a simplified Geo-EAS formatted file.

- **icolx, icoly, icolz,** and **icolvr:** the column numbers for the x, y and z coordinates, and the variable to be simulated. One or two of the coordinate column numbers can be set to zero, which indicates that the simulation is 2D or 1D.

- **tmin** and **tmax:** all values strictly less than **tmin** and strictly greater than **tmax** are ignored.

- **idbg:** an integer debugging level between 0 and 3. The higher the debugging level the more output.

- **dbgfl:** a file for the debugging output.

- **outfl:** the output grid is written to this file. The output file will contain the grid, cycling fastest on x, then y, then z, and last per simulation.

- **nsim:** the number of simulations to generate.

- **nx, xmn, xsiz:** definition of the grid system (x axis).

- **ny, ymn, ysiz:** definition of the grid system (y axis).

- **nz, zmn, zsiz:** definition of the grid system (z axis).

- **seed:** random number seed (a large odd integer).

- **nst** and **c0:** the number of variogram structures and the isotropic nugget constant.

- For each of the **nst** nested structures one must define **it**, the type of structure; **cc**, the c parameter; **ang1,ang2,ang3**, the angles defining the geometric anisotropy; $\mathbf{aa}_{hmax}$, the maximum horizontal range; $\mathbf{aa}_{hmin}$, the minimum horizontal range; and $\mathbf{aa}_{vert}$, the vertical range. A detailed description of these parameters is given in Section II.3. The semivariogram model is that of the normal scores. The kriging variance is directly interpreted as the variance of the conditional distribution; consequently, the nugget constant **c0** and c (sill) parameters should add to 1.0. Recall that the power model is not a legitimate model for a multi-Gaussian phenomenon and it is not allowed in lusim.

V.7.2 Sequential Gaussian Simulation sgsim

The sequential Gaussian algorithm (sGs) is presented in Section V.2.3. The dimensioning parameters in $\boxed{\text{sgsim.inc}}$ should be customized to each simulation study. The grid size should be set explicitly and enough storage should be allocated for the covariance lookup table. The reproduction of long-range covariance structures may be poor if the covariance lookup table is too small, see related note in Section V.11.

The GSLIB library and sgsim program should be compiled following the instructions in Appendix B. The parameters required for the main program sgsim are listed below and shown in Figure V.14:

```
                    Parameters for SGSIM
                    ********************

START OF PARAMETERS:
../data/cluster.dat            \file with data
1  2  0  3  5  0               \  columns for X,Y,Z,vr,wt,sec.var.
-1.0          1.0e21           \  trimming limits
1                              \transform the data (0=no, 1=yes)
sgsim.trn                      \  file for output trans table
0                              \  consider ref. dist (0=no, 1=yes)
histsmth.out                   \  file with ref. dist distribution
1  2                           \  columns for vr and wt
0.0     15.0                   \  zmin,zmax(tail extrapolation)
1       0.0                    \  lower tail option, parameter
1      15.0                    \  upper tail option, parameter
1                              \debugging level: 0,1,2,3
sgsim.dbg                      \file for debugging output
sgsim.out                      \file for simulation output
1                              \number of realizations to generate
50    0.5    1.0               \nx,xmn,xsiz
50    0.5    1.0               \ny,ymn,ysiz
1     0.5    1.0               \nz,zmn,zsiz
69069                          \random number seed
0    8                         \min and max original data for sim
12                             \number of simulated nodes to use
1                              \assign data to nodes (0=no, 1=yes)
1    3                         \multiple grid search (0=no, 1=yes),num
0                              \maximum data per octant (0=not used)
10.0  10.0  10.0               \maximum search radii (hmax,hmin,vert)
 0.0   0.0   0.0               \angles for search ellipsoid
4     0.60   1.0               \ktype: 0=SK,1=OK,2=LVM,3=EXDR,4=COLC
../data/ydata.dat              \  file with LVM, EXDR, or COLC variable
4                              \  column for secondary variable
1    0.1                       \nst, nugget effect
1    0.9  0.0   0.0   0.0      \it,cc,ang1,ang2,ang3
          10.0  10.0  10.0     \a_hmax, a_hmin, a_vert
```

Figure V.14: An example parameter file for **sgsim**.

- **datafl:** the input data in a simplified Geo-EAS formatted file. If this file does not exist, then an unconditional simulation will be generated.

- **icolx, icoly, icolvr, icolwt**, and **icolsec:** the column numbers for the x, y and z coordinates, the variable to be simulated, the declustering weight, and the secondary variable (e.g., for external drift if used). One or two of the coordinate column numbers can be set to zero, which indicates that the simulation is 2D or 1D. For equal weighting, set **icolwt** to zero.

- **tmin** and **tmax:** all values strictly less than **tmin** and strictly greater than **tmax** are ignored.

- **itrans:** if set to 0, then no transformation will be performed; the variable is assumed already standard normal (the simulation results will also be left unchanged). If **itrans=1**, transformations are performed.

- **transfl:** output file for the transformation table if transformation is required (**itrans=0**).

- **ismooth:** if set to 0, then the data histogram, possibly with declustering weights is used for transformation; if set to 1, then the data are transformed according to the values in another file (perhaps from histogram smoothing).

- **smthfl:** file with the values to use for transformation to normal scores (if **ismooth** is set to 1).

- **icolvr** and **icolwt:** columns in **smthfl** for the variable and the declustering weight (set to 1 and 2 if **smthfl** is the output from histsmth).

- **zmin** and **zmax** the minimum and maximum allowable data values. These are used in the back-transformation procedure.

- **ltail** and **ltpar** specify the back-transformation implementation in the lower tail of the distribution: **ltail=1** implements linear interpolation to the lower limit **zmin**, and **ltail=2** implements power model interpolation, with $\omega = $ **ltpar**, to the lower limit **zmin**.
  The middle class interpolation is linear.

- **utail** and **utpar** specify the back-transformation implementation in the upper tail of the distribution: **utail=1** implements linear interpolation to the upper limit **zmax**, **utail=2** implements power model interpolation, with $\omega = $ **utpar**, to the upper limit **zmax**, and **utail=4** implements hyperbolic model extrapolation with $\omega = $ **utpar**. The hyperbolic tail extrapolation is limited by **zmax**.

- **idbg:** an integer debugging level between 0 and 3. The larger the debugging level, the more information written out.

- **dbgfl:** the file for the debugging output.

- **outfl:** the output grid is written to this file. The output file will contain the results, cycling fastest on x, then y, then z, then simulation by simulation.

- **nsim:** the number of simulations to generate.

- **nx, xmn, xsiz:** definition of the grid system (x axis).

- **ny, ymn, ysiz:** definition of the grid system (y axis).

- **nz, zmn, zsiz:** definition of the grid system (z axis).

- **seed:** random number seed (a large odd integer).

- **ndmin** and **ndmax:** the minimum and maximum number of original data that should be used to simulate a grid node. If there are fewer than **ndmin** data points, the node is not simulated.

- **ncnode:** the maximum number of previously simulated nodes to use for the simulation of another node.

- **sstrat:** if set to 0, the data and previously simulated grid nodes are searched separately: The data are searched with a super block search, and the previously simulated nodes are searched with a spiral search (see Section II.4). If set to 1, the data are relocated to grid nodes and a spiral search is used and the parameters **ndmin** and **ndmax** are not considered.

- **multgrid:** a multiple grid simulation will be performed if this is set to 1 (otherwise a standard spiral search for previously simulated nodes is considered).

- **nmult:** the number of multiple grid refinements to consider (used only if **multgrid** is set to 1).

- **noct:** the number of original data to use per octant. If this parameter is set ≤ 0, then it is not used; otherwise, it overrides the **ndmax** parameter and the data are partitioned into octants and the closest **noct** data in each octant are retained for the simulation of a grid node.

- **radius$_{hmax}$, radius$_{hmin}$,** and **radius$_{vert}$** : the search radii in the maximum horizontal direction, minimum horizontal direction, and vertical direction (see angles below).

- **sang1, sang2,** and **sang3:** the angle parameters that describe the orientation of the search ellipsoid. See the discussion on anisotropy specification associated with Figure II.4.

- **ktype:** the kriging type (0 = simple kriging, 1 = ordinary kriging, 2 = simple kriging with a locally varying mean, 3 = kriging with an external drift, or 4 = collocated cokriging with one secondary variable) used throughout the loop over all nodes. SK is required by theory; only in cases where the number of original data found in the neighborhood is large enough can OK be used without the risk of spreading data values beyond their range of influence [87].

- **rho** and **varred:** correlation coefficient and variance reduction factor to use for collocated cokriging (used only if **ktype** = 4). The variance reduction factor modifies the kriging variance after collocated cokriging. The default should be 1.0 (unchanged); however, depending on the continuity of the secondary variable realization, the variance of the resulting model can be too high. A reduction factor of about 0.6 may be required. A better understanding of this and a more automatic correction is being studied.

- **secfl:** the file for the locally varying mean, the external drift variable, or the secondary variable for collocated cokriging (the secondary variable must be gridded at the same resolution as the model being constructed by **sgsim**)

- **nst** and **c0:** the number of semivariogram structures and the isotropic nugget constant.

- For each of the **nst** nested structures one must define **it**, the type of structure; **cc**, the c parameter; **ang1,ang2,ang3**, the angles defining the geometric anisotropy; $\mathbf{aa}_{hmax}$, the maximum horizontal range; $\mathbf{aa}_{hmin}$, the minimum horizontal range; and $\mathbf{aa}_{vert}$, the vertical range. A detailed description of these parameters is given in Section II.3. The semivariogram model is that of the normal scores. The kriging variance is directly interpreted as the variance of the conditional distribution; consequently, the nugget constant **c0** and c (sill) parameters should add to 1.0. Recall that the power model is not a legitimate model for a multi-Gaussian phenomenon.

V.7.3   Multiple Truncations of a Gaussian Field gtsim

The GSLIB library and gtsim program should be compiled following the instructions in Appendix B. The parameters required for the main program gtsim are listed below and shown in Figure V.15:

- **gausfl:** input data file with the Gaussian realizations that will be truncated to create categorical variable realizations.

- **outfl:** output data file for categorical variable realizations.

- **nsim:** number of realizations.

```
                       Parameters for GTSIM
                       *********************

START OF PARAMETERS:
sgsim.out                    \file with input Gaussian realizations
gtsim.out                    \file for output categorical realizations
1                            \number of realizations
50   50   1                  \nx,ny,nz
3                            \number of categories
1     0.25                   \   cat(1)   global proportion(1)
2     0.25                   \   cat(2)   global proportion(2)
3     0.50                   \   cat(3)   global proportion(3)
0                            \proportion curves (0=no, 1=yes)
propc01.dat                  \   file with local proportion (1)
1                            \   column number for proportion
propc02.dat                  \   file with local proportion (2)
1                            \   column number for proportion
propc03.dat                  \   file with local proportion (3)
1                            \   column number for proportion
```

Figure V.15: An example parameter file for **gtsim**.

- **nx, ny,** and **nz:** size of the simulation grid.

- **ncat:** number of categories.

- **cat()** and **proportion():** the integer codes for each category and its corresponding global proportion.

- **pcurve:** if set to 0, then there are no proportion curves and the global proportions are used to establish the thresholds for truncation. If set to 1, then proportion curve files are considered for locally varying thresholds. **ncat** − 1 proportion files are needed:

- **propfl:** file containing a regular 3D grid (same size as the input Gaussian realizations) containing the local proportions of the **ncat** − 1 categories.

- **icolprop:** column with locally varying proportion.

V.8 Sequential Indicator Simulation Programs

V.8.1 Indicator Simulation sisim

The sequential indicator simulation algorithms (sis) for categorical and continuous variables are presented in Sections V.3.1 and V.3.2. The sisim program is for the simulation of either integer-coded categorical variables or continous variables with indicator data defined from a cdf. The GSLIB library and sisim program should be compiled following the instructions in Appendix B. The parameters required for the main program sisim are listed below and shown in Figure V.16:

- **vartype:** the variable type (1=continuous, 0=categorical).

```
                        Parameters for SISIM
                        ********************

START OF PARAMETERS:
1                                \1=continuous(cdf), 0=categorical(pdf)
5                                \number thresholds/categories
0.5    1.0    2.5    5.0    10.0  \    thresholds / categories
0.12   0.29   0.50   0.74   0.88  \    global cdf / pdf
../data/cluster.dat              \file with data
1    2    0    3                  \    columns for X,Y,Z, and variable
direct.ik                        \file with soft indicator input
1    2    0    3  4  5  6  7      \    columns for X,Y,Z, and indicators
0                                \    Markov-Bayes simulation (0=no,1=yes)
0.61   0.54   0.56   0.53   0.29  \    calibration B(z) values
-1.0e21      1.0e21              \trimming limits
0.0    30.0                      \minimum and maximum data value
1      0.0                        \    lower tail option and parameter
1      1.0                        \    middle      option and parameter
1      30.0                       \    upper tail option and parameter
cluster.dat                      \    file with tabulated values
3    0                            \    columns for variable, weight
0                                \debugging level: 0,1,2,3
sisim.dbg                        \file for debugging output
sisim.out                        \file for simulation output
1                                \number of realizations
50    0.5    1.0                  \nx,xmn,xsiz
50    0.5    1.0                  \ny,ymn,ysiz
1     1.0    10.0                 \nz,zmn,zsiz
69069                            \random number seed
12                               \maximum original data  for each kriging
12                               \maximum previous nodes for each kriging
1                                \maximum soft indicator nodes for kriging
1                                \assign data to nodes? (0=no,1=yes)
0     3                          \multiple grid search? (0=no,1=yes),num
0                                \maximum per octant    (0=not used)
20.0   20.0   20.0               \maximum search radii
 0.0    0.0    0.0               \angles for search ellipsoid
0     2.5                        \0=full IK, 1=median approx. (cutoff)
0                                \0=SK, 1=OK
1      0.15                       \One   nst, nugget effect
1      0.85 0.0    0.0    0.0     \      it,cc,ang1,ang2,ang3
            10.0   10.0   10.0    \      a_hmax, a_hmin, a_vert
1      0.10                       \Two   nst, nugget effect
1      0.90 0.0    0.0    0.0     \      it,cc,ang1,ang2,ang3
            10.0   10.0   10.0    \      a_hmax, a_hmin, a_vert
1      0.10                       \Three nst, nugget effect
1      0.90 0.0    0.0    0.0     \      it,cc,ang1,ang2,ang3
            10.0   10.0   10.0    \      a_hmax, a_hmin, a_vert
1      0.10                       \Four  nst, nugget effect
1      0.90 0.0    0.0    0.0     \      it,cc,ang1,ang2,ang3
            10.0   10.0   10.0    \      a_hmax, a_hmin, a_vert
1      0.15                       \Five  nst, nugget effect
1      0.85 0.0    0.0    0.0     \      it,cc,ang1,ang2,ang3
            10.0   10.0   10.0    \      a_hmax, a_hmin, a_vert
```

Figure V.16: An example parameter file for sisim.

- **ncat:** the number of thresholds or categories.

- **cat:** the threshold values or category codes (there should be **ncat** values on this line of input).

- **pdf:** the global cdf or pdf values (there should be **ncat** values on this line of input)

- **datafl:** the input data in a simplified Geo-EAS file.

- **icolx, icoly, icolz,** and **icolvr:** the column numbers for the x, y, and z coordinates and the variable to be simulated. One or two of the coordinate column numbers can be set to zero, which indicates that the simulation is 2D or 1D.

- **directik:** already transformed indicator values are read from this file. Missing values are identified as less than **tmin**, which would correspond to a constraint interval. Otherwise, the cdf data should steadily increase from 0 to 1 and soft categorical probabilities must be between 0 to 1 and sum to 1.0.

- **icolsx, icolsy, icolsz,** and **icoli:** the columns for the x, y, and z coordinates, and the indicator variables.

- **imbsim:** set to 1 if considering Markov-Bayes option for cokriging with soft indicator data, otherwise, set to 0.

- **b(z):** if **imbsim** is set to 1, then the $B(z)$ calibration values are needed.

- **tmin** and **tmax:** all values strictly less than **tmin** and strictly greater than **tmax** are ignored.

- **zmin** and **zmax:** minimum and maximum attribute values when considering a continuous variable.

- **ltail** and **ltpar** specify the extrapolation in the lower tail: **ltail**=1 implements linear interpolation to the lower limit **zmin**; **ltail**=2 power model interpolation, with $\omega = $ **ltpar**, to the lower limit **zmin**; and **ltail** = 3 implements linear interpolation between tabulated quantiles (only for continuous variables).

- **middle** and **midpar** specify the interpolation within the middle of the distribution: **middle** = 1 implements linear interpolation; **middle** = 2 implements power model interpolation, with $\omega = $ **midpar**; and **middle**=3 allows for linear interpolation between tabulated quantile values (only for continuous variables).

- **utail** and **utpar** specify the extrapolation in the upper tail of the distribution: **utail**=1 implements linear interpolation to the upper limit **zmax**, **utail**=2 implements power model interpolation, with $\omega = $

utpar, to the upper limit **zmax**, **utail**=3 implements linear interpolation between tabulated quantiles, and **utail**=4 implements hyperbolic model extrapolation with $\omega = $ **utpar**. The hyperbolic tail extrapolation is limited by **zmax** (only for continuous variables).

- **tabfl:** If linear interpolation between tabulated values is the option selected for any of the three regions, then this simplified Geo-EAS format file is opened to read in the values. One legitimate choice is exactly the same file as the conditioning data, i.e., **datafl**. Note that **tabfl** specifies the tabulated values for all classes.

- **icolvrt** and **icolwtt:** the column numbers for the values and declustering weights in **tabfl**. Note that declustering weights can be used but are not required - just set the column number less than or equal to zero.

  If declustering weights are not used, then the class probability is split equally between the subclasses defined by the tabulated values.

- **idbg:** an integer debugging level between 0 and 3. The larger the debugging level, the more information written out.

- **dbgfl:** the file for the debugging output.

- **outfl:** the output grid is written to this file. The output file will contain the results, cycling fastest on x, then y, then z, then simulation by simulation.

- **nsim:** the number of simulations to generate.

- **nx, xmn, xsiz:** definition of the grid system (x axis).

- **ny, ymn, ysiz:** definition of the grid system (y axis).

- **nz, zmn, zsiz:** definition of the grid system (z axis).

- **seed:** random number seed (a large odd integer).

- **ndmax:** the maximum number of original data that will be used to simulate a grid node.

- **ncnode:** the maximum number of previously simulated nodes to use for the simulation of another node.

- **maxsec:** the maximum number of soft data (at node locations) that will be used for the simulation of a node. This is particularly useful to restrict the number of soft data when an exhaustive secondary variable informs all grid nodes.

- **sstrat:** if set to 0, the data and previously simulated grid nodes are searched separately: The data are searched with a super block search and the previously simulated nodes are searched with a spiral search (see Section II.4). If set to 1, the data are relocated to grid nodes and a spiral search is used; the parameters **ndmin** and **ndmax** are not considered.

- **multgrid:** a multiple grid simulation will be performed if this is set to 1 (otherwise a standard spiral search will be considered).

- **nmult:** the target number of multiple grid refinements to consider (used only if **multgrid** is set to 1).

- **noct:** the number of original data to use per octant. If this parameter is set ≤ 0, then it is not used; otherwise, the closest **noct** data in each octant are retained for the simulation of a grid node.

- **radius$_{hmax}$, radius$_{hmin}$,** and **radius$_{vert}$** : the search radii in the maximum horizontal direction, minimum horizontal direction, and vertical direction (see angles below).

- **sang1, sang2,** and **sang3:** the angle parameters that describe the orientation of the search ellipsoid. See the discussion.

- **mik** and **mikcat:** if **mik** is set to 0, then a full indicator kriging is performed at each grid node location to establish the conditional distribution. If **mik** is set to 1, then the median approximation is used; i.e., a single variogram is used for all categories; therefore, only one kriging system needs to be solved and the computer time is significantly reduced. The variogram corresponding to category **mikcat** will be used.

- **ktype:** the kriging type (0 = simple kriging, 1 = ordinary kriging) used throughout the loop over all nodes. SK is required by theory, only in cases where the number of original data found in the neighborhood is large enough can OK be used without the risk of spreading data values beyond their range of influence [87]. The global pdf values (specified with each category) are used for simple kriging.

The following set of parameters are required for each of the **ncat** categories:

- **nst** and **c0:** the number of semivariogram structures and the isotropic nugget constant.

- For each of the **nst** nested structures one must define **it**, the type of structure; **cc**, the c parameter; **ang1,ang2,ang3**, the angles defining the geometric anisotropy; **aa$_{hmax}$**, the maximum horizontal range; **aa$_{hmin}$**, the minimum horizontal range; and **aa$_{vert}$**, the vertical range. A detailed description of these parameters is given in Section II.3. Each semivariogram model refers to the corresponding indicator transform. A Gaussian variogram with a small nugget constant is not a legitimate variogram model for a discontinuous indicator function. There is no need to standardize the parameters to a sill of one since only the relative shape affects the kriging weights.

Indicator Simulation - Gridded Secondary Variable

The sisim program is inefficient in handling a gridded secondary variable in a Markov-Bayes simulation context. sisim considers the secondary variable to be irregularly spaced and the prior ccdfs have to be stored at all locations and for all primary thresholds considered. The idea behind the sisim_gs program is to store the secondary variable as a regular array and store a calibration table instead of the prior ccdfs at each threshold.

The parameters for sisim_gs are very similar to the sisim parameter file shown on Figure V.16. The parameters related to direct input of the indicator coded data are removed and the following parameters included:

```
ydata.dat              \file with gridded secondary variable
4                      \   column for secondary variable
bicalib.cal            \file with calibration table
```

The grid size of the secondary data must be the same as that being simulated. The file with the calibration table originating from bicalib contains the secondary data thresholds, the prior cdf table, and the $B(z)$ calibration values.

Indicator Simulation - Prior Mean

The sisim program does not allow incorporation of prior cdf or pdf mean values. Prior cdf or pdf information can be a valuable vehicle to incorporate soft knowledge regarding the phenomenon under study. They can also be used to incorporate in the simulation prior guesses for the lithofacies proportions, for example, stemming from geological interpretation or seismic imaging. These prior guesses could be then updated via IK in a Bayesian framework, i.e. the prior local ccdfs are updated to account for the hard data available in their neighborhoods. The sisim_lm program is included in GSLIB to provide this functionality.

The parameters for sisim_lm are very similar to the sisim parameter file shown on Figure V.16. The parameters related to direct input of indicator coded data are removed and a single new parameter, a file containing the prior mean values, is needed:

```
ydataprop.dat          \file with gridded indicator prior mean
```

The gridded prior mean values must be on the same grid as that being simulated. The file should contain only the prior means for the thresholds specified; the column numbers are taken to be 1 through *ncut* and are not read from the parameter file.

```
                    Parameters for PFSIM
                    ********************

START OF PARAMETERS:
1                            \1=continuous(cdf), 0=categorical(pdf)
1                            \1=indicator ccdfs, 0=Gaussian (mean,var)
5                            \   number thresholds/categories
0.5  1.0  2.5  5.0  10.0     \   thresholds / categories
cluster.dat                  \Within-class details: file with global dist
3    0     -1.0   1.0e21     \   ivr,   iwt
0.0      30.0                \   minimum and maximum Z value
1    1.0                     \   lower tail: option, parameter
1    1.0                     \   middle    : option, parameter
1    2.0                     \   upper tail: option, parameter
kt3d.out                     \file with input conditional distributions
1    2                       \   columns for mean, var, or ccdf values
-1.0e21    1.0e21            \   trimming limits
sgsim.out                    \file with input p-field realizations
1                            \   column number in p-field file
0                            \   0=Gaussian, 1=uniform [0,1]
pfsim.out                    \file for output realizations
1                            \number of realizations
50    50    1                \nx, ny, nz
```

Figure V.17: An example parameter file for `pfsim`.

V.8.2 p-Field Simulation `pfsim`

The `pfsim` program is for simulation of categorical or continuous variables with p-field simulation. The parameters required are documented below, and the format of the parameter file is illustrated in Figure V.17:

- **vartype:** the variable type (1=continuous, 0=categorical).

- **idist:** if set to 1, then the conditional distributions are taken to be defined by indicator thresholds. If set to 0, then the conditional distributions are Gaussian, requiring a mean and variance.

- **ncat:** number of thresholds or categories (**idist**=1 or **vartype**=0).

- **cat:** the threshold values or category codes (there should be **ncat** values on this line of input)

- **datafl:** the data file containing z data that provides the details between the IK thresholds. This file is used only if table lookup values are called for in one of the interpolation/extrapolation options.

- **icolvr** and **icolwt:** when **ltail, middle,** or **utail** is 3, then these variables identify the column for the values and the column for the (declustering) weight in **datafl** that will be used for the global distribution.

- **zmin** and **zmax:** minimum and maximum data values allowed. Even when not using linear interpolation it is safe to control the minimum and maximum values entered. The lognormal or hyperbolic upper tail is constrained by **zmax**.

- **ltail** and **ltpar** specify the extrapolation in the lower tail: **ltail**=1 implements linear interpolation to the lower limit **zmin**; **ltail**=2 implements power model interpolation, with ω =**ltpar**, to the lower limit **zmin**; and **ltail**=3 implements linear interpolation between tabulated quantiles.

- **middle** and **midpar** specify the interpolation within the middle of the distribution: **middle**=1 implements linear interpolation; **middle**=2 implements power model interpolation, with ω =**midpar**; and **middle**=3 allows for linear interpolation between tabulated quantile values.

- **utail** and **utpar** specify the extrapolation in the upper tail of the distribution: **utail**=1 implements linear interpolation to the upper limit **zmax**, **utail**=2 implements power model interpolation, with ω =**utpar**, to the upper limit **zmax**; **utail**=3 implements linear interpolation between tabulated quantiles; and, **utail**=4 implements hyperbolic model extrapolation with ω =**utpar**.

- **distfl:** input data file with the conditional distributions.

- **imean** and **ivar:** columns for the Gaussian ccdf means and variances (**idist**=0) or the ccdf values (**idist**=1).

- **tmin** and **tmax:** all values strictly less than **tmin** and strictly greater than **tmax** are ignored.

- **pfile:** file containing the input p-field probability realizations.

- **icolp:** column number in **pfile** for the p-field probability value.

- **pflag:** if set to 0, then the p-field probability values are taken to be Gaussian deviates that need to be converted to cumulative probability values. If set to 1, then the p-field probability values are already cumulative distribution values between 0 and 1.

- **outfl:** the output grid is written to this file. The output file will contain the results, cycling fastest on x, then y, then z, then simulation by simulation.

- **nsim:** number of realizations.

- **nx, ny,** and **nz:** size of the simulation grid.

V.9   A Boolean Simulation Program `ellipsim`

This program provides a very simple example of Boolean simulation: Ellipsoids of various sizes and anisotropies are dropped at random until a target proportion of points within-ellipse is met. The ellipsoids are allowed to overlap each other.

```
                    Parameters for ELLIPSIM
                    ***********************

START OF PARAMETERS:
ellipsim.out               \file for output realizations
1                          \number of realizations
100   0. 1.0               \nx,xmn,xsiz
100   0. 1.0               \ny,ymn,ysiz
1     0. 1.0               \nz,zmn,zsiz
69069                      \random number seed
0.25                       \target proportion (in ellipsoids)
 10.0   10.0    4.0  0.0 0.0 0.0    1.0 \radius[1,2,3],angle[1,2,3],weight
  2.0    2.0    2.0  0.0 0.0 0.0    1.0 \radius[1,2,3],angle[1,2,3],weight
  1.0    1.0    1.0  0.0 0.0 0.0    1.0 \radius[1,2,3],angle[1,2,3],weight
```

Figure V.18: An example parameter file for `ellipsim`.

The GSLIB library and `ellipsim` program should be compiled following the instructions in Appendix B. The parameters required for the main
program `ellipsim` are listed below and shown in Figure V.18:

- **outfl:** the simulated realizations are written to this file. The output
  file will contain the unconditional simulated indicator values (1 if in an
  ellipse, zero otherwise) cycling fastest on x, then y, then z, and then
  simulation by simulation.

- **nsim:** the number of simulations to generate.

- **nx, xsiz:** definition of the grid system (x axis).

- **ny, ysiz:** definition of the grid system (y axis).

- **nz, zsiz:** definition of the grid system (z axis).

- **seed:** random number seed (a large odd integer).

- **targprop:** the target proportion of ellipse volume.

- For the different types of ellipsoids to be added: **radiusx, radiusy,
  radiusz, angle1, angle2, angle3,** and **weight**, specify the ellipsoid
  size, orientation, and relative proportion (counting).

V.10   Simulated Annealing Programs

V.10.1   Simulated Annealing `sasim`

Program `sasim`, presented in Section V.6, allows conditional simulations of a
continuous variable honoring any combination of the following input statistics: histogram, variogram, indicator variograms, correlation coefficient with
a secondary variable, or conditional distributions with a secondary variable.

```
                    Parameters for SASIM
                    *********************

START OF PARAMETERS:
1  1  1  0  0                      \components: hist,varg,ivar,corr,cpdf
1  1  1  1  1                      \weight:    hist,varg,ivar,corr,cpdf
1                                  \0=no transform, 1=log transform
1                                  \number of realizations
50      0.5    1.0                 \grid definition: nx,xmn,xsiz
50      0.5    1.0                 \                 ny,ymn,ysiz
 1      0.5    1.0                 \                 nz,zmn,zsiz
69069                             \random number seed
4                                  \debugging level
sasim.dbg                          \file for debugging output
sasim.out                          \file for simulation output
1                                  \schedule (0=automatic,1=set below)
0.0   0.05  10   3   5   0.001     \   schedule: t0,redfac,ka,k,num,Omin
10.0  0.1                          \   maximum perturbations, reporting
100                                \   maximum number without a change
0                                  \conditioning data:(0=no, 1=yes)
../data/cluster.dat                \   file with data
1    2    0    3                   \   columns: x,y,z,attribute
-1.0e21     1.0e21                 \   trimming limits
1                                  \file with histogram:(0=no, 1=yes)
../data/cluster.dat                \   file with histogram
3    5                             \   column for value and weight
99                                 \   number of quantiles for obj. func.
1                                  \number of indicator variograms
2.78                               \   indicator thresholds
../data/seisdat.dat                \file with gridded secondary data
1                                  \   column number
1                                  \   vertical average (0=no, 1=yes)
0.60                               \correlation coefficient
../data/cal.dat                    \file with paired data
2    1    0                        \   columns for primary, secondary, wt
-0.5    100.0                      \   minimum and maximum
5                                  \   number of primary thresholds
5                                  \   number of secondary thresholds
51                                 \Variograms: number of lags
1                                  \   standardize sill (0=no,1=yes)
1    0.1                           \   nst, nugget effect
1    0.9  0.0   0.0   0.0          \   it,cc,ang1,ang2,ang3
          10.0  10.0  10.0         \   a_hmax, a_hmin, a_vert
1    0.1                           \   nst, nugget effect
1    0.9  0.0   0.0   0.0          \   it,cc,ang1,ang2,ang3
          10.0  10.0  10.0         \   a_hmax, a_hmin, a_vert
```

Figure V.19: An example parameter file for sasim.

The GSLIB library and `sasim` program should be compiled following the instructions in Appendix B. The parameters required for the main program `sasim` are listed below and shown in Figure V.19:

- **objhist, objvarg, objivar, objcorr, objcpdf:** binary flags that specify whether or not a component objective function will be considered (set to 1 if yes, 0 if not).

- **sclhist, sclvarg, sclivar, sclcorr, sclcpdf:** user-imposed scaling factors that will multiply the scaling factors that the program automatically calculates (see Section V.6).

- **ilog:** if set to 1, the logarithm (base 10) of the variable will be considered. This is appropriate for positively skewed variables with large variance.

- **nsim:** the number of realizations to generate.

- **nx, xmn, xsiz:** definition of the grid system (x axis).

- **ny, ymn, ysiz:** definition of the grid system (y axis).

- **nz, zmn, zsiz:** definition of the grid system (z axis).

- **seed:** random number seed (a large odd integer).

- **idbg:** an integer debugging level between 0 and 3.

- **dbgfl:** the file for the debugging output.

- **outfl:** the output grid is written to this file. The output file will contain the results, cycling fastest on x, then y, then z, then simulation by simulation.

- **isas:** the annealing schedule (next set of parameters) can be set explicitly or it can be set automatically (0=automatic, 1=then use the next three lines of input).

- **t0, redfac, ka, k, num, omin:** the annealing schedule: initial temperature, the reduction factor, the maximum number of perturbations at any one given temperature, and the target number of acceptable perturbations at a given temperature, the stopping number (maximum number of times that the **ka** is reached), and a low objective function value indicating convergence.

- **maxpert** and **report:** the maximum number of perturbations (will be scaled by **nx · ny · nz**). After a fixed number of perturbations (**report** scaled by **nx · ny · nz**) the objective function is written to the screen and the debugging file.

- **maxnochange:** the simulated annealing will be halted after **maxnochange** perturbations without a change.

- **icond:** set to 1 if there is conditioning data (0 implies no conditioning data).

- **condfl:** an input data file with the conditioning data (simplified Geo-EAS format). If this file does not exist, then an unconditional simulation is generated.

- **icolx, icoly, icolz,** and **icolvr:** the column numbers for the $x, y,$ and z coordinates and the variable to be simulated. One or two of the coordinate column numbers can be set to zero, which indicates that the simulation is 2D or 1D.

- **tmin** and **tmax:** all values strictly less than **tmin** and strictly greater than **tmax** are ignored.

- **ihist:** set to 1 if the histogram should be taken from the following file (set to 0 if not).

- **histfl:** an input data file with the histogram in simplified Geo-EAS format.

- **icolhvr** and **icolhwt:** the column numbers for the variable to be simulated and a declustering weight.

- **nquant:** number of quantiles for the histogram objective function.

- **nivar:** number of indicator variograms to consider. The threshold values (in units of the primary variable) are input next and the variograms are input directly after the direct variogram of the primary variable with the same format.

- **ithreshold:** the threshold values (in units of the primary variable) for the indicator variograms.

- **secfl:** an input data file with the secondary variable model (needed if cosimulation is being performed).

- **icolsec:** the column number for the secondary variable in **secfl.**

- **vertavg:** if set to 1, then the correlation applies to the secondary variable and a vertical average of the variable being simulated. Otherwise, the secondary variable is considered at each grid node location.

- **corrcoef:** the correlation coefficient (used if the fourth component objective function is turned on).

- **bivfl:** an input data file with the bivariate data to define the conditional distributions (simplified Geo-EAS format).

- **icolpri** and **icolsec:** the column numbers for the primary and secondary variables in **bivfl**.

- **tmin** and **tmax:** all values strictly less than **tmin** and strictly greater than **tmax** are ignored.

- **npricut:** number of thresholds to define the conditional distributions of the primary variable within a class of the secondary variable.

- **nseccut:** number of thresholds to define the classes of secondary variable.

- **nlag:** the number of variogram lags to consider in the objective function. The closest **nlag** lags, measured in terms of variogram distance, are retained.

- **istand:** a flag specifying whether or not to standardize the sill of the semivariogram to the variance of the univariate distribution (**istand=1** will standardize).

  It is essential that the variance of the values of the initial random image matches the spatial (dispersion) variance implied by the variogram model. That dispersion variance should be equal to the total sill if it exists (i.e., a power model has not been used) and the size of the field is much larger than the largest range in the variogram model. Otherwise, the dispersion variance can be calculated from traditional formulae ([109], pp. 61-67).

- **nst** and **c0:** the number of variogram structures and the isotropic nugget constant.

- For each of the **nst** nested structures one must define **it**, the type of structure; **cc**, the c parameter; **ang1,ang2,ang3**, the angles defining the geometric anisotropy; $\mathbf{aa}_{hmax}$, the maximum horizontal range; $\mathbf{aa}_{hmin}$, the minimum horizontal range; and $\mathbf{aa}_{vert}$, the vertical range. A detailed description of these parameters is given in Section II.3.

- Indicator variograms follow the definition of the direct primary variable variogram.

V.10.2   An Annealing Postprocessor anneal

This program requires an integer-coded training image, a series of initial integer-coded realizations (possibly only one) to postprocess, and information about what components of a two-point histogram are to be reproduced. The integer coding could be derived naturally from a categorical variable or it could be derived from the discretization of a continuous variable.

The GSLIB library and anneal program should be compiled following the instructions in Appendix B. The parameters required for the main program anneal are listed below and shown in Figure V.20:

```
                        Parameters for ANNEAL
                        **********************

START OF PARAMETERS:
sisim.out                                \file with input image(s)
ellipsim.out                             \file with training image
3    10                                  \debug level, reporting interval
postsim.dbg                              \file for debug output
postsim.out                              \file for output simulation
1                                        \number of realizations
 25    25   1                            \nx, ny, nz
69069                                    \random number seed
5    0.000001                            \maximum iterations, tolerance
4                                        \number of directions
1   0   0   10                           \ixl, iyl, izl, nlag
0   1   0   10                           \ixl, iyl, izl, nlag
1   1   0    5                           \ixl, iyl, izl, nlag
1  -1   0    5                           \ixl, iyl, izl, nlag
```

Figure V.20: An example parameter file for **anneal**.

- **datafl:** a file that contains the input realizations to postprocess. The realizations must be integer-coded. The maximum number of different integer codes is specified in **anneal.inc**. The input file has one integer per line cycling fastest on x, then y, then z, then by realization.

- **trainfl:** a file with the training image. The training image *must* have the same integer code values as the realizations to postprocess and have the same size and discretization.

- **idbg** and **report:** an integer debugging level between 0 and 3, and a reporting interval. After a fixed number of iterations (**report**) the realization is written to the output file.

- **dbgfl:** file for the debugging information.

- **outfl:** the output file that will contain the postprocessed realizations.

- **nsim:** the number of initial realizations to postprocess.

- **nx, ny,** and **nz:** definition of the grid system.

- **seed:** random number seed (a large odd integer).

- **maxit** and **tol:** the maximum number of iterations over the entire grid network. Typically 99% of the improvement is reached in 5 iterations and the remaining 1% may be reached by 10 iterations. More than 10 iterations is not recommended. The initial objective function is normalized to one; when it is lowered to **tol**, the simulation is stopped.

- **ndir:** the number of directions for the two-point histogram control. For each direction the following parameters are needed:

  ixl: the offset in the x direction for a unit lag.

iyl: the offset in the y direction for a unit lag.

izl: the offset in the z direction for a unit lag.

nlag: the number of lags.

V.11   Application Notes

- The ACORN random number generator [189] is implemented in all simulation programs; Marsaglia's random number generator [131] is available in the first edition and in the utility directory on the diskettes.

- All output files are currently written in simple ASCII format to facilitate machine independence and transfer of output to postprocessing files (gray/color-scale plotting, statistics, etc.). It is strongly advised that each installation consider some file compression scheme. More advanced approaches would consider a consistent packed format that would allow access by other programs without decompressing the entire file. A straightforward approach would be to apply a compression algorithm once the simulation is complete. With fast computers there is no need to store any realization. The realizations are completely specified by the data, the algorithm, and the random number seed. Any time a realization is needed for display or input to a transfer function it could be recreated dynamically with the specified data and algorithm.

- Since each realization is unequivocally identified by its random number seed, all realizations of the same algorithm are equiprobable inasmuch as the random number generator draws independent seed values uniformly distributed in $[0, 1]$, see related discussion in Section II.1.5.

- The procedure of drawing from a cdf constructed with indicator kriging (as implemented in `sisim`) involves simply generating a uniform random number between 0 and 1 and retaining the corresponding conditional quantile as the simulated value; see program `draw` in Section VI.2.4. In fact, the entire conditional distribution is not needed; one only needs to know the details near the drawn cdf value, i.e., the two bounding cutoffs or the tail to consider. Therefore, an efficient algorithm would draw the random number first and then focus on the z interval of interest thereby reducing the number of kriging systems that must be solved to $\log_2(ncut)$. This implementation [28] has not been distributed with GSLIB since, theoretically, order relation problems must be corrected prior to drawing from the distribution; correcting order relations requires knowledge of the complete distribution.

- The statistics (cdf, variogram, etc.) of each realization will fluctuate around the model statistics; see Section V.1.5. This fluctuation will be especially pronounced if the domain being simulated has either a small number of nodes or is small with respect to the range of correlation.

Indicator methods tend to present more fluctuations than Gaussian methods. Depending on the goal of the study, fluctuations can be reduced by postprocessing the realizations by programs such as `trans` (Section VI.2.7) or `anneal` (Section V.10.2) so that histograms and variograms are honored more closely.

- The question of which kriging method (OK or SK) to apply in the sequential Gaussian and sequential indicator methods is often asked. Theoretically and practically, simple kriging yields the most correct and the best realizations, in the sense of cdf and variogram reproduction. If the local mean data value is seen to vary significantly over the study area OK with moving data neighborhoods might be considered [87, 112]. Also the local mean values might be input directly if that additional information is available.

- Virtually all the simulation programs work with a limited data search neighborhood. The spatial structure beyond the limited neighborhood or search radius is uncontrolled and therefore random. In practice, there may be long-range spatial structures, such as involved in zonal anisotropy, that are important. The solution to this problem is to use the *multiple-grid concept*, whereby a coarse grid is simulated first [70, 180]. That coarse grid is then used as conditioning for a second, finer, grid simulation. This grid refinement can be performed multiple times, leading to a better reproduction of the long range variogram structure. The multiple-grid concept has been implemented in `sgsim` and `sisim`.

- Aligning the grid with the major directions of anisotropy and/or using a large covariance lookup table (which may call for redimensioning the array in the "include" file) will help simulated realizations reproduce very strong anisotropies such as a zonal anisotropies.

- The number of fixed cutoffs used by the indicator simulation algorithms must not be too great to reduce order relation problems. A common rule of thumb is that the number of data used in any indicator kriging should be three to five times the number of fixed cutoffs. This constraint makes it more likely that there are data within each class; when there are no data within a class, order relation problems are more likely. An alternative to fixing the cutoffs is to consider as many cutoffs as there are different data values, which requires interpolating the indicator variogram parameters from the models given at fixed cutoffs.

  When too many data are used with fixed cutoffs, negative kriging weights due to screening become significant, again causing order relation problems.

  The median IK model (IV.32) allows a large number of cutoffs to be considered without incurring the related order relation problems.

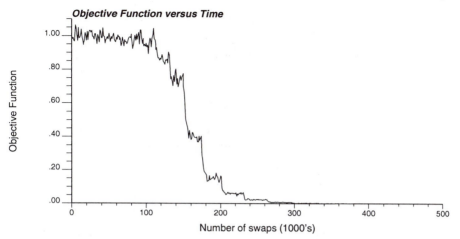

Figure V.21: An example plot of the objective function versus the number of swaps. The sudden vertical drops are when the temperature is lowered.

- Indicator kriging does not provide resolution beyond the cutoff values retained for kriging. The additional class resolution can be obtained by smoothing the resulting cdf (see program `histsmth` in Section IV.2.3), or by interpolating within the original classes using, e.g., the parameters **ltail, middle,** and **utail** in `sisim`. An alternative, equally subjective, approach consists of replacing the hard indicator data seen as step functions of the cutoff z by continuous kernel functions of z; see [123, 165] and Figure IV.2.

- When working with the annealing simulation program `sasim`, useful diagnostics are provided by the plot of the objective function versus the number of swaps; see Figure V.21. The vertical drops occur when the temperature is decreased. The magnitude of the drops is a function of the multiplicative factor λ; see Section V.6. The frequency of the drops is essentially a function of the parameter K; the parameter K_A does not play a role in the early part of the annealing procedure. The parameter K can be reduced if there is a long plateau before the objective function starts a fast decrease (as in Figure V.21). If the objective function does not decrease to near zero, then K and λ should be increased.

V.12 Problem Set Six: Simulation

The goals of this problem set are to experiment with stochastic simulation techniques, to judge the important parameters, and to gain an appreciation of the applicability of the various algorithms. Most runs are based on the data sets presented in Section II.5.

Part A: Gaussian Simulation

Consider the normal scores variogram computed in Problem Set 2 or the following isotropic semivariogram model:

$$\gamma(\mathbf{h}) = 0.30 + 0.70 \mathrm{Sph}(\frac{|\mathbf{h}|}{10.0})$$

Use the 140 clustered data in $\boxed{\text{cluster.dat}}$ and this variogram model to answer the following questions:

1. Describe briefly the sequential Gaussian simulation algorithm and the corresponding **sgsim** code with the aid of a simple flowchart.

2. Simulate point $x = 41, y = 29$ with a high debugging level and check that the solution is as expected. Use the data file $\boxed{\text{parta.dat}}$, defined in Problem Set 3, for the conditioning data and set up a parameter file from the template provided in **sgsim.par**. Comment on the results.

3. Create two conditional simulations using the 140 data with **sgsim** (use declustering weights and **nx**=50, **xmn**=0.5, **xsiz**=1.0, **ny**=50, **ymn**=0.5, **ysiz**=1.0, **nz**=1, **zmn**=0.5, **zsiz**=1.0). Use your documented judgment to choose the other parameters. Generate and plot the following graphics to aid your interpretation:

   - Gray- or color-scale maps of the simulations. Use **pixelplt**.
   - A histogram of the simulated values. Use **histplt**.
   - A Q-Q plot of the declustered data distribution and the distribution of simulated points. Use **qpplt**.
   - Run the normal scores variogram of the simulated values in one direction. Use **gam** and **vmodel** to compute the experimental variogram and the model, respectively. The program **nscore** may be used to retrieve the normal scores from the already back transformed realization.

   Comment on the results.

4. Repeat question 3 without using declustering weights. Comment on the results.

5. Build your own semivariogram model, alter the number of conditioning data, alter the input parameters significantly, and run **sgsim** to observe the consequences of your change. Comment on the sensitivity.

6. One common use of the LU approach is to build local block distributions. Discretize each block by a number of points (say 4 by 4 = 16 nodes), simulate these points simultaneously using the **lusim** program (with the surrounding data values as conditioning points), and obtain a

simulated block value by averaging the simulated point values. This is repeated, say, 100 times to obtain 100 simulated block values. Considering the same block as in the block kriging example (see Figure IV.12 in Problem Set 3) and the lusim program, build the block probability distribution.

7. Describe the truncated Gaussian simulation algorithm for simulating categorical variables and the corresponding gtsim code with the aid of a simple flowchart. Describe how to accomplish data conditioning.

8. Use gtsim to truncate an sgsim realization into three categories with global proportions of 50%, 25%, and 25%.

Part B: Indicator Simulation

The goal of this assignment is to gain an appreciation for sequential indicator simulation by experimenting with the same 140 data.

The three quartile cutoffs ($z_1=0.34$, $z_2=1.20$, $z_3=2.75$) of the declustered histogram of the 140 sample data are considered (see Figure A.5). The corresponding isotropic and standardized indicator semivariogram models[21]:

$$\gamma_I(\mathbf{h}; z_1) = 0.30 + 0.70\text{Sph}(|\mathbf{h}|/9.0)$$
$$\gamma_I(\mathbf{h}; z_2) = 0.25 + 0.75\text{Sph}(|\mathbf{h}|/10.0)$$
$$\gamma_I(\mathbf{h}; z_3) = 0.35 + 0.65\text{Sph}(|\mathbf{h}|/8.0)$$

When performing a simulation study it is good practice to select one or two representative locations and have a close look at the simulation procedure, i.e., the kriging matrices, the kriging weights, the conditional distribution, and the drawing from the distribution. It is also a good idea to evaluate the sensitivity of the realizations to critical assumptions such as the models used for extrapolation in the tails. In addition, once enough realizations have been generated, one should also check for a reasonable (not exact) reproduction of the univariate distribution and indicator variograms; see the related discussion on ergodicity in Section V.1.5.

1. Briefly describe the sequential indicator simulation algorithm and the corresponding sisim code with the aid of a simple flowchart.

2. The original 140 clustered data with an enlargement of the area around the point $x = 41$ and $y = 29$ are shown in Figure IV.9. Simulate point $x = 41, y = 29$ with a high debugging level and check that the solution is as expected. Use the data file boxed(parta.dat) for the conditioning data and set up a parameter file from the template provided in sisim. Comment on the results.

---

[21] Modeled from the 2500 exhaustive data values.

3. Create two conditional simulations using the 140 data with `sisim` (use **nx** = 50, **xmn** = 0.5, **xsiz** = 1.0, **ny** = 50, **ymn** = 0.5, **ysiz** = 1.0, **nz** = 1, **zmn** = 0.5, **zsiz** = 1.0). Use the data in cluster.dat for the distribution within all four classes and the cutoffs and semivariogram models specified earlier; use your own documented judgement to choose all other parameters. Generate and plot the following graphics to aid your interpretation:

   - Gray-scale maps of the simulations. Use `pixelplt`.

   - A histogram of the simulated values. Use `histplt`.

   - A Q-Q plot of the declustered 140 data distribution and the distribution of simulated points. Use `qpplt`.

   - The indicator variograms of simulated values in one direction. Use `gam` and `vmodel` to compute the experimental variogram and the model, respectively.

   Comment on the results.

4. Repeat question 3 using linear interpolation in the lower tail and a hyperbolic model in the upper tail. Comment on the results and compare to question 3.

5. The indicator simulation algorithm is extremely flexible in that it can handle different anisotropy at each cutoff. In this question simply expand the EW range to 90 (10:1 anisotropy) for the first cutoff and expand the NS range to 80 (10:1 anisotropy) for the last cutoff. Create two realizations, plot the gray-scale maps, and comment.

6. Build your own variogram model(s), increase the number of cutoffs, alter the number of conditioning data, possibly including soft data such as constraint intervals, then run `sisim` and comment on the consequences of your changes.

   `sisim` allows a considerable amount of flexibility and the program will go ahead and attempt to reproduce whatever spatial structure is specified. Comment on what you consider practical limitations. For example, it may not be possible to impose a strong anisotropy and have the direction of continuity switch by 90 degrees at alternating cutoffs.

Part C: Markov-Bayes Indicator Simulation

The goal of this assignment is to gain an appreciation for situations when the Markov-Bayes extension to the sequential indicator simulation algorithm is warranted.

The secondary data in ydata.dat are needed for the following questions:

1. Construct the preposterior distributions and the calibration parameter $B(z)$ values needed for sisim using the 140 primary and secondary data in cluster.dat and program bicalib.

2. Run sisim to create two alternate realizations. Compare them to realizations previously obtained in part B.

3. Everything else being equal, reduce drastically the number of hard primary data. Run sisim again and compare the realizations to the ones obtained in question 2. Comment on the impact of the soft data.

Part D: p-Field Simulation

The goal of this assignment is to gain an appreciation for situations when the p-field simulation program may be appropriate.

1. Describe the probability field simulation concept and draw a flowchart for the pfsim program included in GSLIB. Document the characteristics of a fast probability field simulation program.

2. Use ik3d to perform an indicator kriging at each of the 2500 nodes of the reference grid (see Problem Set 5). The program returns the $K = 9$ ccdf values at each node with order relation problems corrected.

3. Calculate and model the semivariogram of the uniform transform of the 140 data in cluster.dat. Renormalize the sill of this variogram to 1.0 for generation of the unconditional Gaussian realizations.

4. Construct 5 unconditional Gaussian realizations with sgsim. Note that there are much faster algorithms for unconditional simulation, e.g., spectral methods and turning bands (see program tb3d in the first edition of GSLIB).

5. Run pfsim to jointly draw 5 realizations from the ik3d ccdfs. Compare them to realizations previously obtained in part B.

6. Try various combinations of class interpolation options; comment on the results.

Part E: Categorical Variable Simulation

The goal of this assignment is to experiment with the Boolean and simulated annealing simulation programs. The goal is to generate 2D realizations of a binary variable that represents presence or absence of, say, shale. The shale bodies are approximated by ellipses elongated in the $90°$ direction with a 5:1 aspect ratio. All the shales are about 20 feet long in their major axis.

1. Use the Boolean simulation program `ellipsim` to simulate a 2D field with 25% elliptical shales. Consider a 100 by 100 foot field discretized into unit square feet.

   The variogram for circles (or ellipses after a geometric transformation) positioned at random in a matrix can be calculated analytically [40, 132, 166]. This semivariogram model, also called the *circular bombing model variogram*, is written:

   $$\gamma(\mathbf{h}) = p\left(1 - p^{\text{Circ}(\mathbf{h})}\right)$$

   where $\mathbf{h}$ is the separation vector of modulus $|\mathbf{h}|$, a is the diameter of the circles, p is the fraction outside the circles, $(1 - p)$ is the fraction within the circles, $p(1 - p)$ is the variance of the indicator $I(\mathbf{u})$, and $\text{Circ}(\mathbf{h})$ is the circular semivariogram model ([125], p. 116):

   $$\text{Circ}(\mathbf{h}) = \begin{cases} \frac{2}{\pi a^2}\left[|h|\sqrt{a^2 - |h|^2} + a^2 \sin^{-1}\left(\frac{|h|}{a}\right)\right], & |h| \leq a \\ 1, & |h| \geq a \end{cases}$$

   where a is the diameter of the circles. This indicator variogram model $\gamma(\mathbf{h})$ has the familiar $p(1 - p)$ variance form with the exponent corresponding to the hyperspherical variogram model for the dimension of the space being considered. In 1D the exponent becomes the linear variogram up to the range a; in 2D the exponent is the circular variogram model; and in 3D the exponent becomes the spherical variogram model.

   This variogram model can be written with the familiar geometric anisotropy if the circles become ellipses. Furthermore, if there are multiple ellipse sizes, then the overall variogram can be written as a nested sum of elementary bombing model variograms.

2. Use that theoretical variogram (or consider an approximate model directly inferred from one of the Boolean realizations) and `sisim` to generate two realizations.

3. Postprocess these latter two realizations with `anneal` so that they share more of the features of the elliptical training image. Use the two-point histogram for 5 lags in the two orthogonal directions aligned with the sides of the training image.

Part F: Simulated Annealing Simulation

The goal of this part is to experiment with the simulated annealing simulation program `sasim`. Consider the following normal scores semivariogram model:

$$\gamma(\mathbf{h}) = 0.10 + 0.90 \, \text{Sph}(\frac{|\mathbf{h}|}{10.0})$$

1. Create two standard normal unconditional simulations. Consider 24 lags for conditioning and use your documented judgment for the other parameters.

   Generate and plot the following graphics to aid your interpretation:

   - Gray-scale maps of the simulations. Use `pixelplt`.
   - A histogram of the simulated values. Use `histplt`.
   - Run the variogram of the simulated values in two directions. Use `gam` and `vmodel` to compute the experimental semivariogram and the model, respectively.

   Comment on the results.

2. Try conditioning to lesser (say 8) and greater (say 100) variogram lags and note the change in the realizations and the impact on CPU time.

3. Experiment with conditional simulation using the weighted nonparametric distribution corresponding to cluster.dat . Consider a one- then a two-part objective function and comment on the results.

Chapter VI

Other Useful Programs

This chapter presents utility programs that depart from the mainstream geostatistical algorithms presented in previous chapters, yet they are useful in producing the graphs and figures shown in the text. Many of the utilities documented in this chapter are available in statistical or numerical analysis software packages and textbooks.

Section VI.1 presents a set of programs that compute classical statistics and generate PostScript graphical displays. Facilities are available for histograms, normal probability plots, Q-Q/P-P plots, scatterplots, semivariogram plots, gray-scale and color-scale maps.

Section VI.2 presents utility programs for spatial declustering, univariate data transformation, solving linear systems of equations, preparation for Markov-Bayes simulation, and postprocessing of IK and simulation results.

VI.1   PostScript Display

The utility programs in this section generate graphics in the PostScript page description language [3, 4, 122]. The PostScript output from all the programs are elementary plot units that are approximately 3 inches high by 4 inches wide. In the first edition, these elementary plot units required a header and footer before they could be printed on a PostScript laser printer or displayed with a PostScript previewer. In this edition, however, the PostScript output may be directly printed or previewed. The elementary plot files may still be combined on a single 8.5 by 11 inch sheet of paper.[1]

Figure VI.1 shows the possible arrangements of elementary plot units on a page. The plot units displayed on one page can be created from any of the utility programs.

The PostScript commands for the necessary header and footer are contained in template files named according to Figure VI.1; e.g., the template

---

[1]The plotting utilities work equally well with A4 paper.

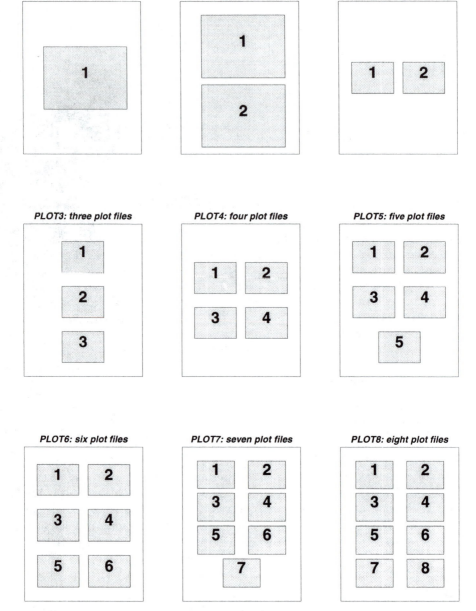

Figure VI.1:   Arrangement of the elementary plot files (shaded areas) on 8.5 by 11 inch paper achieved by the scripts PLOT1 through PLOT8.

file for PLOT3 is named $\boxed{\text{plot3.tem}}$ in the `bin` directory on the diskette. An editor may be used to read the elementary plot files into the template file. The location for each file is noted by comments like "PLOT ONE STARTS HERE:", etc.

Special C-shell script commands have been provided for machines running under the UNIX operating system. The script commands are also named according to Figure VI.1; e.g., the command for PLOT3 is named $\boxed{\text{plot3}}$ in the `bin` directory on the diskette. The names of the elementary plot files are specified as command line arguments and finally the output file, e.g.,

> `plot3 histplt.out probplt.out scatplt.out plot3.ps`

The three input files are $\boxed{\text{histplt.out}}$, $\boxed{\text{probplt.out}}$, and $\boxed{\text{scatplt.out}}$; the output file is $\boxed{\text{temp.ps}}$.

The scripts check that enough file names have been specified; e.g., PLOT3 must have four arguments - three input plot files and the output plot file name.

The motivation behind these PostScript display programs is to give initial users a way of getting quick graphical and statistical summaries of their data or results of other GSLIB subroutines. Inevitably, users will have to learn a little of the PostScript page description language if they want to customize the program output. We strongly recommend using familiar in-house or commercial statistics and graphics packages. The previous utilities have been included because they are useful to prepare report quality figures, while remaining in the GSLIB environment.

VI.1.1 Location Maps `locmap`

For reporting small 2D data sets it is often convenient to plot a location map where the magnitude of the data values are expressed with a gray- or color-scale; see Figure II.10.

The GSLIB library and `locmap` program should be compiled following the instructions in Appendix B. The parameters required for the main program `locmap` are listed below and shown in Figure VI.2:

- **datafl:** the data file in a simplified Geo-EAS format.

- **icolx, icoly,** and **icolvr:** The columns for the x variable (horizontal axis), y variable (vertical axis), and the variable (for coding the gray/color-scale of the symbol).

- **tmin** and **tmax:** all values strictly less than **tmin** and strictly greater than **tmax** are ignored.

- **outfl:** file for PostScript output.

- **xmn** and **xmx:** the limits in the x direction.

```
                    Parameters for LOCMAP
                    *********************

START OF PARAMETERS:
../data/cluster.dat            \file with data
1    2    3                     \    columns for X, Y, variable
-990.    1.0e21                 \    trimming limits
locmap.ps                       \file for PostScript output
0.0      50.                    \xmn,xmx
0.0      50.                    \ymn,ymx
0                               \0=data values, 1=cross validation
1                               \0=arithmetic,  1=log scaling
1                               \0=gray scale,  1=color scale
0                               \0=no labels,   1=label each location
0.01    10.0    1.              \gray/color scale: min, max, increm
0.5                             \label size: 0.1(sml)-1(reg)-10(big)
Locations of Clustered Data    \Title
```

Figure VI.2: An example parameter file for `locmap`.

- **ymn** and **ymx:** the limits in the y direction.

- **idata:** when data values are being posted (**idata**=0), then a circular symbol is used; when the results of cross validation are being posted (**idata**−1), then plus or minus signs are posted and filled in by the absolute magnitude of the error value.

- **ilog:** =0 then an arithmetic scaling of the gray/color-scale is used, =1 then a logarithmic scaling (base 10) is used.

- **igray:** =0 then a gray-scale is used, =1 then a color-scale is used.

- **ilabel:** =0 then no labels will be displayed; =1 then each location will be annotated with the attribute value.

- **gmin, gmax**, and **ginc:** minimum and maximum limits for the variable axis (gray- or color-scale plotting) and increment for labeling a legend bar. These do not have to be compatible with **tmin** and **tmax**. Values that are less than **gmin** or greater than **gmax** will appear just beyond either end of the gray/color spectrum.

- **labsiz:** size of the labels (0.1 is for small labels, 1 is the default size, and 10 is for large labels).

- **title:** a 40-character title for the top of the plot.

VI.1.2   Gray- and Color-Scale Maps `pixelplt`

For plotting 2D slices through gridded 3D data sets, e.g., as generated by kriging or simulation algorithms, one can use gray- or color-scale maps, see, for example, the gray-scale maps of the frontispiece.

```
                    Parameters for PIXELPLT
                    ************************

START OF PARAMETERS:
../data/true.dat                  \file with gridded data
1                                 \ column number for variable
-1.0e21  1.0e21                   \ data trimming limits
pixelplt.ps                       \file with PostScript output
1                                 \realization number
50    0.5     1.0                 \nx,xmn,xsiz
50    0.5     1.0                 \ny,ymn,ysiz
1     0.0     1.0                 \nz,zmn,zsiz
1                                 \slice orientation: 1=XY, 2=XZ, 3=YZ
1                                 \slice number
2-D Reference Data                \Title
East                              \X label
North                             \Y label
1                                 \0=arithmetic, 1=log scaling
1                                 \0=gray scale, 1=color scale
0                                 \0=continuous, 1=categorical
0.01  100.0   2.0                 \continuous:  min, max, increm.
4                                 \categorical: number of categories
1      3      Code_One            \category(), code(), name()
2      1      Code_Two
3      6      Code_Three
4     10      Code_Four


Color Codes for Categorical Variable Plotting:
       1=red, 2=orange, 3=yellow, 4=light green, 5=green, 6=light blue,
       7=dark blue, 8=violet, 9=white, 10=black, 11=purple, 12=brown,
       13=pink, 14=intermediate green, 15=gray
```

Figure VI.3: An example parameter file for `pixelplt`.

The GSLIB library and `pixelplt` program should be compiled following the instructions in Appendix B. The parameters required for the main program `pixelplt` are listed below and shown in Figure VI.3:

- **datafl:** The data file containing the 2D/3D grids with x cycling fastest, then y, then z, then per simulation.

- **icol:** column number in the input file to consider. This is typically one (1).

- **tmin** and **tmax:** all values strictly less than **tmin** and strictly greater than **tmax** are ignored.

- **outfl:** file for PostScript output.

- **ireal:** the grid or realization number to consider.

- **nx, xmn,** and **xsiz:** the number of nodes in the x direction, the origin, and the grid node separation.

- **ny, ymn,** and **ysiz:** the number of nodes in the y direction, the origin, and the grid node separation.

- **nz, zmn,** and **zsiz:** the number of nodes in the z direction, the origin, and the grid node separation.

- **slice:** the type of slice through the 3D grid to consider (1=horizontal XY slice, 2=vertical XZ slice, 3=vertical YZ slice). The axes increase to the right and to the top.

- **sliceno:** the slice number to consider (if **slice**=1 then **sliceno** must be between 1 and **nz**; if **slice**=2, then **sliceno** must be between 1 and **ny**; if **slice**=3, then **sliceno** must be between 1 and **nx**.

- **title:** a 40-character title for the top of the plot.

- **xlab:** a 40-character label that will appear centered below the plot.

- **ylab:** a 40-character label that will appear vertically centered to the left of the plot.

- **ilog:** =0, then an arithmetic scaling of the gray/color-scale is used; =1, then a logarithmic scaling (base 10) is used.

- **igray:** =0, then a gray-scale is used; =1, then a color-scale is used.

- **icont:** =0 for a continuous variable, =1 for a categorical variable.

- **gmin, gmax,** and **ginc:** minimum, maximum, and labeling increment for a continuous variable. These do not have to be compatible with **tmin** and **tmax**. Values that are less than **gmin** or greater than **gmax** will appear just beyond either end of the gray/color spectrum.

- **ncat:** number of categories.

  For each category:

- **code, colorcode,** and **name:** the integer code of the category, the color code to plot the category, and a name to put beside the color on the legend. The color codes have been chosen arbitrarily as follows: 1=red, 2=orange, 3=yellow, 4=light green, 5=green, 6=light blue, 7=dark blue, 8=violet, 9=white, 10=black, 11=purple, 12=brown, 13=pink, 14=intermediate green, and 15=gray.

VI.1.3   Histograms and Statistics `histplt`

`histplt` has been custom written to generate some univariate statistical summaries and a visual output that is compatible with a PostScript display device. The input data are a variable from a simplified Geo-EAS input file where the variable in another column can act as a weight (declustering weight, specific gravity, thickness for 2D data,...). These weights do not need to add up to one. Minimum and maximum trimming limits can be set to remove missing values and outliers. The program will automatically scale the histogram.

```
                   Parameters for HISTPLT
                   **********************

START OF PARAMETERS:
../data/cluster.dat             \file with data
3    0                          \   columns for variable and weight
-1.0e21   1.0e21                \   trimming limits
histplt.ps                      \file for PostScript output
 0.01      100.0                \attribute minimum and maximum
-1.0                            \frequency maximum (<0 for automatic)
25                              \number of classes
1                               \0=arithmetic, 1=log scaling
0                               \0=frequency,  1=cumulative histogram
200                             \   number of cum. quantiles (<0 for all)
2                               \number of decimal places (<0 for auto.)
Clustered Data                  \title
1.5                             \positioning of stats (L to R: -1 to 1)
-1.1e21                         \reference value for box plot
```

Figure VI.4: An example parameter file for `histplt`.

The user can choose to set the minimum and maximum histogram limits, number of classes, and whether or not to use a logarithmic scale.

If the histogram plot appears too erratic because of too few data, one may consider smoothing that histogram; see Section VI.2.2.

The GSLIB library and `histplt` program should be compiled following the instructions in Appendix B. The parameters required for the main program `histplt` are listed below and shown in Figure VI.4:

- **datafl:** the data file in a simplified Geo-EAS format.

- **ivr** and **iwt:** column number for the variable and the weight. If **iwt** $\leq$ 0, equal weighting is considered.

- **tmin** and **tmax:** all values strictly less than **tmin** and strictly greater than **tmax** are ignored.

- **outfl:** file for PostScript output.

- **hmin** and **hmax:** minimum and maximum plotting limits for the variable axis (abscissa). These do not have to be compatible with **tmin** and **tmax**. Values that are less than **hmin** or greater than **hmax** will appear just beyond either end of the histogram plot. They contribute to the statistics calculated. Setting **hmin** $\geq$ **hmax** will cause the program to set the limits to the minimum and maximum data values encountered.

- **fmax:** maximum frequency value for plotting vertical axis. The maximum will be determined automatically if **fmax** is set less than 0.0. This option is used to force the same histogram axis for different sets of data.

- **ncl:** the number of classes.

- **ilog:** =0, then an arithmetic scaling of the variable scale is used; =1, then a logarithmic scaling (base 10) is used.

- **icum:** =0, then a frequency histogram will be constructed; =1, then a cumulative histogram will be constructed.

- **ncum:** the number of quantiles to plot when constructing a cumulative histogram.

- **ndec:** the number of decimal places to use for the statistics annotated on the histogram plot (if set to 0, then the number will be automatically determined).

- **title:** a 40-character title for the top of the plot.

- **pos:** positioning of statistics. At times it is desirable to move the statistics from right to left to avoid overwriting the histogram bars (**pos**=-1 is the left and **pos**=1 is the right).

- **refval:** reference value for a box plot under the histogram. If this value is within the allowable range specified by **tmin** and **tmax**, then a label on the horizontal X axis will not be plotted and, instead, a box plot will be shown (outside whiskers at the 95% probability interval, inside box at the 50% probability interval, vertical solid line at the median, and black dot at the reference value specified by **refval**).

VI.1.4   Normal Probability Plots `probplt`

`probplt` generates either a normal or a lognormal probability plot. This plot displays all the data values on a chart that illustrates the general distribution shape and the behavior of the extreme values. Checking for normality or lognormality is a secondary issue. The input data are a variable from a simplified Geo-EAS input file where the variable in another column can act as a weight. The user must specify the plotting limits - only those data within the limits are shown.

The GSLIB library and `probplt` program should be compiled following the instructions in Appendix B. The parameters required for the main program `probplt` are listed below and shown in Figure VI.5:

- **datafl:** the data file in a simplified Geo-EAS format.

- **icolvr** and **icolwt:** column number for the variable and the weight. If **icolwt** ≤ 0, equal weighting is considered.

- **tmin** and **tmax:** all values strictly less than **tmin** and strictly greater than **tmax** are ignored.

- **outfl:** file for PostScript output.

```
                    Parameters for PROBPLT
                    **********************

START OF PARAMETERS:
../data/cluster.dat          \file with data
3    5                       \  columns for variable and weight
0.0    1.0e21                \  trimming limits
probplt.ps                   \file for PostScript output
99                           \number of points to plot (<0 for all)
1                            \0=arithmetic, 1=log scaling
0.01   30.0    5.0           \min,max,increment for labeling
Clustered Data               \title
```

Figure VI.5: An example parameter file for **probplt**.

```
                    Parameters for QPPLT
                    ********************

START OF PARAMETERS:
../data/cluster.dat          \file with first set of data (X axis)
3    0                       \  columns for variable and weight
../data/data.dat             \file with second set of data (Y axis)
3    0                       \  columns for variable and weight
0.0          1.0e21          \  trimming limits
qpplt.ps                     \file for PostScript output
1                            \0=Q-Q plot, 1=P-P plot
2300                         \number of points to plot (<0 for all)
0.0        20.0              \X minimum and maximum
0.0        20.0              \Y minimum and maximum
1                            \0=arithmetic, 1=log scaling
Small Data Set Versus Clustered Data   \Title
```

Figure VI.6: An example parameter file for **qpplt**.

- **ilog:** $=0$, then an arithmetic scaling of the variable scale is used; $=1$, then a logarithmic scaling (base 10) is used.

- **pmin, pmax**, and **pinc:** minimum, maximum plotting limits, and increment for the variable axis.

- **title:** a 40-character title for the top of the plot.

VI.1.5 Q-Q and P-P plots qpplt

qpplt takes univariate data from two data files and generates either a Q-Q or a P-P plot. Q-Q plots compare the quantiles of two data distributions, and P-P plots compare the cumulative probabilities of two data distributions. These graphical displays are used to compare two different data distributions, e.g., compare an original data distribution to a distribution of simulated points.

The GSLIB library and qpplt program should be compiled following the instructions in Appendix B. The parameters required for the main program qpplt are listed below and shown in Figure VI.6:

- **datafl1:** The data file for the first variable (x axis) in a simplified Geo-EAS format.

- **ivr1** and **iwt1:** column number for the variable and the weight. If **iwt1** ≤ 0, equal weighting is considered.

- **datafl2:** The data file for the second variable (y axis) in a simplified Geo-EAS format. Note: It is possible that **datafl2** is the same as **datafl1**.

- **ivr2** and **iwt2:** column number for the variable and the weight. If **iwt2** ≤ 0, equal weighting is considered.

- **tmin** and **tmax:** all values strictly less than **tmin** and strictly greater than **tmax** are ignored.

- **outfl:** file for PostScript output.

- **qqorpp:** The flag to tell the program whether to construct a Q-Q plot (**qqorpp** $= 0$) or a P-P plot (**qqorpp** $= 1$).

- **npoints:** number of points to use on the Q-Q or P-P plot (should not exceed the smallest number of data in **datafl1** / **datafl2**).

- **xmin** and **xmax:** minimum and maximum limits for x axis.

- **ymin** and **ymax:** minimum and maximum limits for y axis.

- **ilog:** $=0$, then an arithmetic scaling of the axes is used; $=1$, then a logarithmic scaling (base 10) is used. Note that the log scaling only applies to the Q-Q plot; the axis for the P-P plot is between 0 and 1.

- **title:** a 40-character title for the top of the plot.

VI.1.6   Bivariate Scatterplots and Analysis `scatplt`

A scatterplot is the simplest yet often most informative way to compare pairs of values. `scatplt` displays bivariate scatterplots and some statistical summaries for PostScript display. The summary statistics are weighted, except for the rank order (Spearman) correlation coefficient. If the scatterplot appears too erratic because of too few data, one may consider smoothing that bivariate distribution; see Section VI.2.3. Note that the plot may become quite crowded if there are too many data pairs.

The GSLIB library and `scatplt` program should be compiled following the instructions in Appendix B. The parameters required for the main program `scatplt` are listed below and shown in Figure VI.7:

- **datafl:** The data file in a simplified Geo-EAS format.

- **icolx, icoly, icolwt,** and **icol3vr:** The columns for the x variable (horizontal axis), y variable (vertical axis), weight (if less than or equal to zero then equal weighting is used; otherwise, the weight is used to

```
                    Parameters for SCATPLT
                    ***********************

START OF PARAMETERS:
../data/cluster.dat                 \file with data
4    3    5    0                     \ columns for X, Y, wt, third var.
-999.  1.0e21                        \ trimming limits
scatplt.ps                           \file for Postscript output
0.0      20.0      0                 \X min and max, (0=arith, 1=log)
0.0      20.0      0                 \Y min and max, (0=arith, 1=log)
1                                    \plot every nth data point
1.0                                  \bullet size: 0.1(sml)-1(reg)-10(big)
0.0       2.0                        \limits for third variable gray scale
Primary versus Secondary             \title
```

Figure VI.7: An example parameter file for scatplt.

calculate the summary statistics), and third variable (if less than or equal to zero, then no third variable is considered; otherwise, each bullet is gray-scale coded according to this third variable).

- **tmin** and **tmax:** all values strictly less than **tmin** and strictly greater than **tmax** are ignored.

- **outfl:** file for PostScript output.

- **xmin, xmax**, and **xlog:** minimum and maximum plotting limits for the x axis. These do not have to be compatible with the trimming limits. Pairs within the trimming limits, but outside the plotting limits, are still used for computing the correlation coefficient. Setting **xmin** $\geq$ **xmax** will cause the program to use the minimum and maximum data values encountered. If **xlog**=0, then an arithmetic scaling of the X axis is used; if **xlog**=1, then a logarithmic scaling is used.

- **ymin, ymax**, and **ylog:** minimum and maximum plotting limits for the y axis. Setting **ymin** $\geq$ **ymax** will cause the program to use the minimum and maximum data values encountered. If **ylog**=0, then an arithmetic scaling of the Y axis will be used; if **ylog**=1, then a logarithmic scaling will be used.

- **nthpoint:** plot every **nthpoint**'th data point. This should normally be set to 1, however, it may be used to subset large data sets.

- **dotsiz:** size of the black dot marking each data (0.1 is small, 1.0 is the default, and 10 is big).

- **gmin** and **gmax:** minimum and maximum gray-scale limits: used if a third variable is considered to shade each data point

- **title:** a 40-character title for the top of the plot.

```
                    Parameters for BIVPLT
                    *********************

START OF PARAMETERS:
txperm.dat                 \file with data
3    5    0                 \   columns for X variable, Y variable, and wt
1    1                      \   log scaling for X and Y (0=no, 1=yes)
-0.1    1.0e21             \   trimming limits
histsmth.por               \file with smoothed X distribution
1    2                      \   columns for variable, weight
histsmth.per               \file with smoothed Y distribution
1    2                      \   columns for variable, weight
scatsmth.out               \file with smoothed bivariate distribution
1    2    3                 \   columns for X, Y, and weight
bivplt.ps                  \file for PostScript output
1    1                      \pixel map? (0=no, 1=yes), (0=B&W, 1=color)
0.0    0.001  0.0001       \   probability min, max, and labeling increm
30                         \number of histogram classes (marginals)
Porosity-Permeability Cross Plot    \Title
```

Figure VI.8: An example parameter file for `bivplt`.

VI.1.7　Smoothed Scatterplot Display `bivplt`

The `histsmth` and `scatsmth` programs given in Section VI.2.3 smooth univariate and bivariate distributions. The `bivplt` program documented here displays the results. The plot also shows the original data pairs and the original unsmoothed data histogram.

The GSLIB library and `bivplt` program should be compiled following the instructions in Appendix B. The parameters required for the main program `bivplt` are listed below and shown in Figure VI.8:

- **datafl:** The data file with the paired data in a simplified Geo-EAS format.

- **icolx, icoly,** and **icolwt:** The columns for the x variable (horizontal axis), y variable (vertical axis), and weight (if less than or equal to zero then equal weighting is used; otherwise, the weight is used to calculate the summary statistics).

- **ilogx** and **ilogy:** set to 1 if log scaling has been used for the x or y axis, respectively.

- **tmin** and **tmax:** all values strictly less than **tmin** and strictly greater than **tmax** are ignored.

- **smoothx:** the data file with the smoothed x distribution.

- **icolxvr** and **icolxwt:** the column location for the variable and the weight (probability).

- **smoothy:** the data file with the smoothed y distribution.

- **icolyvr** and **icolywt:** the column location for the variable and the weight (probability).

- **smoothxy:** the data file with the smoothed bivariate distribution.

- **icolx, icoly**, and **icolxywt:** the column locations for the x and y variables and the weight (probability).

- **outfl:** file for PostScript output.

- **imap** and **icolor:** specify whether a pixel map of the bivariate probabilities will be created (**imap=1** will create one) and whether it will be color (**icolor=1**) or gray-scale (**icolor=0**).

- **pmin, pmax**, and **pinc:** minimum and maximum plotting limits for the pixel map.

- **title:** a 40-character title for the top of the plot.

VI.1.8 Variogram Plotting `vargplt`

The program `vargplt` takes the special output format used by the variogram programs and creates graphical displays for PostScript display devices. This program is a straightforward display program and does not provide any facility for variogram calculation or model fitting. Good interactive variogram fitting programs are available in the public domain [28, 58, 62].

The GSLIB library and `vargplt` program should be compiled following the instructions in Appendix B. `vargplt` is the only program in GSLIB that works interactively. A file name will be requested that contains the variogram results (from `gam`, `gamv`, `bigaus`, or `vmodel`). The scaling of the distance and the variogram axes may be set automatically or manually. Multiple variograms may be displayed on a single plot or a new elementary PostScript plot file may be started at any time.

VI.2 Utility Programs

This section presents some useful programs that complement the mainstream programs presented earlier. Once again, these are intended to be starting points for initial users. There are programs for coordinate and data transformation, declustering, smoothing, and some postprocessing.

VI.2.1 Coordinates `addcoord` and `rotcoord`

Two utility programs have been provided. The first adds 3D coordinates to gridded points in the ordered GSLIB output files. The second is a straightforward 2D coordinates conversion (translation and rotation).

```
              Parameters for ADDCOORD
              *************************

START OF PARAMETERS:
addcoord.dat                     \file with data
addcoord.out                     \file for output
1                                \realization number
4      5.0       10.0            \nx,xmn,xsiz
4      2.5       5.0             \ny,ymn,ysiz
2      0.5       1.0             \nz,zmn,zsiz
```

Figure VI.9: An example parameter file for **addcoord**.

```
              Parameters for ROTCOORD
              *************************

START OF PARAMETERS:
../data/cluster.dat    \file with data
1    2                 \  columns for original X and Y coordinates
rotcoord.out           \file for output with new coordinates
0.0           0.0      \origin of rotated system in original coordinates
30.0                   \rotation angle (in degrees clockwise)
0                      \0=convert to rotated coordinate system
                       \1=convert from rotated system to original system
```

Figure VI.10: An example parameter file for **rotcoord**.

The GSLIB library and **addcoord** program should be compiled following the instructions in Appendix B. The parameters required for the main program **addcoord** are listed below and shown in Figure VI.9:

- **datafl:** the input gridded data file.

- **outfl:** file for output with X, Y, and Z coordinates.

- **ireal:** the grid or realization number to consider.

- **nx, xmn,** and **xsiz:** the number of nodes in the x direction, the origin, and the grid node separation.

- **ny, ymn,** and **ysiz:** the number of nodes in the y direction, the origin, and the grid node separation.

- **nz, zmn,** and **zsiz:** the number of nodes in the z direction, the origin, and the grid node separation.

The GSLIB library and **rotcoord** program should be compiled following the instructions in Appendix B. The parameters required for the main program **rotcoord** are listed below and shown in Figure VI.10:

- **datafl:** the input data file.

- **icolx** and **icoly:** The columns for the x coordinate and y coordinate.

- **outfl:** file for output with rotated (or back-transformed) coordinates.

```
                        Parameters for DECLUS
                        *********************

START OF PARAMETERS:
../data/cluster.dat             \file with data
1    2    0    3                \ columns for X, Y, Z, and variable
-1.0e21       1.0e21            \ trimming limits
declus.sum                      \file for summary output
declus.out                      \file for output with data & weights
1.0    1.0                      \Y and Z cell anisotropy (Ysize=size*Yanis)
0                               \0=look for minimum declustered mean (1=max)
24   1.0   25.0                 \number of cell sizes, min size, max size
5                               \number of origin offsets
```

Figure VI.11: An example parameter file for declus.

- **xorigin** and **yorigin:** the origin of the rotated system in the original coordinate system.

- **angle:** the rotation angle (in degrees clockwise).

- **switch:** if **switch**=0, then rotate to the new coordinate system (the default case). If **switch**=1, however, then the input x and y coordinates are considered to be already rotated and they are back transformed to the original system.

VI.2.2   Cell Declustering declus

Data are often spatially clustered; yet, we may need to have a histogram that is representative of the entire area of interest. To obtain a representative distribution, one approach is to assign declustering weights whereby values in areas/cells with more data receive less weight than those in sparsely sampled areas. The program declus [41] provides an algorithm for determining 3D declustering weights in cases where the clusters are known to be clustered preferentially in either high- or low-valued-areas. In other cases, polygon-type declustering weights might be considered whereby the weight is made proportional to the sample area (polygon) of influence.

The declustering weights output by declus are standardized so that they sum to the number of data. A weight greater than 1.0 implies that the sample is being overweighted, and a weight lesser than 1.0 implies that the sample is being downweighted (it is clustered with other samples).

For a more detailed discussion on declustering algorithms, see [41, 89, 95]. The GSLIB library and declus program should be compiled following the instructions in Appendix B. The parameters required for the main program declus are listed below and shown in Figure VI.11:

- **datafl:** a data file in simplified Geo-EAS format containing the variable to be declustered.

- **icolx, icoly, icolz,** and **icolvr:** the column numbers for the x, y, z coordinates and the variable.

- **tmin** and **tmax:** all values strictly less than **tmin** and strictly greater than **tmax** are ignored.

- **sumfl:** the output file containing all the attempted cell sizes and the associated declustered mean values.

- **outfl:** file for PostScript output.

- **anisy** and **anisz:** the anisotropy factors to consider rectangular cells. The cell size in the x direction is multiplied by these factors to get the cell size in the y and z directions; e.g., if a cell size of 10 is being considered and **anisy**=2 and **anisz**=3, then the cell size in the y direction is 20 and the cell size in the z direction is 30.

- **minmax:** an integer flag that specifies whether a minimum mean value (**minmax**=0) or maximum mean value (**minmax**=1) is being looked for.

- **ncell, cmin,** and **cmax:** the number of cell sizes to consider, the minimum size, and the maximum size. Note that these sizes apply directly to the x direction and the **anis** factors adjust the sizes in the y and z directions.

- **noff:** the number of origin offsets. Each of the **ncell** cell sizes are considered with **noff** different original starting points. This avoids erratic results caused by extreme values falling into specific cells. A good number is 4 in 2D and 8 in 3D.

VI.2.3   Histogram and Scattergram Smoothing

Sample histograms and scattergrams tend to be erratic when the data are too few. Sawtoothlike fluctuations are usually not representative of the underlying population and they disappear as the sample size grows. The process of smoothing not only removes such fluctuations; it also allows increasing the class resolution and extending the distribution(s) beyond the sample minimum and maximum values. Geostatisticians have developed their own smoothing routines [43, 193] rather than relying on traditional kernel-type smoothing [165] since traditional methods do not honor sample statistics, such as the mean and variance, and simultaneously important quantile-type information such as the median and zero probability of getting negative values. The more flexible annealing approach [43] has been retained for smoothing histograms and scatterplots, programs histsmth and scatsmth.

Histograms

Consider the problem of assigning N probability values $p_i, i = 1, \ldots, N$, to evenly spaced z values between given minimum z_{min} and maximum z_{max}.

The equal spacing of the z_i values is

$$\Delta z = \frac{1}{N} \left(z_{max} - z_{min} \right) \tag{VI.1}$$

with $p_i \geq 0, \forall i = 1, \ldots, N$, $p_1 = 0, z_1 = z_{min}$ and $p_N = 0, z_N = z_{max}$. The idea is to choose N large (100-500) so that the resulting distribution can be reliably used for geostatistical simulation.

The final set of smoothed probabilities is established from an initial set of probabilities by successive perturbations. The perturbation mechanism consists of selecting at random two indices i and j such that $i \neq j$ and $i, j \in]1, N[$. The probability values at i and j are perturbed as follows:

$$p_i^{(new)} = p_i + \Delta p$$

$$p_j^{(new)} = p_j - \Delta p$$

Note that j and i are chosen such that $p_j \leq p_i$ and the incremental change Δp is calculated as:

$$\Delta p = 0.1 \, U \, p_j$$

where 0.1 is a constant chosen to dampen the magnitude of the perturbation (found by trial and error) and U is a random number in $[0, 1]$. There is no need to check that $p_i \leq 1.0$; it must be ≤ 1.0 since $p_i^{(new)}$ is greater than zero.

If the initial $p_i, i = 1, \ldots, N$, values are legitimate (i.e., $p_i \in [0, 1]$, $\forall i$ and $\sum_{i=1}^{N} p_i = 1$), then any set of probabilities derived from multiple applications of this perturbation mechanism is also legitimate.

For simplicity, all perturbations that lower the global objective function (see below) are accepted, and all perturbations that increase the objective function are rejected.

The following component objective functions have been considered:

- For the mean:

$$O_m = (\bar{z} - m_z)^2 \tag{VI.2}$$

where m_z is the target mean (from the data or specified by the user) and $\bar{z}$ is the average from the smoothed distribution:

$$\bar{z} = \sum_{i=1}^{N} p_i \, z_i \tag{VI.3}$$

- For the variance:

$$O_v = \left(s^2 - \sigma^2 \right)^2 \tag{VI.4}$$

where σ^2 is the target variance (from the data or specified by the user) and s^2 is the variance from the smoothed distribution:

$$s^2 = \sum_{i=1}^{N} p_i \, z_i^2 - \bar{z}^2 \tag{VI.5}$$

- For a number of quantiles:

$$O_q = \sum_{i=1}^{n_q} [cp_i - F(z_i)]^2 \qquad (VI.6)$$

where n_q is the number of imposed quantiles, cp_i is the smoothed cdf value for threshold z_i, and $F(z_i)$ is the target cumulative probability (from the data). For example, the median porosity value of 12.5 would be specified as $F(z_1) = 0.5$ and $z_1 = 12.5$. The cumulative cp probability value associated to any z threshold can be calculated by summing the p_i values until the corresponding z_i value exceeds the threshold z, i.e., first establish the index k threshold value:

$$k = \text{integer portion of } \frac{z - z_{min}}{\Delta z}$$

then the p value is computed as

$$cp = \sum_{i=1}^{k} p_i \qquad (VI.7)$$

- The smoothness of the set of probabilities $p_i, i = 1, \ldots, N$, can be measured by summing the squared difference between each p_i and a smooth $\hat{p}_i$ defined as an average of the values surrounding i, that is:

$$O_s = \sum_{i=1}^{N} (p_i - \hat{p}_i)^2 \qquad (VI.8)$$

where the $p_i, i = 1, \ldots, N$, are the smoothed probability values and $\hat{p}_i, i = 1, \ldots, N$, are local averages of the p_i values:

$$\hat{p}_i = \frac{1}{2 n_0} \sum_{k=-n_0, k \neq i}^{n_0} p_{i+k}, \quad , i = 1, \ldots, N \qquad (VI.9)$$

where n_0 is the number of values in the smoothing window (say, 5-10), $p_i = 0 \ \forall i \leq 1$ and $i \geq N$

The global objective function is defined as the sum of these four components:

$$O = \nu_m\, O_m + \nu_v\, O_v + \nu_q\, O_q + \nu_s\, O_s \qquad (VI.10)$$

the weights ν_m, ν_v, ν_q, and ν_s are computed such that each component has an equal contribution to the global objective function (see Section V.6.1). Note that additional components could be added as necessary.

The GSLIB library and `histsmth` program should be compiled following the instructions in Appendix B. The parameters required for the main program `histsmth` are listed below and shown in Figure VI.12:

```
                    Parameters for HISTSMTH
                    ************************
START OF PARAMETERS:
../data/data.dat              \file with data
3    0                        \   columns for variable and weight
-1.0e21   1.0e21              \   trimming limits
Primary Variable             \title
histsmth.ps                  \file for PostScript output
 30                          \   number of hist classes (for display)
histsmth.out                 \file for smoothed output
100      0.01   100.0        \smoothing limits: number, min, max
1                            \0=arithmetic, 1=log scaling
750      50     0.0001   69069  \maxpert, reporting interval,min Obj,Seed
1  1   1   1                 \1=on,0=off: mean,var,smth,quantiles
1.  1.  2.  2.               \weighting : mean,var,smth,quantiles
   5                         \size of smoothing window
-999.0 -999.0               \target mean and variance (-999=>data)
5                            \number of quantiles: defined from data
0                            \number of quantiles: defined by user
0.5      1.66                \   cdf, z
```

Figure VI.12: An example parameter file for `histsmth`.

- **datafl:** the data file with the raw (perhaps declustered) data.

- **icolvr** and **icolwt:** the column location for the variable and the declustering weight (0 if none available).

- **tmin** and **tmax:** all values strictly less than **tmin** and strictly greater than **tmax** are ignored.

- **title:** a 40-character title for the top of the PostScript plot.

- **psfl:** name for the PostScript output file.

- **nhist:** number of histogram classes for the PostScript output file. **nhist** is typically set less than **nz** (see below) to obtain a reasonable histogram display superimposed on the smoothed distribution.

- **outfl:** output file containing the smoothed distribution (evenly spaced z values and variable p values).

- **nz, zmin**, and **zmax:** the number N of evenly spaced z values for the smoothed histogram and the limits for the evenly spaced z values

- **ilog:** $=0$, then an arithmetic scaling is used; $=1$, then a logarithmic scaling (base 10) is used.

- **maxpert, report, omin**, and **seed:** after **maxpert·nz** perturbations the program is stopped. After **report·nz** perturbations the program reports on the current objective function(s). When the normalized objective function reaches **omin**, the program is stopped. The random number seed **seed** should be a large odd integer.

- **imean, ivari, ismth**, and **iquan:** flags for whether closeness to a target mean, closeness to a target variance, smoothness, and closeness to specified quantiles will be considered (1 = yes, 0 = no).

- **sclmean, sclvari, sclsmth**, and **sclquan:** user-imposed weights that scale the weights that the program automatically calculates.

- **nsmooth:** half of the smoothing window size.

- **mean** and **variance:** target mean and variance (if set less than -999, then they will be calculated from the input data).

- **ndq:** number of quantiles defined from the data (evenly spaced cumulative probability values).

- **nuq:** number of quantiles defined by the user (**nuq** lines must follow with a cdf and a z value). The user-defined quantiles allows the user more control over "peaks" and "valleys" in the smoothed distribution. The user should choose these quantiles to be consistent with the quantiles defined from the data.

Scatterplots

Consider the problem of finding a set of smooth bivariate probabilities $p_{i,j}$, $i = 1, \ldots, N_1$, $j = 1, \ldots, N_2$ applied to attribute z_1 (say porosity) and attribute z_2 (say permeability). There are many unknowns in the bivariate case; for example, if N_1 and N_2 are both 250, then the number of bivariate probabilities are $N_1 \cdot N_2 = 62,500$. Given a *smoothed* set of bivariate probabilities the marginal univariate probabilities are retrieved by row and column sums; that is, the univariate distribution of z_1 is given by the set of probabilities:

$$p_i = \sum_{j=1}^{N_2} p_{i,j}, \quad i = 1, \ldots, N_1$$

and the z_2 distribution by:

$$p_j = \sum_{i=1}^{N_1} p_{i,j}, \quad j = 1, \ldots, N_2$$

The approach taken here is to impose the already-smoothed univariate distributions on the smooth bivariate distribution model.

The final set of smoothed bivariate probabilities is established from an initial set of probabilities by successive perturbations. The perturbation mechanism consists of selecting a bivariate index (i', j') and considering the perturbation:

$$p_{i',j'}^{(\text{new})} = p_{i',j'} + \Delta p$$

with Δp chosen such that the candidate probability value $(p_{i',j'}^{(\text{new})})$ is positive.

$$\Delta p = 0.1 \, (U - 0.5) \, p_{i',j'}$$

where 0.1 is a constant chosen to dampen the magnitude of the perturbation (found by trial-and-error) and U is a random number in $[0, 1]$.

The physical constraint on the sum of the bivariate probabilities

$$\sum_{i=1}^{i=N_1} \sum_{j=1}^{j=N_2} p_{i,j} = 1$$

is imposed indirectly through the constraint of consistency with both marginal distributions; see below.

For simplicity, all perturbations that lower the global objective function are accepted and all perturbations that increase the objective function are rejected.

The following component objective functions have been considered in the smoothing of bivariate distributions:

- For consistency with the two marginal distributions:

$$O_m = \sum_{i=1}^{N_1} \left[\left(\sum_{j=1}^{N_2} p_{i,j} \right) - p_i^* \right]^2 + \sum_{j=1}^{N_2} \left[\left(\sum_{i=1}^{N_1} p_{i,j} \right) - p_j^* \right]^2 \qquad (\text{VI.11})$$

where p_i^* is the target smoothed marginal probability of z_{1_i} and p_j^* is the target smoothed marginal probability of z_{2_j}.

- For the correlation coefficient:

$$O_c = (\rho - \rho^*)^2 \qquad (\text{VI.12})$$

where ρ is the correlation coefficient from the bivariate probabilities and ρ^* is the target correlation coefficient (from the data or specified by the user).

- For a number of bivariate quantiles:

$$O_q = \sum_{i=1}^{n_q} [cp_i - F(z_{1_i}, z_{2_i})]^2 \qquad (\text{VI.13})$$

where n_q is the number of imposed quantiles and $F(z_{1_i}, z_{2_i})$ is the i th target cumulative probability. The cumulative cp probability value associated with any z_{1_i}, z_{2_i} pair can be calculated by summing the $p_{i,j}$ values for all z_1, z_2 threshold values less that the threshold values z_{1_i}, z_{2_i}.

- The smoothness of the set of probabilities $p_{i,j}, i = 1, \ldots, N_1, j = 1, \ldots,$ N_2 may be quantified by summing the squared difference between each $p_{i,j}$ and a smooth $\hat{p}_{i,j}$ defined as an average of the values surrounding i, that is:

$$O_s = \sum_{i=1}^{N_1} \sum_{j=1}^{N_2} (p_{i,j} - \hat{p}_{i,j})^2 \qquad\qquad (VI.14)$$

where the $p_{i,j}, i = 1, \ldots, N_1, j = 1, \ldots, N_2$, are the smoothed probability values and $\hat{p}_{i,j}, i = 1, \ldots, N_1, j = 1, \ldots, N_2$, are local averages of the $p_{i,j}$ values. These local averages are defined within an elliptical window. The major axis and the anisotropy of the elliptical window are based on the target correlation, i.e.,

$$\text{angle from } Z_1 \text{ axis} \quad = \quad \tan^{-1}\left(\frac{\sum_{i=1}^{n}(z_{1_i} - \overline{z_1})^2}{\sum_{i=1}^{n}(z_{2_i} - \overline{z_2})^2}\right)$$

$$\text{anisotropy} \quad = \quad \sqrt{1 - \rho^{*2}}$$

where $z_{1_i}, z_{2_i}, i = 1, \ldots, n$, are the available data pairs, $\overline{z_1}$ is the corresponding average of z_1, $\overline{z_2}$ is the average of z_2, and ρ^* is the target correlation coefficient. The number of points in the smoothing ellipse is an input parameter.

The GSLIB library and scatsmth program should be compiled following the instructions in Appendix B. The parameters required for the main program scatsmth are listed below and shown in Figure VI.13:

- **datafl:** the data file with the raw (perhaps declustered) paired observations.

- **icolx, icoly,** and **icolwt:** the column location for the z_1 variable, the z_2 variable, and the declustering weight (0 if none available).

- **smoothx:** the data file with the smoothed z_1 (X) distribution.

- **icolxvr** and **icolxwt:** the column location for the z_1 variable and the weight (probability).

- **smoothy:** the data file with the smoothed z_2 (Y) distribution.

- **icolyvr** and **icolywt:** the column location for the z_2 variable and the weight (probability).

- **dbgfl:** name for the output file for debugging information.

- **finalxfl:** name for the output file of the final X distribution - to check reproduction of the marginal distribution.

- **finalyfl:** name for the output file of the final Y distribution - to check reproduction of the marginal distribution.

```
                    Parameters for SCATSMTH
                    ************************

START OF PARAMETERS:
txperm.dat                      \file with data
3    5    0                     \  columns for X, Y, weight
histsmth.por                    \file with smoothed X distribution
1    2                          \  columns for variable, weight
histsmth.per                    \file with smoothed Y distribution
1    2                          \  columns for variable, weight
0    0                          \log scaling flags
scatsmth.dbg                    \file for debug information
scatsmth.xr                     \file for final X distribution
scatsmth.yr                     \file for final Y distribution
scatsmth.out                    \file for output
150.  1.0   0.0001   69069      \maxpert, report, min obj, seed
1  1  1  1                      \1=on,0=off: marg,corr,smth,quan
1  1  2  2                      \weighting : marg,corr,smt,hquan
 25                             \smoothing Window Size (number)
-999.                           \correlation (-999 take from data)
9  9                            \number of X and Y quantiles
0                               \points defining envelope (0=none)
   0.0      0.0                 \   x(i)    y(i)
   0.0    999.0                 \   x(i)    y(i)
 999.0    999.0                 \   x(i)    y(i)
 999.0      0.0                 \   x(i)    y(i)
```

Figure VI.13: An example parameter file for `scatsmth`.

- **outfl:** the output file. This file will contain the N_1 (nx) by N_2 (ny) X, Y, and probability values.

- **maxpert, report, omin**, and **seed:** after **maxpert·nz** perturbations the program is stopped. After **report·nz** perturbations the program reports on the current objective function(s). When the normalized objective function reaches **omin**, the program is stopped. The random number seed **seed** should be a large odd integer.

- **imarg, icorr, ismth**, and **iquan:** flags for whether closeness to the marginal histograms, closeness to a target correlation coefficient, smoothness, and closeness to specified quantiles are considered (1 = yes, 0 = no).

- **sclmarg, sclcorr, sclsmth**, and **sclquan:** user imposed weights that scale the weights that the program automatically calculates (default is 1.0 for all weights).

- **nsmooth:** size of the smoothing window.

- **correlation:** target correlation coefficient.

- **ndqx** and **ndqy:** number of quantiles defined from the data (evenly spaced cumulative probability values in x and y).

- **nenvelope:** number of points defining the envelope containing the bivariate distribution (may be set to zero so that all points in the bivariate distribution are allowable).

- **x(),y():** points defining the allowable bivariate envelope. This "envelope" is a practical feature that restricts the nonzero bivariate probability values to fall within some user-specified region.

VI.2.4   Random Drawing draw

The core of all stochastic simulation programs is a drawing from a distribution function, e.g., from a ccdf modeling the uncertainty at the location **u** being simulated. In all GSLIB simulation programs this is done by reading the quantile value $z = q(p)$ corresponding to a random number p uniformly distributed in $[0, 1]$; see [157]:

$$z = q(p) = F^{-1}(p), \text{ such that } F(z) = F(q(p)) = p \in [0, 1]$$

The cumulative histogram of the series of values $z^{(l)} = q(p^{(l)}), l = 1, \ldots, L,$ reproduces the target cdf $F(z)$ if the $p^{(l)}$'s are drawn independently.

Drawing from any K-variate cdf is done through a sequence of K drawings from conditional univariate distributions; see Section V.1.3. Consider the task of drawing L pairs of values $(z_1^{(l)}, z_2^{(l)}), l = 1, \ldots, L,$ from a target bivariate cdf $F(z_1, z_2) = \text{Prob}\{Z_1 \leq z_1, Z_2 \leq z_2\}$ with marginal cdf's $F(z_1) - \text{Prob}\{Z_1 \leq z_1\}$ and $F(z_2) = \text{Prob}\{Z_2 \leq z_2\}$, and conditional cdf's $F_{1|2}(z_1|k) = \text{Prob}\{Z_1 \leq z_1|Z_2 \in (z_{2_{k-1}}, z_{2_k}]\}$. The procedure consists of:

1. Two independent random numbers $p_1^{(l)}$, $p_2^{(l)}$ are drawn uniformly distributed in $[0, 1]$.

2. The $p_2^{(l)}$ quantile of the Z_2 distribution gives the simulated value $z_2^{(l)} = F_2^{-1}(p_2^{(l)})$. Let k be the class to which $z_2^{(l)}$ belongs: $z_2^{(l)} \in (z_{2_{k-1}}, z_{2_k}]$.

3. The $p_1^{(l)}$ quantile of the Z_1 conditional distribution $F_{1|2}(z_1|k)$ gives the corresponding simulated value $z_1^{(l)} = F_{1|2}^{-1}(p_1^{(l)}|k)$.

Steps 1 to 3 are repeated L times to draw L independent pairs $(z_1^{(l)}, z_2^{(l)}), l = 1, \ldots, L$. The cumulative bivariate scattergram of these L pairs reproduces the target bivariate cdf $F(z_1, z_2)$.

The **draw** program, documented below, reads as input a K-variate probability density function, computes the corresponding K-variate cdf and draws a user-specified number L samples from that distribution. The input probability density function could be the equal weighted data distribution, the declustered distribution, a smoothed version of the data distribution (e.g., the output of **histsmth** or **scatsmth**), or any user-specified distribution.

The GSLIB library and **draw** program should be compiled following the instructions in Appendix B. The parameters required for the main program **draw** are listed below and shown in Figure VI.14:

- **datafl:** a data file in simplified Geo-EAS format containing the K-variate distribution.

```
                    Parameters for DRAW
                    *******************

START OF PARAMETERS:
../data/cluster.dat          \file with data
3                            \   number of variables
1   2   3                    \   columns for variables
5                            \   column for probabilities (0=equal)
-1.0e21     1.0e21           \   trimming limits
69069     100                \random number seed, number to draw
draw.out                     \file for realizations
```

Figure VI.14: An example parameter file for **draw**.

- **nvar:** the number of variables (K) making up the distribution being drawn from.

- **icolvr(1,...,nvar):** the column numbers for the **nvar** variables of interest.

- **icolwt:** the column number for the probability values (if **icolwt**=0, then the samples will be equal weighted).

- **tmin** and **tmax:** all values strictly less than **tmin** and strictly greater than **tmax** are ignored.

- **seed** and **num:** the random number seed (a large odd integer) and the number $L = $ **num** of simulated values to draw.

- **outfl:** the output file containing the simulated values.

VI.2.5 Normal Score Transformation `nscore`

The Gaussian-based simulation programs `lusim` and `sgsim` work with normal scores of the original data. The simulation is performed in the normal space, then the simulated values are back transformed. This section provides details of the normal score transformation step. All Gaussian simulation programs accept data that have already been transformed.

Consider the original data $z_i, i = 1, ..., n$, each with a specified probability $p_i, i = 1, ..., n$ (with $\sum_{i=1}^{n} p_i = 1.0$) to account for clustering. If clustering is not considered important, then all the p_i's can be set equal to $1/n$. Tied z data values are randomly ordered. When there is a large proportion of z data in a tie, these tied values should be ranked prior to `nscore` with a procedure based on the physics of the data, or the ties may be broken according to the data neighborhood averages.

Because of the limited sample size available in most applications, one should consider a nonzero probability for values less than the data minimum or greater than the data maximum. Thus some assumptions must be made about the (unsampled) tails of the attribute distribution. Such assumptions should account for the specific features of the phenomenon under study.

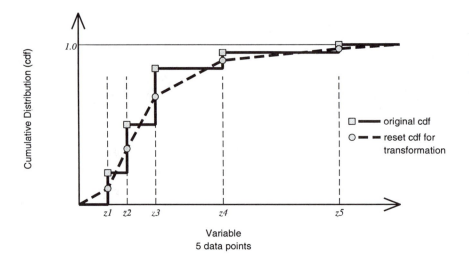

Figure VI.15:   An illustration of how the sample cumulative probability values are adjusted for the transformation. The step increases are all the same unless some declustering has been used.

One common solution ([109], p. 479) is to standardize all previous probabilities p_i to a sum slightly less than one, e.g., to $n/n+1$ if there are n data. This solution is sensitive to the number of data (sample size), and it does not offer any flexibility in modeling the distribution tails.

To avoid the problem of sensitivity to sample size, the cumulative probability associated with each data value is reset to the average between its cumulative probability and that of the next lowest datum. This allows finite probabilities to be lower than the data minimum and greater than the data maximum. Figure VI.15 provides an illustration of this tail adjustment procedure.

For notation, let c_i be the cumulative probability associated with the ith largest data value z_i ($c_i = \sum_{j=1}^{i} p_j$). The normal score transform y_i associated with z_i is then calculated as:

$$y_i = G^{-1}\left(\frac{c_i + c_{i-1}}{2}\right)$$

with $G(y)$ being the standard normal cdf, $y_c = G^{-1}(c)$ being the corresponding standard normal c quantile, and $c_0 = 0.0$. GSLIB utilizes the numerical approximation to $G^{-1}(\cdot)$ proposed by Kennedy and Gentle [116].

The limited sample may have yielded an original sample cdf that is too discrete and consequently deemed not representative of the underlying population. The sample histogram can be smoothed (e.g., by program histsmth); the resulting histogram is then used for the normal score transform (option **ismooth**=1 in program **nscore**).

```
                    Parameters for NSCORE
                    **********************

START OF PARAMETERS:
../data/cluster.dat       \file with data
3    5                    \  columns for variable and weight
-1.0e21    1.0e21         \  trimming limits
0                        \1=transform according to specified ref. dist.
../histsmth/histsmth.out \  file with reference dist.
1    2                   \  columns for variable and weight
nscore.out               \file for output
nscore.trn               \file for output transformation table
```

Figure VI.16: An example parameter file for nscore.

If z_i is the lth of L original sample values found in the kth class $(z_{k-1}, z_k]$ of the smoothed histogram, then the associated cumulative probability is:

$$c_i = \sum_{j=1}^{k-1} p_j + \frac{l}{L}\, p_k$$

with p_k being the (smoothed) probability of class $(z_{k-1}, z_k]$ The normal score transform y_i associated with z_i is then:

$$y_i = G^{-1}(c_i), \quad i = 1, \ldots, n$$

Note that the probability values p_k already account for declustering if declustering weights were used in the smoothing of the sample histogram. Also the smooth p_k values already account for extrapolation in the two tails.

If the input data contain the $(K+1)$ class bounds, $z_k, k = 0, \ldots, K \gg n$, of a previously smoothed sample histogram, the program **nscore** with option **ismooth**=1 will return a transformation lookup table more reliable and with greater resolution than that obtained from the original n sample values. The normal scores transformation is automatically performed in the programs that need normal data. There are times, however, when the transformed data are required (e.g., for variogram calculation) or when the procedure needs to be customized. The program **nscore** performs only the normal score transform.

The GSLIB library and **nscore** program should be compiled following the instructions in Appendix B. The parameters required for the main program **nscore** are listed below and shown in Figure VI.16:

- **datafl:** a data file in simplified Geo-EAS format containing the variable to be transformed and optionally a declustering weight.

- **icolvr** and **icolwt:** the column numbers for the variable and the weight. If **icolwt** ≤ 0, equal weighting is considered.

- **tmin** and **tmax:** all values strictly less than **tmin** and strictly greater than **tmax** are ignored.

- **ismooth:** =1, then consider a smoothed distribution; =0, then only consider the data in **datafl**.

- **smoothfl:** the smoothed distribution file.

- **icolvr** and **icolwt:** the column numbers for the variable and the weight in **smoothfl**.

- **outfl:** the output file containing the normal scores.

- **transfl:** the output file containing the transformation lookup table, i.e., pairs of z and y values with z being the original data values or class bound values and y being the corresponding normal scores values.

VI.2.6   Normal Score Back Transformation `backtr`

The back transformation z_i of the standard normal deviate y_i is given by

$$z_i = F^{-1}(G(y_i))$$

where $F(z)$ is the (declustered) cdf of the original data, possibly smoothed.

Almost always, the value $G(y_i)$ will not correspond exactly to an original sample cdf value F; therefore, some interpolation between the original z values or extrapolation beyond the smallest and largest z value will be required; see Section V.1.6. Linear interpolation is always performed between two known values. A variety of options are available on how the tails of the distribution will be treated.

This back transform aims at exactly reproducing the sample cdf $F(z)$, except for the within-class interpolation and the two extreme class extrapolations. Hence details of the sample cdf that are deemed not representative of the population should be smoothed out prior to using programs `nscore` and `backtr`, e.g., through program `histsmth`.

Back transformation is performed automatically in the programs that normal score transform the data (`sgsim` and `lusim`). There are times when back transformation must be performed outside these programs. The program `backtr` performs only the back transformation.

The GSLIB library and `backtr` program should be compiled following the instructions in Appendix B. The parameters required for the main program `backtr` are listed below and shown in Figure VI.17:

- **datafl:** a data file in simplified Geo-EAS format containing the normal scores.

- **ivr:** the column number for the normal scores.

- **tmin** and **tmax:** all values strictly less than **tmin** and strictly greater than **tmax** are ignored.

- **outfl:** the output file containing the back-transformed values.

```
                    Parameters for BACKTR
                    *********************

START OF PARAMETERS:
../nscore/nscore.out              \file with data
6                                 \  column with Gaussian variable
-1.0e21    1.0e21                 \  trimming limits
backtr.out                        \file for output
../nscore/nscore.trn             \file with input transformation table
0.0 60.0                          \minimum and maximum data value
1    0.0                          \lower tail option and parameter
1   60.0                          \upper tail option and parameter
```

Figure VI.17: An example parameter file for **backtr**.

- **transfl:** an input file containing the transformation lookup table, i.e., pairs of z and y values with z being the original data values or class bound values and y being the corresponding normal scores values. This is an input file that corresponds to the table written out by **nscore**, **sgsim**, ...

- **zmin** and **zmax:** are the minimum and maximum values to be used for extrapolation in the tails.

- **ltail** and **ltpar** specify the back-transformation implementation in the lower tail of the distribution: **ltail** $= 1$ implements linear interpolation to the lower limit **zmin**, and **ltail** $= 2$ implements power model interpolation, with $\omega = ltpar$, to the lower limit **zmin**.

  Linear interpolation is used between known data values.

- **utail** and **utpar** specify the back-transformation implementation in the upper tail of the distribution: **utail** $= 1$ implements linear interpolation to the upper limit **zmax**, **utail** $= 2$ power model interpolation, with $\omega = $ **utpar**, to the upper limit **zmax**, and **utail** $= 4$ implements hyperbolic model extrapolation to the upper limit **zmax** with $\omega = $ **utpar**.

VI.2.7 General Transformation trans

In some applications it is necessary to have a close match between two univariate distributions (histograms or cdf's). For example, a practitioner using a single simulated realization may want its univariate distribution to identify the original data distribution or some target distribution. Program **trans** will transform any set of values so that the distribution of the transformed values identifies the target distribution. This program has the ability to "freeze" the original values and apply the transformation only gradually as one gets further away from the locations of those frozen data values [113]. When data are frozen, the target distribution is only approximately reproduced; the reproduction is excellent if the number of frozen values is small with regard to the number of values transformed.

Note that fluctuations from model statistics are expected when performing stochastic simulation; it is not advisable to transform all simulated realizations systematically so that they match the same sample histogram.

The transformation method is a generalization of the quantile transformation used for normal scores (see Section VI.2.5): The p quantile of the original distribution is transformed to the p quantile of the target distribution. This transform preserves the p quantile indicator variograms of the original values. The variogram (standardized by the variance) will also be stable provided that the target distribution is not too different from the initial distribution.

When "freezing" the original data values, the quantile transform is applied progressively as the location gets further away from the set of data locations. The distance measure used is proportional to a kriging variance at the location of the value being transformed. That kriging variance is zero at the data locations (hence no transformation) and increases away from the data (the transform is increasingly applied). An input kriging variance file must be provided or, as an option, **trans** can calculate these kriging variances using an arbitrary isotropic and smooth (Gaussian) variogram model.

Because not all original values are transformed, reproduction of the target histogram is only approximate. A control parameter, **omega** in $[0, 1]$, allows the desired degree of approximation to be achieved at the cost of generating discontinuities around the data locations. The greater ω, the lesser the discontinuities.

Program **trans** can be applied to either continuous or categorical values. In the case of categorical values a hierarchy or spatial sequencing of the K categories is provided implicitly through the integer coding $k = 1, \ldots, K$ of these categories. Category k may be transformed into category $(k - 1)$ or $(k + 1)$ and only rarely into categories further away.

An interesting side application of program **trans** is in cleaning noisy simulated images [162]. Two successive runs (a "roundtrip") of **trans**, the first changing the original proportions or distribution, the second restituting these original proportions, would clean the original image while preserving data exactitude.

The GSLIB library and **trans** program should be compiled following the instructions in Appendix B. The parameters required for the main program **trans** are listed below and shown in Figure VI.18:

- **vartype:** the variable type (1=continuous, 0=categorical).

- **refdist:** the input data file with the target distribution and weights.

- **ivr** and **iwt:** the column for the values and the column for the (declustering) weight. If there are no declustering weights, then set **iwt**= 0.

- **datafl:** the input file with the distribution(s) to be transformed.

- **ivrd** and **iwtd:** the column for the values and the declustering weights (0 if none).

```
                    Parameters for TRANS
                    ********************

START OF PARAMETERS:
1                            \1=continuous, 0=categorical
../data/true.dat             \file with reference distribution
1    0                       \   columns for variable and weight(0=none)
../data/cluster.dat          \file with original distributions
3    0                       \   columns for variable and weight(0=none)
-1.0e21  1.0e21              \trimming limits
trans.out                    \file for transformed distributions
1                            \number of realizations or "sets" to trans
50   50   1                  \categorical: nx, ny, nz: size of 3-D model
  2    2   0                 \   wx, wy, wz: window size for tie-breaking
1000                         \continuous: number to transform per "set"
0.0    75.0                  \   minimum and maximum values
1      1.0                   \   lower tail: option, parameter
1      75.0                  \   upper tail: option, parameter
0                            \honor local data? (1=yes, 0=no)
kt3d.out                     \   file with estimation variance
2                            \   column number
0.5                          \   control parameter ( 0.33 < w < 3.0 )
69069                        \   random number seed (conditioning cat.)
```

Figure VI.18: An example parameter file for **trans**.

- **tmin** and **tmax:** all values strictly less than **tmin** and strictly greater than **tmax** are ignored.

- **outfl:** output file for the transformed values.

- **nsets:** number of realizations or "sets" to transform. Each set is transformed separately.

- **nx, ny,** and **nz:** size of 3D model (for categorical variable). When transforming categorical variables it is essential to consider some type of tie-breaking scheme. A moving window (of the following size) is considered for tie-breaking when considering a categorical variable.

- **wx, wy,** and **wz:** size of 3D window for categorical variable tie-breaking.

- **nxyz:** the number to transform at a time (when dealing with a continuous variable). Recall that **nxyz** will be considered **nsets** times.

- **zmin** and **zmax:** are the minimum and maximum values that will be used for extrapolation in the tails.

- **ltail** and **ltpar** specify the back-transformation implementation in the lower tail of the distribution: **ltail** = 1 implements linear interpolation to the lower limit **zmin**, and **ltail** = 2 implements power model interpolation, with $\omega = $ **ltpar**, to the lower limit **zmin**.

- **utail** and **utpar** specify the back-transformation implementation in the upper tail of the distribution: **utail** = 1 implements linear interpolation to the upper limit **zmax**, **utail** = 2 implements power model

```
                    Parameters for VMODEL
                    *********************

START OF PARAMETERS:
vmodel.var                      \file for variogram output
2    22                         \number of directions and lags
 0.0    0.0     1.0             \azm, dip, lag distance
90.0    0.0     3.0             \azm, dip, lag distance
2      0.2                      \nst, nugget effect
1      0.4   0.0    0.0   0.0   \it,cc,ang1,ang2,ang3
             10.0   5.0   10.0  \a_hmax, a_hmin, a_vert
1      0.4   0.0    0.0   0.0   \it,cc,ang1,ang2,ang3
             10.0   5.0   10.0  \a_hmax, a_hmin, a_vert
```

Figure VI.19: An example parameter file for **vmodel**.

interpolation, with $\omega = $ **utpar**, to the upper limit **zmax**, and **utail** $= 4$ implements hyperbolic model extrapolation with $\omega = $ **utpar**.

- **transcon:** constrain transformation to honor local data? (1=yes, 0=no)

- **estvfl:** an input file with the estimation variance (must be of size **nxyz**).

- **icolev:** column number in **estvfl** for the estimation variance.

- **omega:** the control parameter for how much weight is given to the original data ($0.33 < \omega < 3.0$).

- **seed:** random number seed used when constraining a categorical variable transformation to local data.

VI.2.8   Variogram Values from a Model vmodel

The details of entering the parameters of a semivariogram model are given in Section II.3. This program will take the semivariogram model and write out a file with the same format as the **gam** program so that it can be plotted with **vargplt**. The primary uses of **vmodel** are to overlay a model on experimental points and also to provide a utility to check the definition of the semivariogram model.

The GSLIB library and **vmodel** program should be compiled following the instructions in Appendix B. The parameters required for the main program **vmodel** are listed below and shown in Figure VI.19:

- **outfl:** the output file that will contain the semivariogram values.

- **ndir** and **nlag:** the number of directions and the number of lags to be considered.

- **azm, dip**, and **lag:** for each of the **ndir** directions a direction must be specified by **azm** and **dip** and a unit lag offset must be specified (**lag**).

```
                    Parameters for BIGAUS
                    **********************

START OF PARAMETERS:
bigaus.out                         \file for output of variograms
3                                  \number of thresholds
0.25    0.50      0.75             \threshold cdf values
1    20                            \number of directions and lags
0.0    0.0     1.0                 \azm(1), dip(1), lag(1)
2     0.2                          \nst, nugget effect
1     0.4   0.0    0.0    0.0      \it,cc,ang1,ang2,ang3
            10.0   5.0   10.0      \a_hmax, a_hmin, a_vert
1     0.4   0.0    0.0    0.0      \it,cc,ang1,ang2,ang3
            10.0   5.0   10.0      \a_hmax, a_hmin, a_vert
```

Figure VI.20: An example parameter file for bigaus.

- **nst** and **c0:** the number of structures and the nugget effect.

- For each of the **nst** nested structures one must define **it**, the type of structure; **cc**, the c parameter; **ang1,ang2,ang3**, the angles defining the geometric anisotropy; aa_{hmax}, the maximum horizontal range; aa_{hmin}, the minimum horizontal range; and aa_{vert}, the vertical range. A detailed description of these parameters is given in Section II.3.

VI.2.9 Gaussian Indicator Variograms bigaus

This program allows checking for bivariate normality; see Section V.2.2. A model of the normal scores variogram is input to bigaus. The corresponding indicator semivariograms for specified cutoffs, lags, and directions are output from the program. The idea is to see how far the experimental indicator variograms deviate from those implied by the bivariate Gaussian distribution.

A numerical integration algorithm is used to compute the Gaussian model-derived indicator variograms; see relation (V.20) and [102, 191].

The GSLIB library and bigaus program should be compiled following the instructions in Appendix B. The parameters required for the main program bigaus are listed below and shown in Figure VI.20:

- **outfl:** the output file for the theoretical Gaussian indicator semivariograms. The format is the same as that created by the gam program; therefore, the program vargplt could be used to plot these indicator variograms.

- **ncut:** the number of thresholds.

- **zc(ncut):** **ncut** threshold values are required. These thresholds are expressed in units of cumulative probability; e.g., the lower quartile is 0.25, the median is 0.50. Note that the results are symmetric: the variogram for the 5th percentile (0.05) is the same as the 95th percentile (0.95).

- **ndir** and **nlag:** the number of directions and the number of lags to be considered.

- **azm, dip**, and **lag:** for each of the **ndir** directions a direction must be specified by **azm** and **dip** and a unit lag offset must be specified (**lag**).

- **nst** and **c0:** the number of structures and the nugget effect.

- For each of the **nst** nested structures one must define **it**, the type of structure; **cc**, the c parameter; **ang1,ang2,ang3**, the angles defining the geometric anisotropy; aa_{hmax}, the maximum horizontal range; aa_{hmin}, the minimum horizontal range; and aa_{vert}, the vertical range. A detailed description of these parameters is given in Section II.3.

VI.2.10   Library of Linear System Solvers

Next to searching for data solving the kriging equations is the most time-consuming aspect of any kriging or simulation program. Depending on the application, a different solution algorithm is appropriate. Discussions are provided in references [66, 69, 83, 154, 163, 164, 186].

All the linear system solvers documented below call for a small numerical tolerance parameter that is used to decide whether a matrix is singular. This parameter has been set between 10^{-10} and 10^{-4}. Ideally, this parameter should be linked to the units of the covariance function; the larger the magnitude of the covariance, the larger the tolerance for a singular matrix. Code is provided for the following algorithms:

- **ksol:** A standard Gaussian elimination algorithm without pivot search is adequate for SK and OK applications. When either the simple or ordinary kriging matrix is built with the covariance or pseudocovariance, the diagonal term is taken as the pivot. The computer savings can be significant.

- **smleq:** Another standard Gaussian elimination algorithm without any pivot search for SK and OK applications. The coding for *smleq* uses double-precision pivot/diagonal elements for greater accuracy.

- **ktsol:** A Gaussian elimination algorithm with partial pivoting is necessary for KT. The pivoting is necessary because of the zero elements on the diagonal arising from the unbiasedness conditions.

- **gaussj:** Gauss-Jordan elimination is a stable approach, but it is less efficient than Gaussian elimination with back substitution. The method is not recommended, but is included for completeness.

- **ludcmp:** An LU decomposition is used for certain simulation algorithms and can be used for any kriging system. This algorithm is general and provides one of the most stable solutions to any nonsingular set of equations. Another advantage is that the right-hand-side vectors are not required in advance.

- **cholfbs:** A Cholesky decomposition, with a forward and back substitution, is quite fast, but can only be used for symmetric positive-definite matrices. One condition for a positive-definite matrix is that it must have nonvanishing positive diagonal elements. The last diagonal element that comes from any unbiasedness constraint implies that the OK and KT kriging matrices do not fulfill that requirement. The Cholesky decomposition algorithm, or any of the methods requiring a positive definite symmetric matrix, will not work with the OK or KT matrices. A small ϵ cannot be used instead of zero because of the further condition that $a_{ik}^2 < a_{ii}a_{kk}$ for all i and k (positive-definite condition). An alternative, which the authors did not investigate, consists of partitioning the OK or KT system and using a Cholesky decomposition only for the strict covariance matrix.

- **cholinv:** The SK covariance matrix can be inverted through a Cholesky decomposition, followed by an inverse of the lower or upper triangular decomposition and finally a matrix multiplication. The SK weights, the OK weights, and the weights for the OK estimate of the mean can be obtained simultaneously [38].

- **gsitrn:** A straightforward Gauss-Seidel iteration algorithm may be used to solve any of the kriging systems. A simple reordering of the equations (which destroys symmetry) is necessary to remove the zero from the diagonal of the OK system. This method seems particularly attractive when considering any of the indicator approaches because the solution at one cutoff provides the logical initial choice for the next cutoff. If the variogram model has not changed much, then the solution will be obtained very quickly. Lacking a previous solution, the Lagrange parameter can be estimated as zero and the weights as the normalized right-hand-side covariances.

To illustrate the relative speed of the algorithms, a system (which system only matters with an iterative approach) was solved 10,000 times. This took *ksol* about a second for a 10 by 10 system on an Apollo 10000. The following table shows the times for the different subroutines with different SK matrix sizes. The time required for the various algorithms does not increase at the same rate as the size of the matrix.

Routine	10 samples	20 samples	40 samples	80 samples
ksol	1.0	4.8	28.4	184.2
smleq	1.0	5.9	39.5	292.1
ktsol	1.4	7.4	47.9	339.5
gaussj	6.2	44.8	337.6	2644.7
ludcmp	2.1	11.6	73.9	521.0
cholfbs	1.3	7.0	43.7	313.2
cholinv	8.7	60.5	447.4	3457.9
gsitrn	2.6	11.0	72.8	309.3

The times in the above table are only illustrative. Different computers, compilers, and coding will result in different numbers. The following comments can be made on the basis of the above results:

- The *ksol* subroutine, which is commonly used in GSLIB, is an efficient direct method to solve either the OK or SK system of equations.

- The *ktsol* subroutine seems the most efficient direct method to solve the KT equations (*ksol* and *smleq* would not work).

- The Gauss-Jordan solution is a stable numerical method to solve a general system of equations. As illustrated above, the method is inefficient for kriging.

- The *ludcmp* subroutine, as it stands in [154], is not an efficient way to solve kriging systems.

- The standard Cholesky LU decomposition with a forward and backward substitution is better than the general LU approach, but is not as fast as the *ksol* subroutine. In principle the Cholesky decomposition should be quicker; after all, it is just a modification of the Gauss algorithm in which the elimination is carried out while maintaining symmetry. The reason for the slowness is the square root operation on all the diagonal elements.

- *cholinv* completes the inverse and solves the system explicitly to get simultaneously the SK estimate, the OK estimate, and the OK estimate of the mean. This method appears attractive, due to its elegance, until speed is considered. On the basis of the tabulated results it would be more efficient to solve two systems (the second one with two right-hand side vectors) with *ksol* to obtain these three estimates.

- *gsitrn* is not an efficient method to solve a system without a good initial solution. Experimentation has shown that the final solution is obtained more quickly than it is obtained by *ksol* when the initial solution is known to the first decimal place and the matrix is larger than 10 by 10. The time also depends on the kriging system and the ordering of the

```
                    Parameters for BICALIB
                    ***********************

START OF PARAMETERS:
../data/ydata.dat                \file with secondary data
4                                \   column for secondary variable
../data/cluster.dat              \file with calibration scatterplot
3  4  5                          \   columns of pri, sec, and weight
-1.0e21    1.0e21                \   trimming limits
bicalib.out                      \file for output data / distributions
bicalib.cal                      \file for output calibration (SISIM)
bicalib.rep                      \file for calibration report
5                                \number of thresholds on primary
0.50 1.00 2.50 5.00 10.0         \   thresholds on primary
5                                \number of thresholds on secondary
0.50 1.00 2.50 5.00 10.0         \   thresholds on secondary
```

Figure VI.21: An example parameter file for bicalib.

data points. Additional programming could result in an efficient iterative solver that optimally (with respect to speed) orders the equations and possibly accelerates convergence through some type of relaxation scheme [163].

VI.2.11   Bivariate Calibration bicalib

Whenever soft data are considered in an indicator formalism, it is necessary to calculate the soft indicator values (the prior distributions). Also, when considering the Markov-Bayes option in the indicator simulation program sisim, the $B(z)$ calibration parameters are called for; see Section IV.1.12 for a more complete discussion. The program bicalib computes the prior distributions and the $B(z)$ calibration parameters. These prior distributions are written to one output file and the $B(z)$ parameters are written to a reporting file from which they must be transferred to the sisim parameter file.

Often in practice the sparse sample size does not provide enough resolution for the scattergram of primary versus secondary variables. Program scatsmth presented in Section VI.2.3 could be used to smooth and add resolution to the sample scattergram. Instead of the original data pairs (size n) the input data file **datafl** to program bicalib could contain the smoothed sample scattergram.

The GSLIB library and bicalib program should be compiled following the instructions in Appendix B. The parameters required for the main program bicalib are listed below and shown in Figure VI.21:

- **datafl:** a data file in simplified Geo-EAS format containing the secondary data values (the output file will contain the corresponding prior distributions for the primary variable).

- **icolsec:** the column number for the secondary variable.

- **calibfl:** a data file in simplified Geo-EAS format containing pairs of primary and secondary data values (the calibration data).

- **ivru, ivrv,** and **iwt:** the column numbers for the primary and secondary variables and declustering weight/bivariate frequency.

- **tmin** and **tmax:** all values strictly less than **tmin** and strictly greater than **tmax** are ignored.

- **outfl:** an output file that contains the prior distributions corresponding to all input secondary data.

- **repfl:** an output file that contains a report of the pairs retained, some summary statistics, a bivariate distribution table, the prior distributions, and the $B(z)$ parameter values.

- **calfl:** an output file that contains the secondary data thresholds, the calibration table, and the $B(z)$ values for the Markov-Bayes model. To be used as input to `sisim_gs`.

- **ncutu:** the number of thresholds applied to the primary variable.

- **cutu(i), i=1,ncutu:** the threshold values applied to the primary variable.

- **ncutv:** the number of thresholds applied to the secondary variable.

- **cutv(i),i=1,ncutv:** the threshold values applied to the secondary variable.

VI.2.12   Postprocessing of IK Results `postik`

The output from `ik3d` requires additional processing before being used. The program `postik` performs order relations corrections, change of support calculations, and various statistical computations according to the following options:

1. Compute the "E-type" estimate, i.e., the mean value of the conditional distribution; see Section IV.1.9 and relation (IV.39).

2. Compute the probability of exceeding a fixed threshold, the average value above that threshold, and the average value below that threshold.

3. Compute the value where a fixed conditional cumulative distribution function (cdf) value p is reached, i.e., the conditional p quantile value.

4. Compute the variance the conditional distribution.

```
                    Parameters for POSTIK
                    *********************
START OF PARAMETERS:
../ik3d/ik3d.out              \file with IK3D output (continuous)
postik.out                    \file for output
3    0.5                      \output option, output parameter
5                             \number of thresholds
0.5  1.0  2.5  5.0  10.0      \the thresholds
0    1       0.75             \volume support?, type, varred
cluster.dat                   \file with global distribution
3    0     -1.0    1.0e21     \   ivr,  iwt,  tmin,  tmax
0.0     30.0                  \minimum and maximum Z value
1    1.0                      \lower tail: option, parameter
1    1.0                      \middle    : option, parameter
1    2.0                      \upper tail: option, parameter
100                           \maximum discretization
```

Figure VI.22: An example parameter file for postik.

Although order relations are corrected by ik3d, they are checked again in postik.

A facility for volume support correction has been provided, although we **strongly** suggest that volume support correction be approached through small-scale simulations; see Section IV.1.13 and [67, 87]. Either the affine correction or the indirect lognormal correction is available; see [67, 89], pp. 468-476.

There are a number of consequential decisions that must be made about how to interpolate between and extrapolate beyond known values of the cumulative distribution function. These are discussed in Section V.1.6. Both the conditional (E-type) mean and the mean above a specified threshold are calculated by summing the z values at regularly spaced ccdf increments. A parameter **maxdis** specifies how many such cdf values to calculate. This can be increased depending on the precision required.

The GSLIB library and postik program should be compiled following the instructions in Appendix B. The parameters required for the main program postik are listed below and shown in Figure VI.22:

- **distfl:** the output IK file (from program ik3d), which is considered as input.

- **outfl:** the output file (output differs depending on output option; see below).

- **iout** and **outpar:** iout=1 computes the E-type estimate and the conditional variance; **iout**=2 computes the probability and means above and below the value **outpar**; **iout**=3 computes the z value (p quantile) corresponding to the cdf value **outpar**=p; **iout**=4 computes the conditional variance.

- **ncut:** the number of thresholds.

- **cut(i), i=1,ncut:** the threshold values used in `ik3d`.

- **ivol, ivoltyp,** and **varred:** if **ivol** = 1, then volume support correction is attempted with the affine correction (**ivoltyp**=1) or the indirect correction through permanence of a lognormal distribution (**ivoltyp**=2). The variance reduction factor (between 0 and 1) is specified as **varred**.

- **datafl:** the data file containing z data that provides the details between the IK thresholds. This file is used only if table lookup values are called for in one of the interpolation/extrapolation options.

- **icolvr** and **icolwt:** when **ltail, middle,** or **utail** is 3 then these variables identify the column for the values and the column for the (declustering) weight in **datafl** that will be used for the global distribution.

- **tmin** and **tmax:** all values strictly less than **tmin** and strictly greater than **tmax** are ignored.

- **zmin** and **zmax:** minimum and maximum data values allowed. Even when not using linear interpolation it is safe to control the minimum and maximum values entered. The lognormal or hyperbolic upper tail is constrained by **zmax**.

- **ltail** and **ltpar** specify the extrapolation in the lower tail: **ltail**=1 implements linear interpolation to the lower limit **zmin**; **ltail**=2 implements power model interpolation, with ω =**ltpar**, to the lower limit **zmin**; and **ltail**=3 implements linear interpolation between tabulated quantiles.

- **middle** and **midpar** specify the interpolation within the middle of the distribution: **middle**=1 implements linear interpolation to the lower limit **zmin**; **middle**=2 implements power model interpolation, with ω =**midpar**, to the lower limit **zmin**; and **middle**=3 allows for linear interpolation between tabulated quantile values.

- **utail** and **utpar** specify the extrapolation in the upper tail of the distribution: **utail**=1 implements linear interpolation to the upper limit **zmax**, **utail**=2 implements power model interpolation, with ω =**utpar**, to the upper limit **zmax**; **utail**=3 implements linear interpolation between tabulated quantiles; and **utail**=4 implements hyperbolic model extrapolation with ω =**utpar**.

- **maxdis:** a maximum discretization parameter. The default value of 50 is typically fine; greater accuracy is obtained with a larger number, and quicker execution time is obtained with a lower number.

```
                    Parameters for POSTSIM
                    **********************

START OF PARAMETERS:
sgsim.out                        \file with simulated realizations
50                               \   number of realizations
-0.001    1.0e21                 \   trimming limits
20    20    1                    \nx, ny, nz
postsim.out                      \file for output array(s)
3    0.0001                      \output option, output parameter
```

Figure VI.23: An example parameter file for `postsim`.

VI.2.13 Postprocessing of Simulation Results `postsim`

`postsim` allows a number of summaries to be extracted from a set of simulated realizations:

1. The "E-type" estimates, i.e., the point-by-point average of the realizations.

2. Compute the variance of the conditional distribution.

3. The probability of exceeding a fixed threshold, the average value above that threshold, and the average value below that threshold.

4. The value where a fixed conditional cumulative distribution function (cdf) value p is reached, i.e., the conditional p quantile value.

The GSLIB library and `postsim` program should be compiled following the instructions in Appendix B. The parameters required for the main program `postsim` are listed below and shown in Figure VI.23:

- **simfl:** the output simulation file that contains all the realizations. This file should contain the simulated values, one value per line, cycling fastest on x, then y, then z, and last per simulation.

- **nsim** specifies the number of realizations.

- **tmin** and **tmax:** all values strictly less than **tmin** and strictly greater than **tmax** are ignored.

- **nx, ny,** and **nz** specify the number of nodes in the x, y, and z direction.

- **outfl:** an output file that contains the E-type estimate, the fraction and mean above threshold, the fraction and mean below threshold, or the conditional p quantile values.

- **iout** and **outpar: iout**=1 computes the E-type estimate,; **iout**=2 computes the probability and mean above and below the value **outpar**, **iout**=3 computes the z value (p quantile) corresponding to the cdf value **outpar**=p; **iout**=4 computes the symmetric probability interval (total width = **outpar**); and **iout**=5 computes an estimate of the conditional variance.

Appendix A

Partial Solutions to Problem Sets

The partial solutions given in this appendix illustrate the use of GSLIB programs, provide base cases for debugging, and serve as references to check program installations. As much as possible, bugs should be reported to the authors using the example data; this would make it simpler to document the problem.

The data, parameter, and output files are provided on the distribution diskettes.

The runs given in this appendix illustrate only some of the available options. It would be too cumbersome to illustrate exhaustively all program options and all solution alternatives. Whenever possible, the results of a program should be checked against analytical results or through an example small enough to allow calculation by hand.

A.1 Problem Set One: Data Analysis

Most summary statistics are affected by preferential clustering. In the case of $\boxed{\text{cluster.dat}}$ the equal-weighted mean and median will be higher since clustering is in high-concentration areas. The variance may also be increased because the clustered data are disproportionately valued high. Note that the effect of clustering is not always so predictable. For example, if the clustered samples are all close to the center of the distribution, then the equal-weighted mean and median may be unbiased but the variance could be too low.

An equal-weighted histogram and lognormal probability plot are shown in Figure A.1. The distribution is positively skewed with a coefficient of variation of 1.5. The parameter files required by histplt and probplt to generate the plots shown on Figure A.1 are presented in Figures A.2 and A.3. These parameter files are provided on the software distribution diskettes.

A quick-and-easy way to decluster these data would be to consider only the first 97 nearly regularly spaced samples. If cell declustering is considered, a reasonable range of square cell sizes would be from 1 mile (no samples are spaced closer than 1 mile) to 25 miles (half of the size of the area of interest). A *natural* cell size would be the spacing of the underlying pseudoregular grid, i.e., $\sqrt{\frac{2500}{97}} \simeq 5.0$. The resulting scatterplot of declustered mean versus cell size is shown in Figure A.4. The minimum declustered mean is found for a cell size of 5.0 miles; the corresponding declustered histogram and lognormal probability plot are shown in Figure A.5.

The parameters required by program declus are shown in Figure A.6. The first 10 lines and the last 10 lines of the declus output file are given in Figure A.7. The scatterplot parameter file is shown in Figure A.8.

A histogram and probability plot of the 2500 exhaustive measurements are shown in Figure A.9. Summary statistics for all 140 data equal weighted, the first 97 data equal weighted, 140 data with declustering weights, and the true exhaustive measurements are shown below:

	n	m	M	σ
Equal weighting (140 data)	140	4.35	2.12	6.70
Equal weighting (97 data)	97	2.21	1.02	3.17
Declustered (140 data)	140	2.52	1.20	4.57
Exhaustive	2500	2.58	0.96	5.15

The Q-Q and P-P plots of the sample data (equal weighted and declustered) versus the exhaustive data are shown in Figure A.10. Identical distributions should plot as a 45° line. When the points fall on any straight line, the shapes of the distributions are the same but with different means and/or variances.

The parameter file for the lower left Q-Q plot in Figure A.10 is shown in Figure A.11 and provided on the software distribution diskettes.

The histogram of the 140 primary data and secondary data were smoothed with histsmth. The logarithmic scaling option should be used when the distribution has significant skewness. The results are shown with the scatsmth bivariate distribution in Figure A.12. The parameter file for histsmth is shown in Figure A.13 and the output file is summarized in Figure A.14. Similar results for scatsmth are shown in Figures A.15-A.16.

2500 independent realizations were drawn from this smoothed histogram using the draw Monte Carlo program. Figure A.17 shows two Q-Q plots comparing the results to the histsmth target input distribution and to the reference distribution in ⟨true.dat⟩. Note that the draw results are very close to the target distribution as expected; however, that target distribution appears to have too low a variance when compared to the reference; that is, the proportion of extreme values are underestimated by the histsmth target distribution.

The parameter file for draw is shown in Figure A.18, and the results

are summarized in Figure A.19 and provided on the software distribution
diskettes.

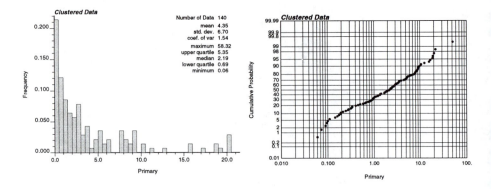

Figure A.1: An equal-weighted histogram and lognormal probability plot of all 140 sample data.

```
                    Parameters for HISTPLT
                    ************************

START OF PARAMETERS:
../../data/cluster.dat      \file with data
3    0                      \   columns for variable and weight
-1.0e21   1.0e21            \   trimming limits
histplt.out                 \file for PostScript output
  0.0       20.0            \attribute minimum and maximum
-1.0                        \frequency maximum (<0 for automatic)
40                          \number of classes
0                           \0=arithmetic, 1=log scaling
0                           \0=frequency,  1=cumulative histogram
200                         \number of cum. quantiles (<0 for all)
2                           \number of decimal places (<0 for auto.)
Clustered Data              \title
1.5                         \positioning of stats (L to R: -1 to 1)
-1.1e21                     \reference value for box plot
```

Figure A.2: Parameter file for the histogram shown in Figure A.1.

```
                    Parameters for PROBPLT
                    **********************

START OF PARAMETERS:
../../data/cluster.dat      \file with data
3    0                      \   columns for variable and weight
-1.0e21   1.0e21            \   trimming limits
probplt.out                 \file for PostScript output
99                          \number of points to plot (<0 for all)
1                           \0=arithmetic, 1=log scaling
0.01   30.0    5.0          \min,max,increment for labeling
Clustered Data              \title
```

Figure A.3: Parameter file for the probability plot shown in Figure A.1.

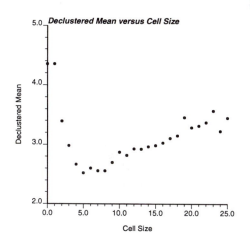

Figure A.4: Scatterplot of the declustered mean versus cell size.

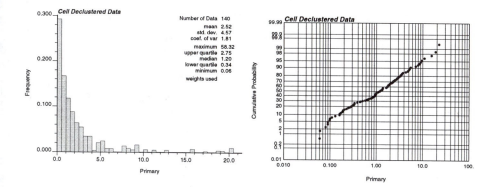

Figure A.5: Declustered histogram and lognormal probability plot of the 140 data.

```
                    Parameters for DECLUS
                    *********************

START OF PARAMETERS:
../../data/cluster.dat       \file with data
1    2    0    3             \  columns for X, Y, Z, and variable
-1.0e21      1.0e21          \  trimming limits
declus.sum                   \file for summary output
declus.out                   \file for output with data & weights
1.0   1.0                    \Y and Z cell anisotropy (Ysize=size*Yanis)
0                            \0=look for minimum declustered mean (1=max)
24   1.0   25.0              \number of cell sizes, min size, max size
5                            \number of origin offsets
```

Figure A.6: Parameter file for the declustering results shown in Figure A.4.

```
Clustered 140 primary and secondary data
   6
Xlocation
Ylocation
Primary
Secondary
Declustering Weight
Weight*nd
39.5 18.5    .06    .22 1.619    1.28130
 5.5  1.5    .06    .27 1.619    1.40010
38.5  5.5    .08    .40 1.416    1.61324
20.5  1.5    .09    .39 1.821    1.79745
27.5 14.5    .09    .24 1.349    1.43016
40.5 21.5    .10    .48  .944    1.08686
15.5  3.5    .10    .21 1.214    1.21589
 6.5 25.5    .11    .36 1.619    1.08344
38.5 21.5    .11    .22 1.146    1.08686
23.5 18.5    .16    .30 1.821    1.79745
 • • •
 2.5 15.5   2.74   4.07  .789    0.68918
 3.5 14.5   3.61   5.03  .497    0.68895
29.5 41.5  58.32  10.26  .450    0.40063
30.5 40.5  11.08   9.31  .396    0.67861
30.5 42.5  21.08  10.26  .326    0.31106
31.5 41.5  22.75   8.21  .427    0.36750
34.5 32.5   9.42   6.76  .413    0.41742
35.5 31.5   8.48  12.78  .419    0.43272
35.5 33.5   2.82   9.21  .271    0.33691
36.5 32.5   5.26  12.40  .252    0.28779
```

Figure A.7: Sample output from **declus** created by the parameter file shown in Figure A.6.

```
              Parameters for SCATPLT
              **********************

START OF PARAMETERS:
declus.sum                        \file with data
1    2    0    0                  \  columns for X, Y, wt, third var.
-1.0e21    1.0e21                 \  trimming limits
scatplt.out                       \file for PostScript output
0.0       25.0        0           \X min and max, (0=arith, 1=log)
2.0        5.0        0           \Y min and max, (0=arith, 1=log)
1                                 \plot every n'th data point
1.0                               \bullet size: 0.1(sml)-1(reg)-10(big)
0.0        2.0                    \limits for third variable gray scale
Declustered Mean versus Cell Size \title
```

Figure A.8: Parameter file for the scatterplot shown in Figure A.4.

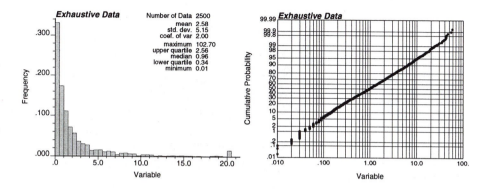

Figure A.9: Histogram and lognormal probability plot of the 2500 exhaustive measurements.

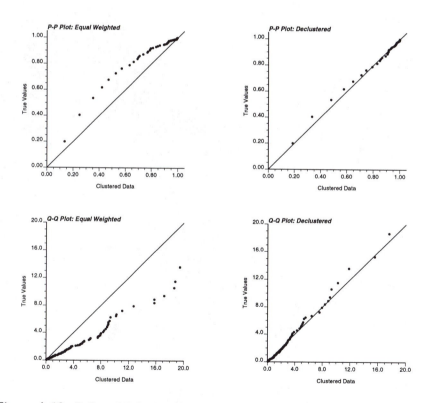

Figure A.10: Q-Q and P-P plots for equal weighted and declustered sample distributions (140 data) versus the exhaustive distribution.

```
                         Parameters for QPPLT
                         ********************

START OF PARAMETERS:
../../data/cluster.dat              \file with first set of data (X axis)
3    0                              \  columns for variable and weight
../../data/true.dat                 \file with second set of data (Y axis)
1    0                              \  columns for variable and weight
0.0           1.0e21                \  trimming limits
qpplt.03                            \file for PostScript output
0                                  \0=Q-Q plot, 1=P-P plot
-1                                 \number of points to plot (<0 for all)
0.0       20.0                     \X minimum and maximum
0.0       20.0                     \Y minimum and maximum
0                                  \0=arithmetic, 1=log scaling
Q-Q Plot: Equal Weighted           \Title
```

Figure A.11: Parameter file for the Q-Q plot shown in the lower left of Figure A.10.

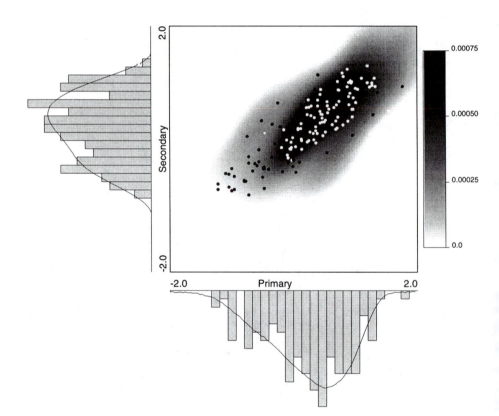

Figure A.12: Smoothed histogram of primary data, secondary data, and scatter-plot.

```
                        Parameters for HISTSMTH
                        ************************

START OF PARAMETERS:
../../data/cluster.dat            \file with data
3    0                            \   columns for variable and weight
-1.0e21   1.0e21                  \   trimming limits
Primary Variable Distribution    \title
histsmth.ps                       \file for PostScript output
 25                               \   number of hist classes (for display)
histsmth.out                      \file for smoothed output
100    0.01   100.0               \smoothing limits: number, min, max
1                                 \0=arithmetic, 1=log scaling
2500    50   0.0001    69069      \maxpert, reporting interval,min Obj,Seed
 1  1  1  1                       \1=on,0=off: mean,var,smth,quantiles
1.  1.  2.  2.                    \weighting : mean,var,smth,quantiles
   5                              \size of smoothing window
-999.0 -999.0                     \target mean and variance (-999=>data)
11                                \number of quantiles: defined from data
0                                 \number of quantiles: defined by user
0.5     1.66                      \   cdf, z
```

Figure A.13: Parameter file for the histogram smoothing results shown in Figure A.12.

```
Smooth Results
3
Z-value
Log10 Z-value
P-value
      0.0100     -2.0000    0.000380
      0.0110     -1.9596    0.000261
      0.0120     -1.9192    0.000539
      0.0132     -1.8788    0.000489
      0.0145     -1.8384    0.000365
      0.0159     -1.7980    0.000573
      0.0175     -1.7576    0.000371
      0.0192     -1.7172    0.000609
      0.0210     -1.6768    0.000522
      0.0231     -1.6364    0.000678
• • •
     43.2876      1.6364    0.000599
     47.5081      1.6768    0.000441
     52.1401      1.7172    0.000442
     57.2237      1.7576    0.000479
     62.8029      1.7980    0.000296
     68.9261      1.8384    0.000598
     75.6463      1.8788    0.000215
     83.0218      1.9192    0.000389
     91.1163      1.9596    0.000216
    100.0000      2.0000    0.000262
```

Figure A.14: Sample output from **histsmth** created by the parameter file shown in Figure A.13.

```
                         Parameters for SCATSMTH
                         ************************

START OF PARAMETERS:
../../data/cluster.dat                 \file with data
3    4    0                            \  columns for X, Y, weight
histsmth.out                           \file with smoothed X distribution
2    3                                 \  columns for variable, weight
histsmth2.out                          \file with smoothed Y distribution
2    3                                 \  columns for variable, weight
1    1                                 \log scaling for X and Y (0=no, 1=yes)
scatsmth.dbg                           \file for debug information
scatsmth.xr                            \file for final X distribution
scatsmth.yr                            \file for final Y distribution
scatsmth.out                           \file for output
250.   1.0    0.0001    69069          \maxpert, report, min obj, seed
1   1   1   1                          \1=on,0=off: marg,corr,smth,quan
1   1   5   2                          \weighting : marg,corr,smt,hquan
100                                    \smoothing Window Size (number)
0.9                                    \correlation (-999 take from data)
5    5                                 \number of X and Y quantiles
0                                      \points defining envelope (0=none)
   0.0      0.0                        \    x(i)    y(i)
   0.0    999.0                        \    x(i)    y(i)
 999.0    999.0                        \    x(i)    y(i)
 999.0      0.0                        \    x(i)    y(i)
```

Figure A.15: Parameter file for the scatterplot smoothing results shown in Figure A.12.

```
Smooth Results
5
X-value
Y-value
10**X-value
10**Y-value
P-value
    -2.0000      -2.0000       0.0100       0.0100    0.000011
    -1.9596      -2.0000       0.0110       0.0100    0.000012
    -1.9192      -2.0000       0.0120       0.0100    0.000012
    -1.8788      -2.0000       0.0132       0.0100    0.000012
    -1.8384      -2.0000       0.0145       0.0100    0.000010
    -1.7980      -2.0000       0.0159       0.0100    0.000009
    -1.7576      -2.0000       0.0175       0.0100    0.000008
    -1.7172      -2.0000       0.0192       0.0100    0.000006
    -1.6768      -2.0000       0.0210       0.0100    0.000005
    -1.6364      -2.0000       0.0231       0.0100    0.000004
• • •
     1.6360       1.9996      43.2516      99.9086    0.000087
     1.6764       1.9996      47.4682      99.9086    0.000087
     1.7168       1.9996      52.0958      99.9086    0.000088
     1.7572       1.9996      57.1745      99.9086    0.000090
     1.7976       1.9996      62.7484      99.9086    0.000091
     1.8380       1.9996      68.8657      99.9086    0.000092
     1.8784       1.9996      75.5793      99.9086    0.000094
     1.9188       1.9996      82.9474      99.9086    0.000093
     1.9592       1.9996      91.0338      99.9086    0.000087
     1.9996       1.9996      99.9086      99.9086    0.000082
```

Figure A.16: Sample output from **scatsmth** created by the parameter file shown in Figure A.15.

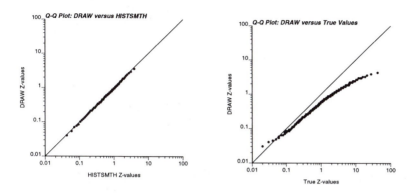

Figure A.17: Results from **draw** program. The Q-Q plot on the left compares the **draw** results with the **histsmth** target input distribution and the Q-Q plot on the right compares the **draw** results with the reference distribution in ⌈ true.dat ⌋.

```
                       Parameters for DRAW
                       ********************

START OF PARAMETERS:
histsmth.out                     \file with data
2                                \   number of variables
1    2                           \   columns for variables
3                                \   column for probabilities (0=equal)
-1.0e21    1.0e21                 \   trimming limits
69069     2500                   \random number seed, number to draw
draw.out                         \file for realizations
```

Figure A.18: Parameter file for the Monte Carlo program **draw**.

```
DRAW output
  2
Z-value
Log10 Z-valu
        2.915         0.4646
        0.2595       -0.5859
        1.668         0.2222
        0.4132       -0.3838
        0.2848       -0.5455
        0.5995       -0.2222
        0.7221       -0.1414
        0.6579       -0.1818
        1.668         0.2222
        0.1963       -0.7071
 • • •
        0.1789       -0.7475
        0.7221       -0.1414
        1.262         0.1010
        0.5462       -0.2626
        0.2595       -0.5859
        0.1353       -0.8687
        0.2154       -0.6667
        1.831         0.2626
        3.853         0.5859
        1.048         0.02020
```

Figure A.19: Sample output from **draw** created by the parameter file shown in Figure A.18.

A.2 Problem Set Two: Variograms

Preferential clustering of data in, say, high grade zones influences not only univariate statistics such as the sample histogram but also all two-point statistics such as the sample variogram.

Unfortunately there is no general declustering algorithm for bivariate and multivariate statistics, similar to the cell-declustering algorithm for univariate statistics[1]. Different measures of spatial variability, as presented in Section III.1, will have different robustness properties with regard to data clustering and sparsity, presence of outliers, or any cause of sampling fluctuations and biases. Therefore, in the presence of sparse and preferentially clustered data, it is good practice to run a few other measures of spatial variability in addition to the traditional sample variogram (III.1). The modeling of the traditional sample variogram can take advantage of features, such as range and anisotropy, better seen on other more robust measures.

Figure A.20 gives seven experimental measures of spatial variability calculated on the 140 clustered sample values in cluster.dat . These measures are listed in Section III.1. The omnidirectional $(0^\circ \pm 90^\circ)$, $NS(0^\circ \pm 22.5^\circ)$, and $EW(90^\circ \pm 22.5^\circ)$ directions are plotted.

Figures A.21 and A.22 give the corresponding input parameter file to program gamv and the first 10 and last 10 lines of the output file. Note that the results are slightly different (more pairs) than those shown in the first edition of GSLIB. The new gamv program allows overlapping lag bins; pairs exactly at the tolerance limit of two lags will be reported in both.

The traditional sample variogram and madogram[2] appear noisy while the other measures of continuity are much more stable, revealing an isotropic range between 10 and 20. The covariance/correlogram measures and relative variograms account for the data mean at each specific lag $\mathbf{h}$; see expressions (III.3) to (III.6). This allows for some correction of the preferential clustering; indeed, pairs with short separation $\mathbf{h}$ tend to be preferentially located in the high-z-valued zones.

The sample traditional variogram can be modeled borrowing ranges and anisotropic features better revealed by other more robust measures of spatial variability.

---

[1] Henning Omre [147] has proposed a variogram declustering algorithm that amounts to weighting each pair of data in the traditional variogram estimate:

$$2\gamma^*(\mathbf{h}) \;=\; \frac{1}{N(\mathbf{h})} \sum_{\alpha=1}^{N(\mathbf{h})} \omega_\alpha(\mathbf{h}) \cdot [\, z(\mathbf{u}_\alpha) \;-\; z(\mathbf{u}_\alpha \,+\, \mathbf{h})]^2$$

The weights $\omega_\alpha(\mathbf{h})$ are such that the marginal histogram built from the bivariate distribution of the $N(\mathbf{h})$ pairs of data $z(\mathbf{u}_\alpha), z(\mathbf{u}_\alpha + \mathbf{h})$ approximates best the prior declustered sample histogram built on all available z data.

[2] The madogram is robust with regard to outlier values, since the influence of such outliers is not squared as it is in the expression of the traditional variogram estimate. The problem in this data set, however, is not so much outlier data but clustered data.

One way around the problem of clustering is to ignore those data that are clustered; in the present case considering only the 97 first data. This is not always possible, particularly if drilling has not been done in recognizable sequences, with the first sequence being on a pseudoregular grid. Also, ignoring data amounts to a loss of information; for example, ignoring clustered data may lead to ignoring most of the short-scale information.

Figure A.23 gives the same eight experimental measures but now calculated on the first 97 data only. Fluctuations are seen to be somewhat reduced. Note the considerable reduction in variance due to the removal of the high-valued clustered data. The sample traditional semivariogram stabilizes around a sill value of about 11, a value consistent with the 97 data sample variance 10, but smaller than the 140 data declustered variance 20.9 and the actual variance 26.5. The naive, equal-weighted, 140 data sample variance is 44.9. These variance deviations stress the difficulty in estimating the correct (ordinate axis) scale for variograms.

Experimental variogram values for the first lag are often extremely unstable as seen in Figure A.20, because of the small number of pairs, preferential locations of close-spaced data, and the large averaging within that first lag.

Yet, with the help of the more robust experimental measures, an isotropic model can be inferred with a range of about 10 distance units and a relative nugget effect between 30 and 50%:

$$\gamma(\mathbf{h}) \; = \; K \; \left[0.3 \, + \, 0.7 \; \text{Sph} \; \left(\frac{|\mathbf{h}|}{10}\right)\right] \;, \quad \forall \, |\mathbf{h}| \, > \, 0 \qquad (A.1)$$

The factor K defining the total sill of the semivariogram model can be identified to the 140 data declustered variance: $K \, = \, 20.9$. Fortunately, such a proportional factor carries no influence on the kriging weights, hence on the kriging estimate. The proportional factor, however, does directly affect the kriging variance value: this is why nonlinear kriging estimates, such as provided by lognormal kriging, which depend on the kriging variance value, are particularly nonrobust; see [152] and discussion in Section IV.1.8.

Figure A.24 gives the same eight measures of spatial variability calculated from all 2500 reference data in the NS and EW directions. The thick continuous curve on the top left graph corresponds to the isotropic model:

$$\gamma(\mathbf{h}) \; = \; 10 \, + \, 16 \; \text{Sph} \; \left(\frac{|\mathbf{h}|}{8}\right) \qquad (A.2)$$

$$\approx \; 26 \; \left[0.38 \, + \, 0.62 \; \text{Sph} \; \left(\frac{|\mathbf{h}|}{10}\right)\right] \;, \quad \forall \, |\mathbf{h}| \, > \, 0$$

It appears that, except for the variance factor K, the model (A.1) inferred from the various sample measures is reasonably close to the exhaustive model (A.2).

Figures A.25 and A.26 give the corresponding input parameter file to program **gam** and the first 10 and last 10 lines of the output file.

Figure A.27 shows a variogram map calculated with the 2500 true values. The variogram map does not reveal any strong anisotropy. Figures A.28 and A.29 give the corresponding input parameter file to program varmap and the first 10 and last 10 lines of the output file.

The first 97 data were normal score transformed using their own equal-weighted histogram. The parameter file and output summary for program nscore are shown in Figures A.30 and A.31. Then the omnidirectional traditional semivariogram was calculated from these 97 normal score data. The results are shown in Figure A.32 together with the fit by the isotropic model:

$$\gamma_Y(\mathbf{h}) = 0.3 + \text{Sph}\left(\frac{|\mathbf{h}|}{12.0}\right) \tag{A.3}$$

Using the model (A.3) for $\gamma_Y(\mathbf{h})$ and assuming a bivariate Gaussian model for $RF\ Y(\mathbf{u})$, the theoretical indicator semivariograms for the three quartile y threshold values were calculated. Program bigaus described in Section VI.2.9 was used. The parameter file and output summary from bigaus are shown in Figures A.33 and A.34. These theoretical isotropic indicator semivariograms are given as the solid lines in Figure A.35. The corresponding 97 data sample indicator semivariograms are also plotted in Figure A.35. The reasonable fits indicate that the sample indicator semivariograms do not invalidate a multivariate Gaussian model for $Y(\mathbf{u})$; see related discussion in Section V.2.2.

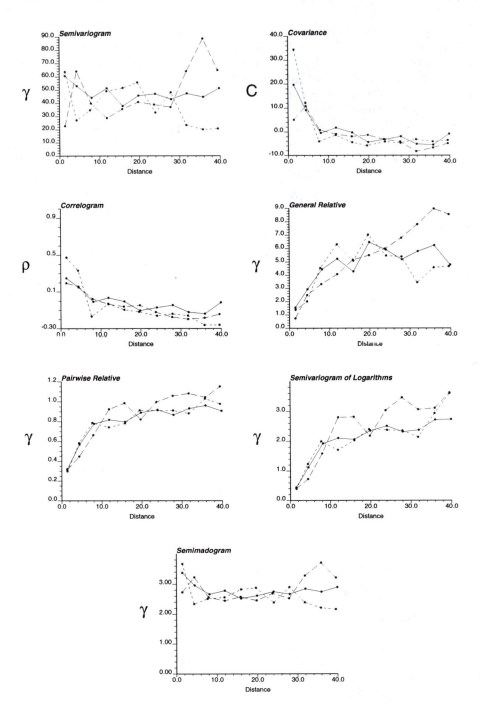

Figure A.20: Seven experimental measures of spatial variability calculated from the 140 sample values of cluster.dat .

```
                        Parameters for GAMV
                        *******************

START OF PARAMETERS:
../../data/cluster.dat              \file with data
1    2    0                         \   columns for X, Y, Z coordinates
1    3                              \   number of varables,column numbers
-1.0e21       1.0e21                \   trimming limits
gamv.out                           \file for variogram output
10                                 \number of lags
4.0                                \lag separation distance
2.0                                \lag tolerance
3                                  \number of directions
   0.0  90.0 50.0    0.0  10.0  10.0  \azm,atol,bandh,dip,dtol,bandv
   0.0  22.5 10.0    0.0  10.0  10.0  \azm,atol,bandh,dip,dtol,bandv
  90.0  22.5 10.0    0.0  10.0  10.0  \azm,atol,bandh,dip,dtol,bandv
0                                  \standardize sills? (0=no, 1=yes)
7                                  \number of variograms
1    1    1                        \tail var., head var., variogram type
1    1    3                        \tail var., head var., variogram type
1    1    4                        \tail var., head var., variogram type
1    1    5                        \tail var., head var., variogram type
1    1    6                        \tail var., head var., variogram type
1    1    7                        \tail var., head var., variogram type
1    1    8                        \tail var., head var., variogram type
```

Figure A.21: Parameter file for **gamv** run.

Semivariogram		tail:Primary		head:Primary	direction 1
1	0.000	0.00000	280	4.35043	4.35043
2	1.412	60.62550	256	8.84773	8.84773
3	4.405	52.95214	886	6.02221	6.02221
4	8.085	44.15183	1438	4.46609	4.46609
5	11.903	51.22849	1798	4.43352	4.43352
6	15.922	37.84546	2042	4.20964	4.20964
7	19.849	45.76223	1844	3.76399	3.76399
8	23.966	47.00244	2220	3.99143	3.99143
9	27.947	42.96206	2168	4.07630	4.07630
10	31.879	47.26698	2046	4.04337	4.04337
• • •					
3	4.307	0.91249	113	5.73292	4.69912
4	7.670	0.93571	171	4.97889	2.86626
5	11.802	0.91542	208	4.53212	3.37067
6	15.926	0.97906	263	4.43011	4.58665
7	19.599	0.99215	247	3.90296	4.05923
8	24.027	0.92365	292	3.10380	3.84740
9	27.892	1.01726	303	3.36109	5.10584
10	31.725	0.94431	257	3.71121	3.66541
11	35.775	0.92067	179	2.66838	3.27084
12	39.491	0.87960	101	4.04752	1.92030

Figure A.22: Output of **gamv** corresponding to the parameter file shown in Figure A.21.

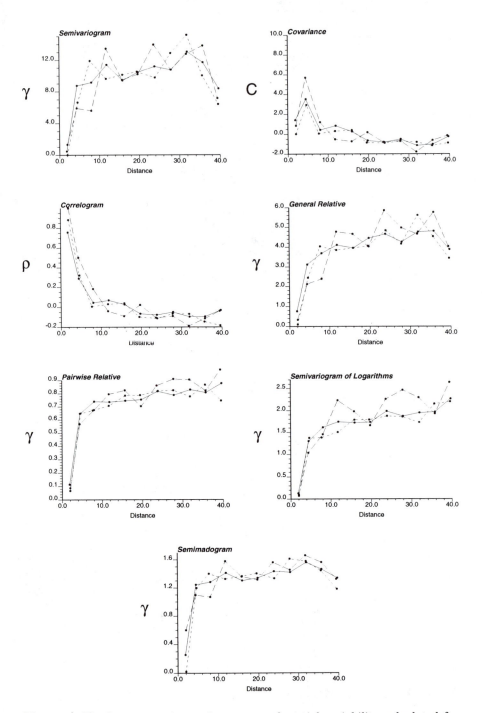

Figure A.23: Seven experimental measures of spatial variability calculated from the first 97 sample values of cluster.dat .

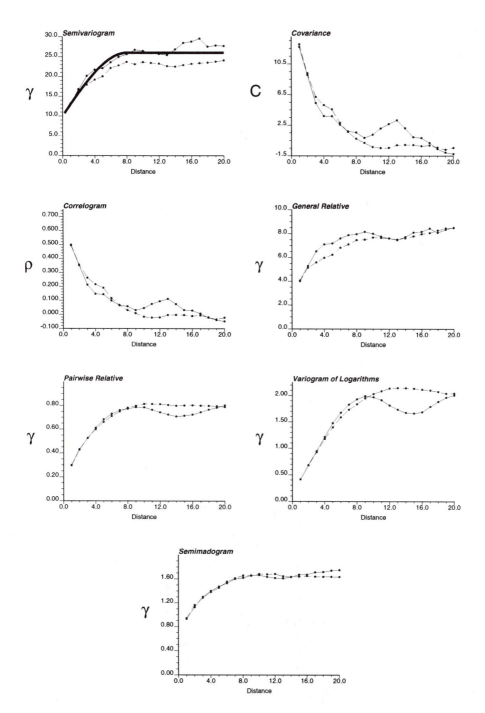

Figure A.24: Seven measures of spatial variability calculated on all 2500 reference data values of $\boxed{\text{true.dat}}$.

```
                              Parameters for GAM
                              *******************

START OF PARAMETERS:
../../data/true.dat           \file with data
1   1                         \   number of variables, column numbers
-1.0e21        1.0e21         \   trimming limits
gam.out                       \file for variogram output
1                             \grid or realization number
50    0.5   1.0               \nx, xmn, xsiz
50    0.5   1.0               \ny, ymn, ysiz
 1    0.5   1.0               \nz, zmn, zsiz
2   20                        \number of directions, number of lags
 1   0   0                    \ixd(1),iyd(1),izd(1)
 0   1   0                    \ixd(2),iyd(2),izd(2)
0                             \standardize sill? (0=no, 1=yes)
7                             \number of variograms
1   1   1                     \tail variable, head variable, variogram type
1   1   3                     \tail variable, head variable, variogram type
1   1   4                     \tail variable, head variable, variogram type
1   1   5                     \tail variable, head variable, variogram type
1   1   6                     \tail variable, head variable, variogram type
1   1   7                     \tail variable, head variable, variogram type
1   1   8                     \tail variable, head variable, variogram type
```

Figure A.25: Parameter file for **gam** run.

```
Semivariogram            tail:primary - Z  head:primary - Z    direction   1
        1      1.000    13.03226    2450      2.53364            2.53909
        2      2.000    16.64981    2400      2.49753            2.52300
        3      3.000    20.16062    2350      2.46945            2.50771
        4      4.000    21.73316    2300      2.45293            2.49312
        5      5.000    22.07451    2250      2.46641            2.49146
        6      6.000    23.60346    2200      2.48491            2.50075
        7      7.000    24.93002    2150      2.50757            2.52079
        8      8.000    25.62192    2100      2.53510            2.53821
        9      9.000    26.70175    2050      2.55013            2.56459
       10     10.000    26.42958    2000      2.57868            2.55968
• • •
       11     11.000     1.67927    1950      2.81330            2.13936
       12     12.000     1.68288    1900      2.80745            2.14270
       13     13.000     1.64556    1850      2.79272            2.11211
       14     14.000     1.63687    1800      2.80217            2.05762
       15     15.000     1.64110    1750      2.84072            2.01293
       16     16.000     1.65112    1700      2.87931            1.94995
       17     17.000     1.64072    1650      2.92879            1.89065
       18     18.000     1.63764    1600      2.99539            1.77568
       19     19.000     1.63833    1550      3.06566            1.67994
       20     20.000     1.63538    1500      3.13565            1.63567
```

Figure A.26: Output of **gam** corresponding to the parameter file shown in Figure A.25.

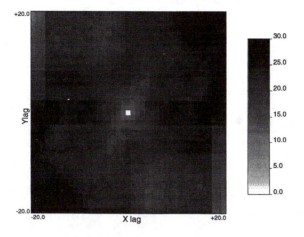

Figure A.27: Variogram map computed with **varmap**.

```
                    Parameters for VARMAP
                    *********************

START OF PARAMETERS:
../../data/true.dat          \file with data
1   1                        \   number of variables: column numbers
-1.0e21      1.0e21          \   trimming limits
1                            \1=regular grid, 0=scattered values
 50    50    1               \if =1: nx,     ny,    nz
1.0  1.0  1.0                \        xsiz, ysiz, zsiz
1    2    0                  \if =0: columns for x,y, z coordinates
varmap.out                   \file for variogram output
 20    20     0              \nxlag, nylag, nzlag
1.0   1.0   1.0              \dxlag, dylag, dzlag
5                            \minimum number of pairs
0                            \standardize sill? (0=no, 1=yes)
1                            \number of variograms
1    1    1                  \tail, head, variogram type
```

Figure A.28: Parameter file for the **varmap** run.

```
Variogram Volume: nx  41 ny  41nz   1
6
variogram
number of pairs
head mean
tail mean
head variance
tail variance
        34.11169         900.         3.99163         1.79398         0.00000         0.00
        33.67606         930.         3.93667         1.79766         0.00000         0.00
        33.65016         960.         3.88683         1.80668         0.00000         0.00
        33.28286         990.         3.87093         1.77759         0.00000         0.00
        32.46981        1020.         3.80836         1.75618         0.00000         0.00
        32.10539        1050.         3.75921         1.72686         0.00000         0.00
        31.21413        1080.         3.69202         1.70562         0.00000         0.00
        30.21509        1110.         3.62503         1.68489         0.00000         0.00
        29.29567        1140.         3.55804         1.66092         0.00000         0.00
        27.78258        1170.         3.51191         1.63003         0.00000         0.00
 • • •
        27.78258        1170.         1.63003         3.51191         0.00000         0.00
        29.29567        1140.         1.66092         3.55804         0.00000         0.00
        30.21509        1110.         1.68489         3.62503         0.00000         0.00
        31.21413        1080.         1.70562         3.69202         0.00000         0.00
        32.10539        1050.         1.72686         3.75921         0.00000         0.00
        32.46981        1020.         1.75618         3.80836         0.00000         0.00
        33.28286         990.         1.77759         3.87093         0.00000         0.00
        33.65016         960.         1.80668         3.88683         0.00000         0.00
        33.67606         930.         1.79766         3.93667         0.00000         0.00
        34.11169         900.         1.79398         3.99163         0.00000         0.00
```

Figure A.29: Output of **varmap** corresponding to the parameter file shown in Figure A.28.

```
                    Parameters for NSCORE
                    *********************

START OF PARAMETERS:
../../data/97data.dat       \file with data
3    0                       \  columns for variable and weight
-1.0e21    1.0e21           \  trimming limits
0                           \1=transform according to specified ref. dist.
../histsmth/histsmth.out   \  file with reference dist.
1    2                       \  columns for variable and weight
97data.nsc                  \file for output
97data.trn                  \file for output transformation table
```

Figure A.30: Parameter file for the **nscore** transformation of the first 97 data.

```
Normal Score Transform:First 97 data of cluster.dat
  6
Xlocation
Ylocation
Primary
Secondary
Declustering Weight
Normal Score value
   39.5000      18.5000        .0600        .2200      4.0000       -2.56528
    5.5000       1.5000        .0600        .2700      4.0000       -2.15800
   38.5000       5.5000        .0800        .4000      3.5000       -1.94690
   20.5000       1.5000        .0900        .3900      4.5000       -1.68089
   27.5000      14.5000        .0900        .2400      3.3330       -1.79808
   40.5000      21.5000        .1000        .4800      2.3330       -1.58309
   15.5000       3.5000        .1000        .2100      3.0000       -1.49843
    6.5000      25.5000        .1100        .3600      4.0000       -1.42333
   38.5000      21.5000        .1100        .2200      2.8330       -1.35550
   23.5000      18.5000        .1600        .3000      4.5000       -1.29339
 • • •
    2.5000       9.5000       6.2600      17.0200       .7220        1.29339
   32.5000      36.5000       6.4100       2.4500       .8610        1.35550
     .5000       8.5000       6.4900      14.9500       .8110        1.42333
   31.5000      45.5000       7.5300      10.2100       .7560        1.49843
    9.5000      29.5000       8.0300       5.2100      1.1000        1.58309
   39.5000      31.5000       8.3400       8.0200       .8430        1.68089
   17.5000      15.5000       9.0800       3.3200      1.1000        1.79808
    2.5000      14.5000      10.2700       5.6700      1.0610        1.94690
   30.5000      41.5000      17.1900      10.1000       .8890        2.15800
   35.5000      32.5000      18.7600      10.7600       .6460        2.56528
```

Figure A.31: Output of **nscore** corresponding to the parameter file shown in Figure A.30.

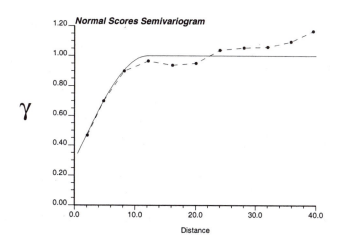

Figure A.32: Omnidirectional normal score semivariogram and its model fit (calculated from 97 data).

```
                      Parameters for BIGAUS
                      **********************

START OF PARAMETERS:
bigaus.out                          \file for output of variograms
3                                   \number of thresholds
0.25    0.50      0.75              \threshold cdf values
1    80                             \number of directions and lags
 0.5   0.0      0.0                 \azm(1), dip(1), lag(1)
1     0.3                           \nst, nugget effect
1     0.7  0.0     0.0     0.0      \it,cc,ang1,ang2,ang3
           12.0   12.0    12.0      \a_hmax, a_hmin, a_vert
```

Figure A.33: Parameter file for **bigaus** run.

```
Model Indicator Variogram: cutoff =    -0.674 Direction
  1     0.000         0.000    1    1.17810      0.00000
  2     1.000         0.114    1    0.46445      0.60576
  3     2.000         0.126    1    0.38598      0.67237
  4     3.000         0.137    1    0.31552      0.73217
  5     4.000         0.147    1    0.25229      0.78585
  6     5.000         0.156    1    0.19589      0.83373
  7     6.000         0.164    1    0.14621      0.87589
  8     7.000         0.171    1    0.10332      0.91230
  9     8.000         0.177    1    0.06740      0.94279
 10     9.000         0.181    1    0.03870      0.96715
• • •
 71    70.000         0.188    1    0.00000      1.00000
 72    71.000         0.188    1    0.00000      1.00000
 73    72.000         0.188    1    0.00000      1.00000
 74    73.000         0.188    1    0.00000      1.00000
 75    74.000         0.188    1    0.00000      1.00000
 76    75.000         0.188    1    0.00000      1.00000
 77    76.000         0.188    1    0.00000      1.00000
 78    77.000         0.188    1    0.00000      1.00000
 79    78.000         0.188    1    0.00000      1.00000
 80    79.000         0.188    1    0.00000      1.00000
```

Figure A.34: Output of **bigaus** corresponding to the parameter file shown in Figure A.33.

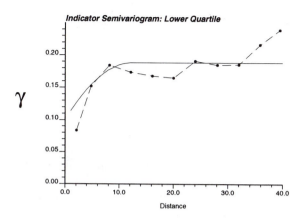

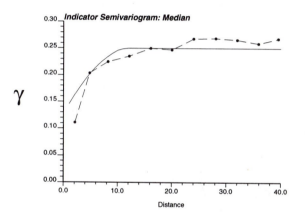

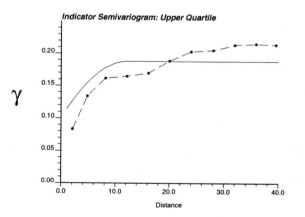

Figure A.35: Experimental indicator semivariograms (97 data) and their fits by Gaussian RF model-derived theoretical curves. The solid continuous line is the model.

A.3   Problem Set Three: Kriging

This solution set documents kriging results with some comments. The numerical results are less interesting than the interpretation.

Part A: Variogram Parameters

1. **Relative nugget effect:** When the relative nugget effect is small, the data redundancy and the proximity to the point being estimated become important. This implies that close samples are weighted more and samples within clusters are treated as less informative. A low relative nugget effect may cause a wider variation in the weights (from negative to greater than one in some cases), which may cause outlier data values to be more problematic. No general comment can be made about the magnitude of the estimate, although a surface estimated with a large nugget effect will be smoother.

2. **Range:** When the range is smaller than the interdistance between any two data points, then the situation is equivalent to a pure nugget effect model. As the range increases, the continuity increases, causing the proximity to the estimated location and data redundancy to become important. As the range becomes very large with a nugget effect, then the estimation is equivalent to a pure nugget system; if there is no nugget effect, then the kriging system behaves as if a linear variogram model were chosen.

3. **Scale:** The scaling of the variogram affects only the kriging (estimation) variance; only the relative shape of the variogram determines the kriging weights and estimate.

4. **Shape:** Both the exponential and spherical variograms have a linear shape near the origin. The exponential model growth from the origin is steeper, causing it to behave in a manner similar to a shorter-range spherical model. The Gaussian model is parabolic near the origin, indicating a spatially very continuous variable. This may cause very large positive and negative weights (which are entirely reasonable given such continuity). If the actual data values are *not* extremely continuous and a Gaussian model is used, then erratic results can arise from the negative weights. Numerical precision problems also arise because of off-diagonal elements that have nearly zero variogram (maximum covariance) values. The problem is not so much the shape of the variogram as the lack of a nugget. The addition of a nugget does not change appreciably the shape of the variogram but improves numerical stability.

5. **Anisotropy:** The anisotropy factor may be applied to the range of the variogram, or equivalently to the coordinates (after rotation to identify

the main directions of continuity). It is instructive to think of geometric anisotropy as a geometric correction to the coordinates followed by application of an isotropic model.

Part B: Data Configuration

1. The proposed configuration is symmetric about the 45° line. As the second point approaches either endpoint, the redundancy with the far point decreases and the redundancy with the near point increases. The rate of increase and decrease of the kriging weight differs depending on the variogram model. In the first case (low nugget effect $C_0 = 0.1$) the increase in redundancy with the near point is less than the decrease in redundancy with the far point; hence, the weight attributed to the second point increases as the point nears an endpoint.

2. With a high nugget effect ($C_0 = 0.9$) the increase in redundancy with the near point is greater than the decrease in redundancy with the far point; hence, the weight decreases as the point nears an endpoint. All the kriging weights in this case are almost equal.

Part C: Screen Effect

1. The first point gets more weight in all cases (unless a hole effect variogram model is considered). The weight applied to the second point decreases as it falls behind the first point. It is important to note that this weight may become negative, in which case the second point actually carries a significant information content. For example, a large negative weight (with an even larger positive weight applied to the closer data) would effectively impose the *trend* of the two data values (e.g., if the close datum is smaller than the far datum then the estimate will be less than either datum).

2. The screen effect is reduced when the nugget is large. Both samples have an almost equal contribution to the estimate.

Part D: Support Effect

1. Discretization with a 1 by 1 point is *equivalent* to point kriging; a greater level of discretization is required to perform block kriging. When the block size increases with respect to the average data spacing, the estimation variance drops considerably because of short scale averaging of errors. The discretization level is not as important as the relative size of the block.

2. The block is effectively larger than the scale of variability and its average smooths out most of the spatial variability. The estimation variance becomes small and the resulting surface (spatial distribution of block estimates) smooth.

Part E: Cross Validation

- **Cross validation with ordinary kriging:** The scatterplots of the true values and kriging estimates exhibit the typical spatial smoothing of kriging. Conditional bias, i.e., overestimation of low values and underestimation of high values, can be observed from the scatterplot.

- **Histogram of errors:** The histogram of the kriging errors shows a fairly symmetric distribution that is typically more "peaked" than a normal distribution. Note that any reasonable interpolation method can achieve global unbiasedness.

- **Map of Errors:** The errors are most significant near the high values but do not show any correlation (a variogram of the errors would confirm this).

- **Kriging variance:** If the kriging variance was a good measure of local accuracy, there should be a positive correlation between the actual absolute errors and the kriging variance. The significant negative correlation found with the sample of 140 is due to the fact that there is clustering in the high values. Near the high values the kriging variance is lowest because extra samples were preferentially located near these highs. Conversely, the actual error happens to be greatest near the high values because the variable is heteroscedastic (its local variance depends on its magnitude). The kriging variance is generally not a good measure of local accuracy because the data values are not taken into account. Further, the SK variance is always less than the OK variance, which in turn is always less than the KT variance. This is due to the additional uncertainty in estimating the mean (surface), which is assumed known in SK.

- **Kriging parameters:** Often the choice of kriging parameters (search radius, minimum and maximum number of data) is based more on data availability than on the physical process underlying the data. Enough data must be found to provide a stable estimate; too many data will unnecessarily smooth the kriging estimates and, possibly, create matrix instability.

- **Simple kriging:** SK requires knowledge of the stationary mean. In this case the mean is given a considerable weight (due to the high nugget effect), causing significant smoothing and conditional bias.

- **Kriging with a trend:** KT allows the locally estimated trend (mean) surface to be fit with monomials of the coordinates. In actuality it behaves much like OK in interpolation conditions and may yield erratic results in extrapolation (unless the trend model is appropriate); see also [112].

Part F: Kriging on a Grid

- **Simple kriging:** the mean and variance of the kriging estimates are considerably less than the true values. The mean is less because kriging does a good job of declustering the original data and the variance is less because of the smoothing effect of kriging. The general trends are reproduced quite well.

- **Minimum number of data:** This effectively prevents the kriging algorithm from estimating in areas where there are too few data. The fewer data retained, the more (artificially) variable the estimate becomes.

- **Maximum number of data:** The more data that are used the smoother the estimates appear. In general, enough data should be used so that the estimate is stable and does not present any artificial discontinuities. Ordinary kriging allows accounting for all data in the search neighborhood, even if they are not correlated with the point being estimated, because the mean is implicitly re-estimated locally. For that reason it may not be appropriate to use a large maximum limit *ndmax* even if the computer can handle it. Note that the computation cost of matrix inversion increases as the number of samples cubed.

- **Search radius:** The search radius has to be large enough to allow a stable estimate. If the search radius is set too large, however, the resulting surface may be too smooth. Note that the search radius and the maximum number of data interact with one another. If *ndmax* is set small enough, it does not matter how large the search radius is set and vice versa. It is generally an error to set a priori the search radius smaller than the correlation range, because in OK remote data contribute to the local re-estimation of the mean.

- **Ordinary kriging:** OK results appear similar to SK except for less smoothing and conditional bias. By accumulating the kriging weights applied to each datum and then normalizing these weights to sum to one, reasonable declustering weights are obtained.

- **Kriging with a trend:** KT results appear similar to OK except that they are more erratic near the borders. In fact there are some negative estimates due to the trend surface being estimated unreasonably low (the kriging systems do not know that contaminant concentration cannot be negative).

- **Block kriging:** The block kriging results are similar to the point kriging results because the block size is small. The kriging variance is less due to the averaging of errors. The smoothing effect of block kriging increases with larger blocks.

- **Trend surface:** The shape of the trend surface in interpolation areas is similar to that indicated by a contour map of block kriging estimates with a large block size.

A.4 Problem Set Four: Cokriging

Cokriging is a form of kriging where the data can relate to attribute(s) different from that being estimated; see Section IV.1.7.

The problem proposed here consists of estimating the 2500 reference z values (see Figure A.36) from the sparse sample of 29 z data in file $\boxed{\text{data.dat}}$ and 2500 related y data in file $\boxed{\text{ydata.dat}}$. The gray-scale map of the y data is given in Figure A.37.

Figure A.38 gives the histograms of the 2500 reference z values to be estimated, the 29 z data, the 2500 secondary y data and the 29 y data collocated with the z data.

Figure A.39 gives the scattergram of the 29 pairs of collocated z and y data. The rank correlation (0.80) is seen to be larger than the linear correlation (0.50). Also heteroscedasticity is observed: The variance of z values varies depending on the class of y values. These features indicate that there is potentially more information to gather from the secondary y information than the mere linear correlation and regression; see [198] and Section IV.1.12.

The variograms were calculated from the 140 collocated $(z\text{-}y)$ data. Figure A.40 gives the two experimental direct semivariograms and the experimental cross semivariogram and their model fits by the linear model of coregionalization (A.4). For each variogram, the two NS and EW directions with tolerance $\pm 22.5^o$ and the omnidirectional curve (tolerance $\pm 90^o$) were considered. The traditional expression (III.1) for the direct semivariograms and (III.2) for the cross semivariogram were used. The range (10) was better picked from the pairwise relative semivariograms (III.6) for both z and y variable, not shown here. The isotropic linear model of coregionalization consists of a nugget effect and a spherical structure of range 10:

$$
\begin{cases}
\gamma_Z(\mathbf{h}) &= 10.00 + 32.0 \text{ Sph } (|\mathbf{h}|/10) \\
\gamma_Y(\mathbf{h}) &= 0.10 + 19.0 \text{ Sph } (|\mathbf{h}|/10) \\
\gamma_{ZY}(\mathbf{h}) &= 0.01 + 16.0 \text{ Sph } (|\mathbf{h}|/10), \quad \text{for } |\mathbf{h}| > 0.
\end{cases}
\tag{A.4}
$$

The determinants of the nugget coefficients and the spherical structure factors are positive; thus this model is legitimate ([89], p. 391):

$$
\begin{vmatrix} 10.0 & 0.01 \\ 0.01 & 0.10 \end{vmatrix} = 1 - 0.0001 > 0 \; ; \quad
\begin{vmatrix} 32.0 & 16.0 \\ 16.0 & 19.0 \end{vmatrix} = 608 - 256 > 0
$$

Ordinary kriging with program kb2d was performed to estimate the 2500 z reference values using only the 29 primary z data and the corresponding γ_Z semivariogram model given in (A.4). There were no negative kriging estimates. The resulting gray-scale estimated map is given in Figure A.41, and the histogram of estimation errors $(z^* - z)$ is given in Figure A.42. The OK estimated map can be compared with the reference gray-scale map of Figure A.36.

Figures A.43 and A.44 give the corresponding input parameter file to program kb2d and the first 10 and last 10 lines of the output file.

Cokriging using both the 29 primary z data and the 2500 secondary data was performed with program cokb3d. The corresponding input parameter file and first 10 and last 10 lines of the output file are given in Figures A.45 and A.46. The traditional ordinary cokriging option was considered with the two unbiasedness conditions (IV.22). Because the y data are much more numerous than the z data, a different search strategy was considered for z and y data. Figure A.47 gives the gray-scale map of the cokriging estimates and Figure A.48 gives the histogram of the cokriging errors $(z^* - z)$. The cokriging map is to be compared with the reference map of Figure A.36 and the OK map of Figure A.41. The improvement brought by the large amount of secondary y data is seen to be significant.

Because the y data were generated as moving averages of z values taken at a different level, the external drift concept is ideally suited to integrate such information. Program kt3d with the external drift option was used; see Section IV.1.5.

Figures A.49 and A.50 provide the input parameter file and the first 10 and last 10 lines of the output file. The resulting set of 2500 z estimates include 65 negative values which were reset to zero.[3] This large number of negative estimates is explained by the fact that kriging with an external drift amounts to extrapolating the y trend to the z values, thus incurring a risk of extrapolating beyond the limit zero z value. Figure A.51 gives the gray-scale map of the z estimates, and Figure A.52 gives the histogram of errors $(z^* - z)$. The results appear significantly better than those provided by kriging using only the 29 primary z data; compare to Figure A.41 and the reference map of Figure A.36.

---

[3]This resetting to zero of negative estimates corresponds to the solution that would be provided by quadratic programming with the condition $z^*(\mathbf{x}) \geq 0$; see [127].

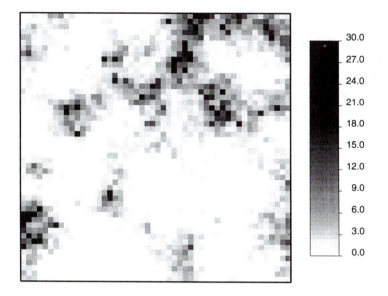

Figure A.36: Reference z values to be estimated.

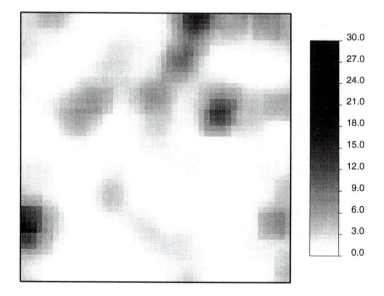

Figure A.37: Secondary information (2500 y values).

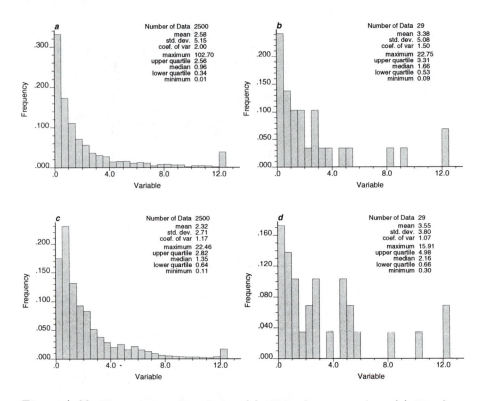

Figure A.38: Histograms and statistics. (a) 2500 reference z values, (a) 29 z data, (c) 2500 secondary y data, (d) 29 y data collocated with the z data.

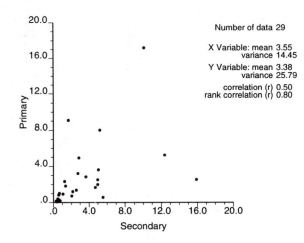

Figure A.39: Scattergram of collocated z primary and y secondary data.

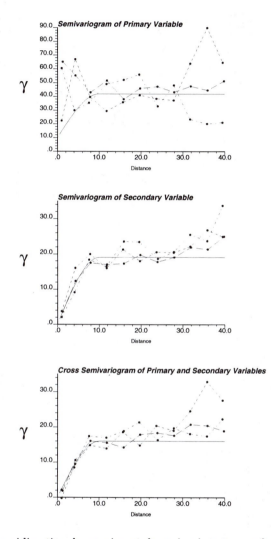

Figure A.40: Omnidirectional experimental semivariograms and cross semivariogram and their model fits (140 collocated z-y data). The thick continuous line is the model.

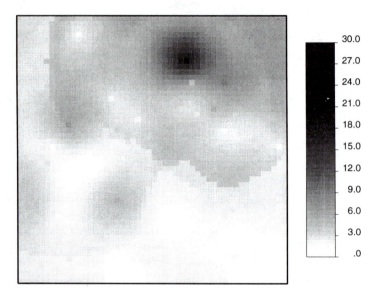

Figure A.41: Gray scale map of OK z estimated values (29 data).

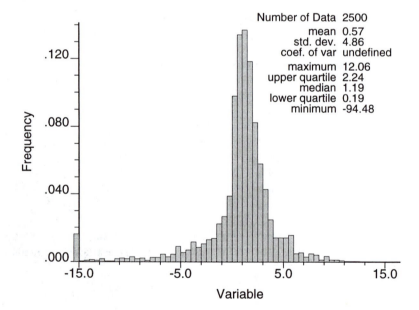

Figure A.42: Histogram of OK kriging errors $(z^* - z)$.

```
                         Parameters for KB2D
                         *******************

START OF PARAMETERS:
../../data/data.dat          \file with data
1   2   3                    \   columns for X, Y, and variable
-1.0e21    1.0e21            \   trimming limits
3                            \debugging level: 0,1,2,3
kb2d.dbg                     \file for debugging output
kb2d.out                     \file for kriged output
50   0.5  1.0                \nx,xmn,xsiz
50   0.5  1.0                \ny,ymn,ysiz
1   1                        \x and y block discretization
4   16                       \min and max data for kriging
25.0                         \maximum search radius
1    2.302                   \0=SK, 1=OK,  (mean if SK)
1  10.0                      \nst, nugget effect
1   32.0   0.0  10.0  10.0   \it, c, azm, a_max, a_min
```

Figure A.43: Parameter file for **kb2d**.

```
KB2D Output
2
Estimate
Estimation Variance
    2.048    42.183
    1.874    42.147
    1.874    42.147
    1.874    42.147
    1.874    42.147
    1.874    42.147
    1.524    42.118
    1.663    42.123
    1.663    42.123
    1.663    42.123
• • •
    6.721    42.191
    5.794    42.160
    6.358    42.191
    6.358    42.191
    6.358    42.191
    6.358    42.191
    6.358    42.191
    6.358    42.191
    6.358    42.191
    6.358    42.191
```

Figure A.44: Output of **kb2d** corresponding to the parameter file shown in Figure A.43.

```
                        Parameters for COKB3D
                        *********************

START OF PARAMETERS:
../../data/ydata.dat                \file with data
2                                   \   number of variables primary+other
1   2   0   3   4                   \   columns for X,Y,Z and variables
-0.01      1.0e21                    \   trimming limits
0                                   \co-located cokriging? (0=no, 1=yes)
../data/ydata.dat                   \   file with gridded covariate
4                                   \   column for covariate
3                                   \debugging level: 0,1,2,3
cokb3d.dbg                          \file for debugging output
cokb3d.out                          \file for output
50   0.5   1.0                      \nx,xmn,xsiz
50   0.5   1.0                      \ny,ymn,ysiz
 1   0.5   1.0                      \nz,zmn,zsiz
 1    1     1                       \x, y, and z block discretization
 1   12     8                       \min primary,max primary,max all sec
25.0  25.0  25.0                    \maximum search radii: primary
10.0  10.0  10.0                    \maximum search radii: all secondary
 0.0   0.0   0.0                    \angles for search ellipsoid
2                                   \kriging type (0=SK, 1=OK, 2=OK-trad)
3.38  2.32  0.00  0.00              \mean(i),i=1,nvar
1    1                              \semivariogram for "i" and "j"
1   10.0                            \   nst, nugget effect
1   32.0  0.0   0.0   0.0           \   it,cc,ang1,ang2,ang3
          10.0  10.0  10.0          \   a_hmax, a_hmin, a_vert
1    2                              \semivariogram for "i" and "j"
1   0.01                            \   nst, nugget effect
1   16.0  0.0   0.0   0.0           \   it,cc,ang1,ang2,ang3
          10.0  10.0  10.0          \   a_hmax, a_hmin, a_vert
2    2                              \semivariogram for "i" and "j"
1   0.1                             \   nst, nugget effect
1   19.0  0.0   0.0   0.0           \   it,cc,ang1,ang2,ang3
          10.0  10.0  10.0          \   a_hmax, a_hmin, a_vert
```

Figure A.45: Parameter file for cokb3d.

```
COKB3D with:All of the secondary data plus 29 primary
2
estimate
estimation variance
        2.6481      38.6260
        2.1062      38.2560
        1.7395      38.2538
        2.1147      37.9623
        1.3728      38.2607
        1.3428      37.9651
        1.5092      38.2491
        1.6393      38.2543
        1.7466      38.2561
        1.6917      37.9352
• • •
        7.2301      38.3237
        7.6823      38.2931
        4.6941      38.0107
        6.7518      38.0107
        7.5002      38.3237
        7.0268      38.3237
        6.6970      38.0107
        7.9967      38.3237
        6.3431      38.3237
        5.5095      38.6741
```

Figure A.46: Output of cokb3d corresponding to the parameter file shown in Figure A.45.

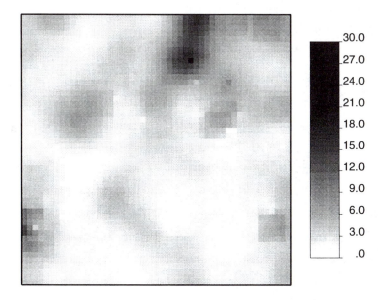

Figure A.47: Gray scale map of cokriging z estimated values.

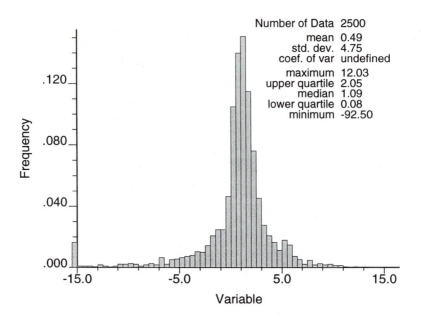

Figure A.48: Histogram of cokriging errors $(z^* - z)$.

```
                    Parameters for KT3D
                    *******************

START OF PARAMETERS:
../../data/data.dat                 \file with data
1   2   0   3   4                   \   columns for X, Y, Z, var, sec var
-0.5        1.0e21                  \   trimming limits
0                                   \option: 0=grid, 1=cross, 2=jackknife
xvk.dat                             \file with jackknife data
1   2   0   3   0                   \   columns for X,Y,Z,vr and sec var
3                                   \debugging level: 0,1,2,3
kt3d.dbg                            \file for debugging output
kt3d.out                            \file for kriged output
50   0.5    1.0                     \nx,xmn,xsiz
50   0.5    1.0                     \ny,ymn,ysiz
1    0.5    1.0                     \nz,zmn,zsiz
1    1      1                       \x,y and z block discretization
1    16                             \min, max data for kriging
0                                   \max per octant (0-> not used)
25.0  25.0  25.0                    \maximum search radii
 0.0   0.0   0.0                    \angles for search ellipsoid
3      2.302                        \0=SK,1=OK,2=non-st SK,3=exdrift
0 0 0 0 0 0 0 0 0                   \drift: x,y,z,xx,yy,zz,xy,xz,zy
0                                   \0, variable; 1, estimate trend
../../data/ydata.dat                \gridded file with drift/mean
4                                   \   column number in gridded file
1    10.0                           \nst, nugget effect
1    32.0   0.0    0.0    0.0       \it,cc,ang1,ang2,ang3
            10.0   10.0   10.0      \a_hmax, a_hmin, a_vert
```

Figure A.49: Parameter file for **kt3d**.

```
KT3D ESTIMATES WITH:  Irregular Spaced 29 primary and secondary data
2
Estimate
EstimationVariance
    2.017    49.702
    1.827    48.212
    1.776    48.353
    1.728    48.576
    1.604    49.552
    1.579    49.821
    1.205    48.106
    1.380    48.321
    1.390    48.245
    1.391    48.236
• • •
    8.206    51.238
    8.563    52.018
    7.147    50.294
    9.203    53.641
    9.574    54.649
    9.559    54.604
    8.902    52.913
    9.404    54.172
    7.835    50.992
    7.247    50.370
```

Figure A.50: Output of **kt3d** corresponding to the parameter file shown in Figure A.49.

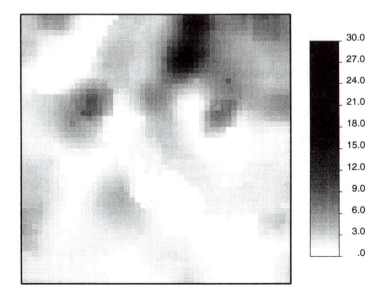

Figure A.51: Gray scale map of kriging with external drift z estimated values.

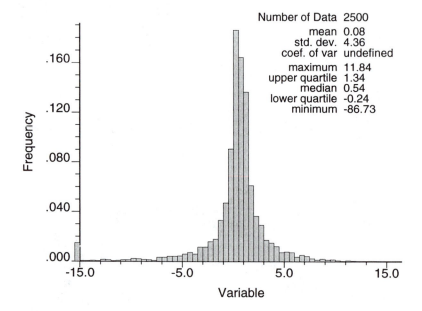

Figure A.52: Histogram of errors of kriging with external drift $(z^* - z)$.

A.5 Problem Set Five: Indicator Kriging

The main objective of indicator kriging is not to provide an indicator value estimate, but to provide a model of uncertainty about the unsampled original value. That model of uncertainty takes the form of a conditional cumulative distribution function (ccdf), recall relation (IV.26):

$$\text{Prob}^* \left\{ Z(\mathbf{u}) \leq z \mid (n) \right\} = E^* \left\{ I(\mathbf{u}; z) \mid (n) \right\}$$

The first column of Table A.1 gives the 9 decile cutoffs of the declustered marginal cdf $F^*(z_k)$ of the 140 z data. The third column gives the corresponding *non*-declustered *cdf* values $\hat{F}(z_k)$ related to the raw indicator data variances:

$$\widehat{\text{Var}} \left\{ I(\mathbf{u}; z_k) \right\} = \hat{F}(z_k) \left[1 - \hat{F}(z_k) \right]$$

Figures A.53 and A.54 give the corresponding 9 omnidirectional sample indicator semivariograms and their model fits. All semivariograms have been standardized by the nondeclustered indicator data variance $\hat{F}(z_k)[1 - \hat{F}(z_k)]$. All nine models are of the same type:

$$\gamma_I(\mathbf{h}; z_k) = C_0(z_k) + C_1(z_k) \, \text{Sph}(|\mathbf{h}|/11) + C_2(z_k) \, \text{Sph}(|\mathbf{h}|/30)$$

$$\text{with} \qquad C_0(z_k) + C_1(z_k) + C_2(z_k) = 1, \ \forall \, k = 1, \dots, 9$$

$$\text{(A.5)}$$

Table A.1 gives the list of the sill parameter values $C_0(z_k), C_1(z_k), C_2(z_k)$. The corresponding model fits should be evaluated globally over all nine cutoffs z_k, because the parameter values were adjusted so that they vary smoothly with cutoff.

Note that, except for the smallest and two largest cutoffs, the model is the same. The lowest sample data (indicator at cutoff z_1) appear more spatially correlated, whereas the highest sample data (indicators at cutoffs z_8 and z_9) appear less correlated. In the absence of ancillary information confirming this different behavior of extreme values and given the uncertainty due to small sample size and preferential data clustering, one could adopt the same model for all relative indicator variograms, i.e. a median IK model; see relation (IV.30) and hereafter.

Indicator kriging using program ik3d was performed with the 9 cutoffs z_k and semivariogram models specified by model (A.5) and Table A.1. The resulting ccdf's are corrected for order relations problems (within ik3d). Figure A.55 gives the input parameter file for the ik3d run and Figure A.56 gives the first and last 10 lines of output.

A correct handling of within-class interpolation and extreme classes extrapolation is critical when using IK. Interpolation using the marginal declustered sample cdf was adopted for all classes.

Program postik provides the E-type estimate for the unsampled z value, i.e., the mean of the calculated *ccdf*. Figure A.57 gives the input parameter file for the postik run; the first 10 and last 10 lines of the postik output

	z_k	$F^*(z_k)$	$\hat{F}(z_k)$	$C_0(z_k)$	$C_1(z_k)$	$C_2(z_k)$
$k=1$	0.159	0.1	0.079	0	0.5	0.5
	0.278	0.2	0.150	0	1.0	0
	0.515	0.3	0.221	0	1.0	0
	0.918	0.4	0.300	0	1.0	0
	1.211	0.5	0.379	0	1.0	0
	1.708	0.6	0.443	0	1.0	0
	2.325	0.7	0.521	0	1.0	0
	3.329	0.8	0.643	0.4	0.6	0
$k=9$	5.384	0.9	0.757	0.9	0.1	0

Table A.1: Parameters for the indicator variogram models.

file are shown in Figure A.58. Figure A.59 gives the gray-scale map of the 2500 E-type estimates and the corresponding histogram of the error (z^*-z). This E-type map compares favorably to the reference map of Figure A.36 although there is the typical smoothing effect of kriging. The location of the conditioning data (see Figure II.10) are apparent on the E-type map; the configuration of the clustered data appears as dark "plus" signs. Figure A.59a can not be compared directly to the ordinary kriging, cokriging, or kriging with an external drift of Problem Set 4 because more data have been used in this case and a different gray-scale has been adopted.

The discontinuities apparent at the data locations are due to the preferential clustering of the 140 data combined with the large nugget effect adopted for the high cutoff values; see Table A.1. One way to avoid such discontinuities is to shift the estimation grid slightly so that the data do not fall on grid node locations. Another solution is to arbitrarily set all nugget constants to zero. Discontinuities are also present but much less apparent in the figures of Problem Set four because only 29 unclustered data were used.

Program postik also provides any p quantile value derived from the ccdf, for example, the 0.5 quantile or conditional median, also called M type estimate: This is the estimated value that has equal chance to be exceeded or not exceeded by the actual unknown value. Figure A.60 provides gray-scale maps for the M type estimate and the 0.1 quantile. Locations with high 0.1 quantile values are almost certainly high in the sense that there is a 90% chance that the actual true values would be even higher.

Last, program postik also provides the probability of exceeding any given threshold value z_0, which need not be one of the original cutoffs used in ik3d. Figure A.61 provides the isopleth maps of the probability of *not* exceeding the low threshold value $z_1 = 0.16$ and the probability of exceeding the high threshold value $z_9 = 5.39$. Dark areas indicate zones that are most likely low z valued on the first map, and zones most likely high z valued in the second map. These dark areas relate well to the corresponding low and high z valued zones of the reference map of the Figure A.36. At the data locations there is no uncertainty and the gray-scale is either white (the data are below 5.39) or black (the data are above 5.39).

Sensitivity to Indicator Variogram Models

The last probability map, that of exceeding the high cutoff $z_9 = 5.39$, was produced again starting from

- a median IK model where all indicator semivariograms are made equal to the median indicator model:

$$\gamma_I(\mathbf{h}; z_k) \ / \ F(z_k)[1 - F(z_k)] \ = \ \mathrm{Sph}(|\mathbf{h}|/11) \qquad (A.6)$$

- a model stating a stronger continuity of high z values with a NW-SE direction of maximum continuity. The model is of type (A.6), except for the last three cutoffs z_7, z_8, z_9. For these cutoffs z_7, z_8, z_9, the minor range in direction SW-NE is kept at 11. The major range is direction NW-SE is increased to 22 for z_7, 44 for z_8, and 88 for z_9. The largest range 88 is greater than the dimension of the field and, for all practical purposes, the corresponding indicator variogram is linear without reaching its sill.

Figure A.62 gives the probability map (of exceeding z_9) using the median IK model. When compared with Figure A.61 this map shows essentially the same features, although with much more contrast. Median IK is much faster CPU-wise.

Figure A.62(b) gives the probability map corresponding to a model with artificial greater continuity of high z values in the diagonal direction NW-SE. The difference with Figure A.61(b) is dramatic: The dark areas are now all tightly connected diagonally.

Figure A.63 gives the probability map corresponding to a model in all points similar to that used for Figure A.62(b) except that the major direction of continuity is now NE-SW. Figures A.62(b) and A.63 are conditional to the same 140 sample data; their differences are strictly due to the "soft" structural information introduced through the indicator variogram models.

As opposed to Gaussian-related models and the median IK model, where only one variogram is available for the whole range of z values, multiple IK with as many different models as cutoff values allows a greater flexibility in modeling different behaviors of different classes of z values. Recall though from relation (IV.38) that a measure of continuum in the variability of indicator variogram model parameters should be preserved from one cutoff to the next.

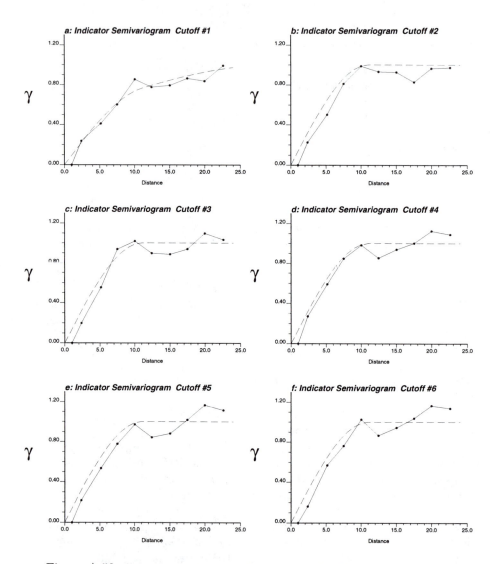

Figure A.53: Sample omnidirectional standardized indicator semivariograms (140 data) and their model fits.

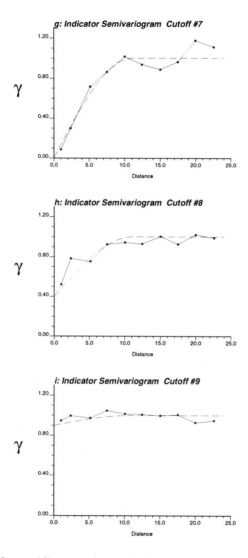

Figure A.54: Sample omnidirectional standardized indicator semivariograms (140 data) and their model fits.

```
                        Parameters for IK3D
                        ********************

START OF PARAMETERS:
1                                       \1=continuous(cdf), 0=categorical(pdf)
9                                       \number thresholds/categories
0.159 0.278 0.515 0.918 1.211 1.708 2.325 3.329 5.384      \thresholds
0.1   0.2   0.3   0.4   0.5   0.6   0.7   0.8   0.9         \global cdf
../../data/cluster.dat                  \file with data
1   2   0   3                           \   columns for X,Y,Z and variable
direct.ik                               \file with soft indicator input
1   2   0   3  4  5  6                   \   columns for X,Y,Z and indicators
-1.0e21   1.0e21                         \trimming limits
1                                       \debugging level: 0,1,2,3
ik3d.dbg                                \file for debugging output
ik3d.out                                \file for kriging output
50    0.5    1.0                        \nx,xmn,xsiz
50    0.5    1.0                        \ny,ymn,ysiz
1     0.0    1.0                        \nz,zmn,zsiz
1   16                                  \min, max data for kriging
25.0  25.0  25.0                        \maximum search radii
 0.0   0.0   0.0                        \angles for search ellipsoid
0                                       \max per octant (0-> not used)
0    2.5                                \0=full IK, 1=Median IK(threshold num)
1                                       \0=SK, 1=OK
2    0.00                               \One   nst, nugget effect
1    0.5  0.0   0.0   0.0               \     it,cc,ang1,ang2,ang3
          11.0  11.0  11.0              \     a_hmax, a_hmin, a_vert
1    0.5  0.0   0.0   0.0               \     it,cc,ang1,ang2,ang3
          30.0  30.0  30.0              \     a_hmax, a_hmin, a_vert
1    0.0                                \Two   nst, nugget effect
1    1.0  0.0   0.0   0.0               \     it,cc,ang1,ang2,ang3
          11.0  11.0  11.0              \     a_hmax, a_hmin, a_vert
1    0.0                                \Three nst, nugget effect
1    1.0  0.0   0.0   0.0               \     it,cc,ang1,ang2,ang3
          11.0  11.0  11.0              \     a_hmax, a_hmin, a_vert
1    0.0                                \Four  nst, nugget effect
1    1.0  0.0   0.0   0.0               \     it,cc,ang1,ang2,ang3
          11.0  11.0  11.0              \     a_hmax, a_hmin, a_vert
1    0.0                                \Five  nst, nugget effect
1    1.0  0.0   0.0   0.0               \     it,cc,ang1,ang2,ang3
          11.0  11.0  11.0              \     a_hmax, a_hmin, a_vert
1    0.0                                \Six   nst, nugget effect
1    1.0  0.0   0.0   0.0               \     it,cc,ang1,ang2,ang3
          11.0  11.0  11.0              \     a_hmax, a_hmin, a_vert
1    0.0                                \Seven nst, nugget effect
1    1.0  0.0   0.0   0.0               \     it,cc,ang1,ang2,ang3
          11.0  11.0  11.0              \     a_hmax, a_hmin, a_vert
1    0.4                                \Eight nst, nugget effect
1    0.6  0.0   0.0   0.0               \     it,cc,ang1,ang2,ang3
          11.0  11.0  11.0              \     a_hmax, a_hmin, a_vert
1    0.9                                \Nine  nst, nugget effect
1    0.1  0.0   0.0   0.0               \     it,cc,ang1,ang2,ang3
          11.0  11.0  11.0              \     a_hmax, a_hmin, a_vert
```

Figure A.55: Input parameter file for **ik3d**.

```
IK3D Estimates with:Clustered 140 primary and secondary data
  9
Threshold:  1 =       0.15900
Threshold:  2 =       0.27800
Threshold:  3 =       0.51500
Threshold:  4 =       0.91800
Threshold:  5 =       1.21100
Threshold:  6 =       1.70800
Threshold:  7 =       2.32500
Threshold:  8 =       3.32900
Threshold:  9 =       5.38400
  0.0000  0.0653  0.0653  0.1058  0.1058  0.1058  0.1058  0.3398  0.5451
  0.0176  0.0768  0.0768  0.1029  0.1029  0.1029  0.1029  0.3389  0.5553
  0.1309  0.1520  0.1520  0.1714  0.1714  0.1714  0.1714  0.3705  0.5656
  0.3823  0.3823  0.3823  0.4021  0.4021  0.4021  0.4021  0.4948  0.6270
  0.6435  0.6435  0.6435  0.6435  0.6435  0.6435  0.6435  0.6435  0.6558
  0.7463  0.7580  0.7580  0.7662  0.7662  0.7662  0.7662  0.7662  0.7662
  0.7458  0.7951  0.7951  0.8187  0.8187  0.8187  0.8187  0.8187  0.8187
  0.7030  0.7979  0.8118  0.8219  0.8219  0.8219  0.8219  0.8219  0.8219
  0.6667  0.8186  0.8406  0.8438  0.8438  0.8438  0.8438  0.8438  0.8438
  0.6182  0.8369  0.8369  0.8415  0.8415  0.8415  0.8415  0.8415  0.8415
• • •
  0.0000  0.0000  0.0000  0.0000  0.0000  0.0000  0.0000  0.4823  0.5805
  0.0000  0.0000  0.0000  0.0000  0.0000  0.0000  0.0006  0.5438  0.6324
  0.0000  0.0000  0.0000  0.0000  0.0000  0.0000  0.0000  0.4964  0.6796
  0.0000  0.0000  0.0000  0.0000  0.0000  0.0000  0.0000  0.4555  0.7303
  0.0000  0.0000  0.0000  0.0000  0.0000  0.0000  0.0000  0.4085  0.7276
  0.0000  0.0000  0.0000  0.0000  0.0000  0.0000  0.0000  0.3119  0.7788
  0.0000  0.0000  0.0000  0.0000  0.0000  0.0000  0.0000  0.2692  0.7773
  0.0000  0.0000  0.0000  0.0000  0.0000  0.0000  0.0000  0.2303  0.7660
  0.0000  0.0000  0.0000  0.0000  0.0000  0.0000  0.0000  0.2085  0.7640
  0.0000  0.0000  0.0000  0.0000  0.0000  0.0000  0.0000  0.1995  0.7621
```

Figure A.56: Output of `ik3d` corresponding to the parameter file shown in Figure A.55.

```
                    Parameters for POSTIK
                    *********************

START OF PARAMETERS:
ik3d.out                          \file with IK3D output (continuous)
postik.out                        \file for output
1    0.5                          \output option, output parameter
9                                 \number of thresholds
0.159 0.278 0.515 0.918 1.211 1.708 2.325 3.329 5.384    \thresholds
0    1       0.75                 \volume support?, type, varred
../../data/cluster.dat            \file with global distribution
3    5     -1.0    1.0e21         \   ivr,   iwt,   tmin,   tmax
0.0      30.0                     \minimum and maximum Z value
3    1.0                          \lower tail: option, parameter
3    1.0                          \middle    : option, parameter
3    1.0                          \upper tail: option, parameter
100                               \maximum discretization

Output options:

  1 --> compute E-type estimate
  2 --> compute probability to exceed a threshold (outpar), the mean
        above the threshold, and the mean below the threshold
  3 --> compute Z value corresponding to the outpar-quantile
  4 --> approximate conditional variance
```

Figure A.57: Input parameter file for `postik`.

```
E-type mean values
1
mean
        6.5358
        6.4705
        6.1640
        4.9358
        3.8376
        2.7830
        2.2125
        2.1903
        1.9543
        1.9724
  • • •
        6.3211
        5.8496
        5.5131
        5.5157
        5.6012
        5.3720
        5.4458
        5.5821
        5.6276
        5.6549
```

Figure A.58: Output of `postik` corresponding to the parameter file shown in Figure A.57.

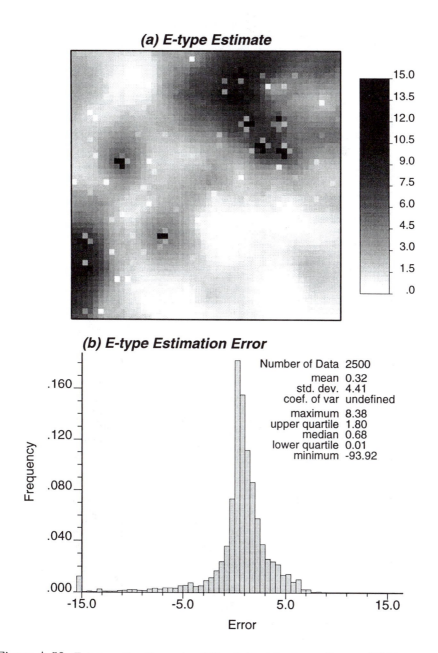

Figure A.59: E-type estimation using 140 z data. (a) gray-scale map, (b) histogram of errors $(z^*\text{-}z)$.

(a) M-type Estimate

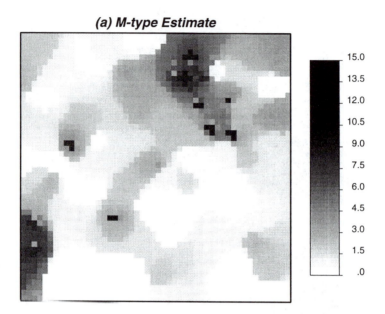

(b) 0.1 Quantile

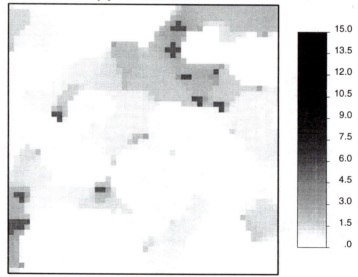

Figure A.60: Conditional quantile maps using 140 z data. (a) 0.5 quantile or M type estimate map, (b) 0.1 quantile map.

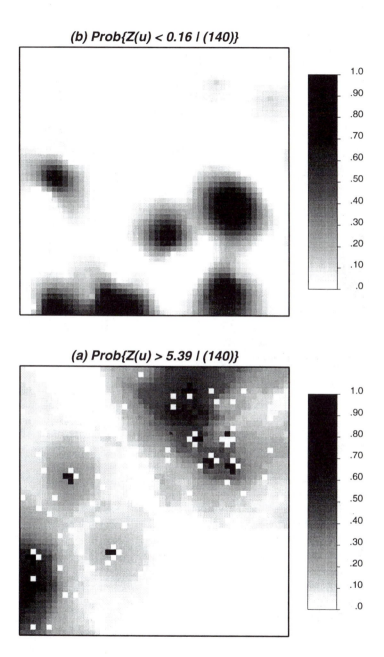

Figure A.61: Probability maps using 140 z data. (a) Probability of not exceeding $z_1 = 0.16$, (b) Probability of exceeding $z_9 = 5.39$.

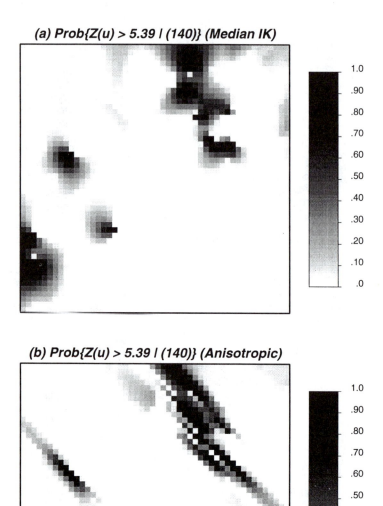

Figure A.62: Probability of exceeding $z_9 = 5.39$, (a) with a median IK model, (b) with a model indicating greater continuity of high z values in direction NW-SE.

Prob{Z(u) > 5.39 | (140)} (Anisotropic)

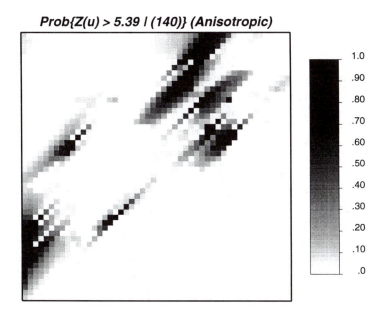

Figure A.63: Probability of exceeding $z_9 = 5.39$ under the same condition as Figure A.62(b) but the direction of extreme-values continuity is NE-SW.

A.6   Problem Set Six: Simulations

There are two related objectives for stochastic simulation. The first is to create realizations that honor, up to statistical fluctuations, the histogram, variogram (or set of indicator variograms), and conditioning data. These realizations overcome the smoothing effect of kriging if used as estimated maps. The second objective is to assess spatial uncertainty through multiple realizations.

The gray-scale used throughout most of this solution set is that considered for the reference images of Figures A.36 and A.37, i.e., between 0-white and 30-black.

Part A: Gaussian Simulation

As discussed in Chapter V, the most straightforward algorithm for simulation of continuous variables is the sequential Gaussian algorithm. The 140 data in cluster.dat are used with declustering weights in sgsim to generate the two realizations shown at the top of Figure A.64. The parameter file for sgsim is shown in Figure A.65, and the first and last 10 lines of the corresponding output file are shown in Figure A.66.

A number of typical checks are also shown in Figure A.64. It is good practice to ensure that the univariate distribution and the normal scores variogram is reproduced within reasonable statistical fluctuations. The input isotropic variogram model is reproduced fairly well. The deviations from the model are explained by both statistical fluctuations and the fact that the input model (A.3) is not deduced from all the 140 conditioning data.

Sensitivity analysis to the input (normal score) variogram model can be done retaining the same 140 conditioning data. Figure A.67 shows three different input variograms corresponding to increasing spatial continuity and an example realization corresponding to each variogram.

A large number of realizations can be generated and used in conjunction with the postsim program to generate quantile maps and maps showing the probability of exceeding a cutoff. Note that these maps can also be generated as a direct result of multi-Gaussian kriging or indicator kriging; see Figure A.61.

The matrix approach to Gaussian simulation is particularly efficient for small-scale simulations to be used for change of support problems; see Section IV.1.13 and V.2.4. The program lusim can be used to simulate a grid of normal score point values within any block of interest; see parameter file and output in Figures A.68 and A.69. These normal score values are back-transformed (with program backtr) and averaged to generate block values. Sixteen point values discretizing a block were simulated 100 times and then averaged to create 100 simulated block values. The histogram of point and block values are shown in Figure A.70. Note that the linear averaging reduces the variance but leaves the overall mean unchanged.

The parameters required by program `gtsim` are shown in Figure A.71. The first 10 lines and the last 10 lines of the `gtsim` output file are given in Figure A.72. The gray-scale realization obtained by truncating an unconditional `sgsim` realization is shown in Figure A.73.

Part B: Indicator Simulation

One advantage of the indicator formalism is that it allows a more complete specification of the bivariate distribution than afforded, e.g., by the multi-Gaussian model. Another advantage is that it allows the straightforward incorporation of secondary information and constraint intervals. An example of two realizations generated with `sisim` are shown in Figure A.74. The parameter file and a portion of the output file are shown in Figures A.75 and A.76. The apparent discontinuity of the very high simulated values is due to the fact that only three cutoffs were used, entailing artificially large within-class noise. The discrepancy between the model and simulated variograms for the first quartile is due to a biased univariate cdf $\hat{F}(z_k)$ (0.31 instead of 0.25) due to clustering. Recall that the variance of an indicator variable is $\hat{F}(z_k)[1.0 - \hat{F}(z_k)]$.

More cutoffs can be used to reduce the within-class noise. The variogram modeling, however, may then become tedious, order relation problems are likely to increase, and the computational requirements increase. As discussed in Section IV.1.9, the median IK approximation is one way to increase the number of cutoffs without any of the disadvantages just mentioned. The price for the median IK approximation is that all indicator variograms are proportional; i.e., the standardized variograms are identical. Figure A.77 shows two realizations generated with `sisim` using 9 cutoffs (same as in Problem Set 5) and the median IK option (`mik=1`).

Once again it is good practice to ensure that the univariate distribution and the indicator variograms are reproduced within reasonable statistical fluctuations. Some of these checks are shown in Figure A.74. Quantile maps and maps of the probability of exceeding a cutoff can be derived if a large number of realizations have been generated.

Part C: Markov-Bayes Indicator Simulation

The Markov-Bayes simulation algorithm coded in program `sisim` allows using soft prior information as provided by a secondary data set. A full indicator cokriging approach to handle the secondary variable would be quite demanding in that a matrix of cross-covariance models must be inferred. This inference is avoided by making the Markov hypothesis; see Section IV.1.12. Calibration parameters from the calibration scatterplot provide the linear rescaling parameters needed to establish the cross covariances.

Indicator cutoffs must be chosen for both the primary and secondary (soft) data. In this case the secondary data have essentially the same units as the primary data; therefore, the same five cutoffs were chosen for both:

0.5, 1.0, 2.5, 5.0, and 10.0. Based on the 140 declustered data, these cut-offs correspond to primary cdf values of 0.30, 0.47, 0.72, 0.87, and 0.96. A parameter file for the calibration program `bicalib` is shown in Figure A.78. The output from `bicalib`, shown in Figure A.79, is directly input to the `sisim` program. The $B(z_k)$ values are obtained from the `bicalib` report file (0.60,0.53,0.53,0.52,0.32 in this case).

Gray-scale representations of two realizations accounting for the soft data via the Markov-Bayes approach are shown in Figure A.80. These realizations must be compared to the true distribution (see Figure A.36) and the distribution of the secondary variable (see Figure A.37).

Another approach to incorporate a secondary variable is the external drift formalism; see Section IV.1.5. The Gaussian simulation program `sgsim` with option **ktype**=3 was used with the secondary variable as an external drift. Two realizations of this simulation procedure are shown in Figure A.81. When using the external drift constraint with Gaussian simulation the kriging variance or standard deviation is always larger than the theoretically correct simple kriging variance. This causes the final back-transformed values also to have too high a variance. In this case, however, the average kriging standard deviation increased by less than one percent. In other situations where the increase is more pronounced, it may be necessary to solve the SK system separately and use the external drift estimate with the SK variance; for a similar implementation; see [94].

Part D: p-Field Simulation

The p-field simulation program `pfsim` in GSLIB implements the p-field concept but does not have an in-built fast unconditional simulation algorithm.

The `sgsim` program was used to generate 5 unconditional Gaussian realizations. The `pfsim` parameters shown in Figure A.82 specifies the drawing from the `ik3d`-derived distributions (Problem Set 4). The first and last 10 lines of output are shown in Figure A.83.

Gray-scale representations of two realizations accounting for the soft data via the Markov-Bayes approach are shown in Figure A.84. These realizations may be compared with the Gaussian and indicator realizations in Figures A.64 and A.74, respectively.

Part E: Categorical Variable Simulation

This section provides only a hint of the potential applications of Boolean and annealing-based simulation techniques.

The parameter file for `ellipsim` is shown in Figure A.85, the first and last 10 lines of output are shown in Figure A.86. The two corresponding simulated images are shown in Figure A.87. In this case the radius, anisotropy ratio, and direction of anisotropy were kept constant. As illustrated in Figure A.88, neither the radius, nor the angle, nor the anisotropy ratio need be kept constant.

The analytical bombing model described in the problem setting was approximated by a spherical variogram model. Program `sisim` was then used to create two realizations; the parameter file is shown in Figure A.89, and the first and last 10 lines of output are shown in Figure A.90. The two realizations, shown in Figure A.91, reproduce the overall proportion of shales and the average aspect ratio. Nevertheless, they contain too much randomness, i.e., the shale shapes are not as well defined as in the `ellipsim` realizations of Figure A.87.

To demonstrate the use of the `anneal` program, consider the postprocessing of the `sisim` realizations of Figure A.91 to honor more of the two-point statistics of the `ellipsim` realizations of Figure A.87. The parameter file for `anneal` is shown in Figure A.92. The control statistics were computed from the first `ellipsim` realization of Figure A.87. The two-point histogram for 10 lags in the two orthogonal directions aligned with the coordinate axes and for 5 lags in the two diagonal directions were retained. The first and last 10 lines of output from `anneal` are shown in Figure A.93. The two postprocessed realizations are shown in Figure A.94. These realizations share more of the characteristics of the original `ellipsim` realizations.

Part F: Annealing Simulation

This section presents two possible realizations provided by the annealing simulation program `sasim`; see the top of Figure A.95. The parameter file for `sasim` is shown in Figure A.96, and the first and last 10 lines of the corresponding output file are shown in Figure A.97.

A number of typical checks are also shown in Figure A.95. The realizations honor the 140 conditioning data from $\boxed{\text{cluster.dat}}$, the univariate distribution, and the variogram model.

Figure A.98 shows two realizations constrained to match the conditioning data, the declustered histogram, the variogram, and the bivariate relationship with the secondary data. The scatterplot reproduction of the first realization is also shown in Figure A.98. Figure A.99 shows how the component objective functions decrease with the number of perturbations. Note the logarithmic scale and the low final objective function values (less than 1% of the starting value).

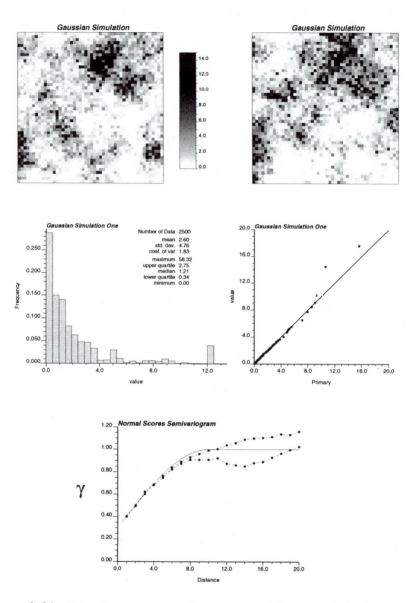

Figure A.64: Two Gaussian realizations generated by sgsim (140 data). The histogram, Q-Q plot, and NS and EW normal scores semivariogram of the first realization. The dotted line is the input isotropic model.

```
                       Parameters for SGSIM
                       ********************

START OF PARAMETERS:
../../data/cluster.dat           \file with data
1   2   0   3   5   0            \  columns for X,Y,Z,vr,wt,sec.var.
-1.0        1.0e21              \  trimming limits
1                               \transform the data (0=no, 1=yes)
sgsim.trn                       \  file for output trans table
0                               \  consider ref. dist (0=no, 1=yes)
histsmth.out                    \  file with ref. dist distribution
1   2                           \  columns for vr and wt
0.0     30.0                    \  zmin,zmax(tail extrapolation)
1       0.0                     \  lower tail option, parameter
1       30.0                    \  upper tail option, parameter
1                               \debugging level: 0,1,2,3
sgsim.dbg                       \file for debugging output
sgsim.out                       \file for simulation output
2                               \number of realizations to generate
50      0.5     1.0             \nx,xmn,xsiz
50      0.5     1.0             \ny,ymn,ysiz
1       0.5     1.0             \nz,zmn,zsiz
69069                           \random number seed
0       8                       \min and max original data for sim
16                              \number of simulated nodes to use
1                               \assign data to nodes (0=no, 1=yes)
0       3                       \multiple grid search (0=no, 1=yes),num
0                               \maximum data per octant (0=not used)
20.0  20.0  20.0                \maximum search radii (hmax,hmin,vert)
 0.0   0.0   0.0                \angles for search ellipsoid
0       0.60    0.75            \ktype: 0=SK,1=OK,2=LVM,3=EXDR,4=COLC
../data/ydata.dat               \  file with LVM, EXDR, or COLC variable
4                               \  column for secondary variable
1       0.3                     \nst, nugget effect
1       0.7  0.0    0.0    0.0  \it,cc,ang1,ang2,ang3
             10.0  10.0  10.0   \a_hmax, a_hmin, a_vert
```

Figure A.65: An example parameter file for sgsim.

```
SGSIM Realizations '
1
value
        1.9692
        0.6191
        0.4640
        0.2296
        0.0827
        0.4379
        0.0154
        0.1000
        0.0699
        0.4034
• • •
       10.3039
        1.1731
        1.3800
        2.9838
        0.6469
        0.1796
        0.3400
        2.1700
        2.2654
        1.9042
```

Figure A.66: Output of sgsim corresponding to the parameter file shown in Figure A.65.

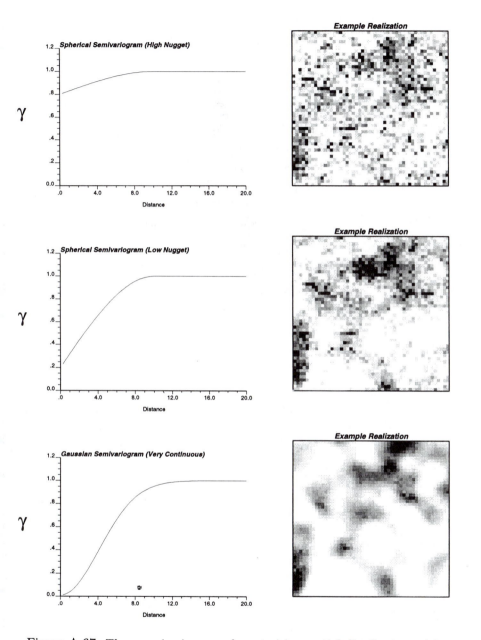

Figure A.67: Three semivariograms characterizing spatial distributions with vary-ing degrees of spatial continuity are shown on the left. Example realizations cor-responding to the variograms are shown on the right. The same 140 conditioning data were retained in all three cases.

```
                        Parameters for LUSIM
                        ********************

START OF PARAMETERS:
../../data/parta.dat                \file with data
1    2    0    4                     \    columns for X,Y,Z, normal scores
-1.0e21     1.0e21                   \    trimming limits
3                                    \debugging level: 0,1,2,3
lusim.dbg                           \file for debugging output
lusim.out                           \file for realization(s)
100                                 \number of realizations
4    40.25    0.5                   \nx,xmn,xsiz
4    28.25    0.5                   \ny,ymn,ysiz
1     0.00    1.0                   \nz,zmn,zsiz
69069                               \random number seed
1    0.3                            \nst, nugget effect
1    0.7    0.0    0.0    0.0       \it,cc,ang1,ang2,ang3
            10.0   10.0   10.0      \a_hmax, a_hmin, a_vert
```

Figure A.68: An example parameter file for lusim.

```
LUSIM Output
1
simulated values
-1.1041
-0.6383
-0.6233
 0.0951
-0.3330
-0.1919
 0.2546
-0.7371
 0.1371
 0.3837
• • •
 0.1391
 0.4089
 0.1707
-0.0820
 0.1986
-0.6004
 0.2300
 0.3736
 1.0573
 0.9094
```

Figure A.69: Output of lusim corresponding to the parameter file shown in Figure A.68.

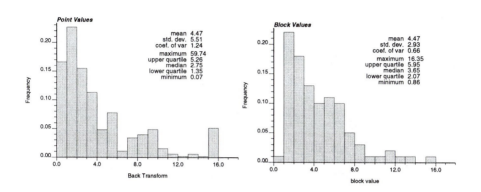

Figure A.70: The distribution of point values and the corresponding distribution of block values for a 2 by 2 unit block centered at $x=41$, $y=29$ generated by lusim.

```
                        Parameters for GTSIM
                        *********************

START OF PARAMETERS:
sgsimp.out                      \file with input Gaussian realizations
gtsim.out                       \file for output categorical realizations
1                               \number of realizations
50    50    1                   \nx,ny,nz
3                               \number of categories
1     0.50                      \cat(1)   global proportion(1)
2     0.25                      \cat(2)   global proportion(2)
3     0.25                      \cat(3)   global proportion(3)
0                               \proportion curves (0=no, 1=yes)
propc01.dat                     \    file with local proportion (1)
1                               \    column number for proportion
propc02.dat                     \    file with local proportion (2)
1                               \    column number for proportion
propc03.dat                     \    file with local proportion (3)
1                               \    column number for proportion
```

Figure A.71: An example parameter file for gtsim.

```
GTSIM outputSGSIM Realizations
1
category
2.00
1.00
1.00
2.00
1.00
1.00
1.00
1.00
1.00
1.00
• • •
3.00
3.00
2.00
2.00
3.00
3.00
2.00
3.00
3.00
1.00
```

Figure A.72: Output of gtsim corresponding to the parameter file shown in Figure A.71.

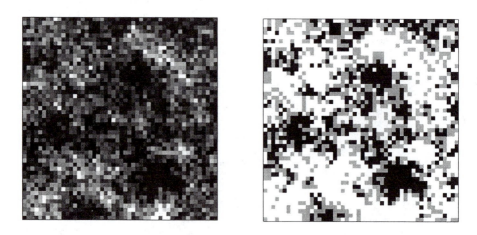

Figure A.73: A continuous sgsim realization and the corresponding categorical gtsim realization.

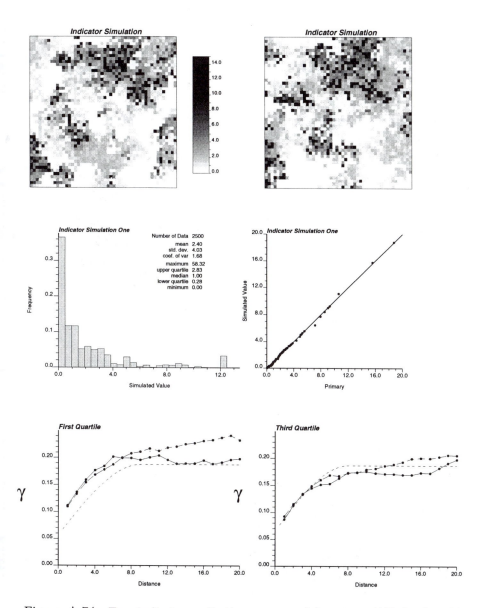

Figure A.74: Two indicator realizations generated by sisim (140 data), a histogram, Q-Q plot, and two directional indicator semivariograms from the first realization. The dotted line is the isotropic model.

```
                          Parameters for SISIM
                          ********************

START OF PARAMETERS:
cluster.dat                          \data file
1    2    0    3                     \column: x,y,z,vr
-1.0e21      1.0e21                  \data trimming limits
0.0    30.0                          \minimum and maximum data value
3      1.0                           \lower tail option and parameter
3      1.0                           \middle       option and parameter
3      1.0                           \upper tail option and parameter
cluster.dat                          \tabulated values for classes
3    5                               \column for variable, weight
direct.ik                            \direct input of indicators
sisim.out                            \output file for simulation
2                                    \debugging level: 0,1,2,3
sisim.dbg                            \output File for Debugging
0                                    \0=standard order relation corr.
69069                                \random number seed
2                                    \number of simulations
50    0.5    1.0                     \nx,xmn,xsiz
50    0.5    1.0                     \ny,ymn,ysiz
1     1.0    10.0                    \nz,zmn,zsiz
1                                    \0=two part search, 1=data-nodes
0                                    \   max per octant(0 -> not used)
20.0                                 \   maximum search radius
  0.0  0.0   0.0  1.0   1.0          \   sang1,sang2,sang3,sanis1,2
0    12                              \   min, max data for simulation
12                                   \number simulated nodes to use
1     1.21                           \0=full IK, 1=med approx(cutoff)
0                                    \0=SK, 1=OK
3                                    \number cutoffs
0.34 0.25    1   0.30                \cutoff, global cdf, nst, nugget
     1     9.0   0.70               \         it, aa, cc
     0.0   0.0   0.0  1.0   1.0     \         ang1,ang2,ang3,anis1,2
1.21 0.50    1   0.25                \cutoff, global cdf, nst, nugget
     1    10.0   0.75               \         it, aa, cc
     0.0   0.0   0.0  1.0   1.0     \         ang1,ang2,ang3,anis1,2
2.75 0.75    1   0.35                \cutoff, global cdf, nst, nugget
     1     8.0   0.65               \         it, aa, cc
     0.0   0.0   0.0  1.0   1.0     \         ang1,ang2,ang3,anis1,2
```

Figure A.75: An example parameter file for sisim.

```
SISIM SIMULATIONS:   Clustered 140 primary and secondary data
1
Simulated Value
 13.1375
  5.1796
 10.2620
   .0834
   .1900
   .1000
   .1900
   .2099
   .2800
   .0874
• • •
   .3666
   .4426
  1.1090
  1.2100
  1.1875
  1.8182
   .5110
   .8897
   .3382
   .3222
```

Figure A.76: Output of `sisim` corresponding to the parameter file shown in Figure A.75.

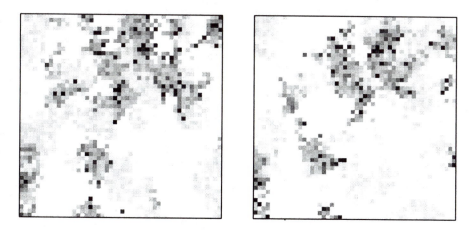

Figure A.77: Two indicator realizations generated with the median IK approximation of `sisim`.

```
                        Parameters for BICALIB
                        **********************

START OF PARAMETERS:
../../data/ydata.dat                 \file with data
4                                    \    column for secondary variable
../../data/cluster.dat               \file with calibration scatterplot
3   4   5                       .    \    columns of pri, sec, and weight
-1.0e21    1.0e21                    \    trimming limits
bicalib.out                          \file for output data / distributions
bicalib.rep                          \file for calibration report
bicalib.cal                          \calibration data for sisim
5                                    \number of thresholds on primary
0.50 1.00 2.50 5.00 10.0             \    thresholds on primary
5                                    \number of thresholds on secondary
0.50 1.00 2.50 5.00 10.0             \    thresholds on secondary
```

Figure A.78: Parameter file for bicalib.

```
All of the secondary data plus 29 primar
   9
Xlocation
Ylocation
Primary
Secondary
Primary Threshold              0.5000
Primary Threshold              1.0000
Primary Threshold              2.5000
Primary Threshold              5.0000
Primary Threshold             10.0000
    .5    .5 -1.00   3.26    0.1144   0.1144   0.4096   0.8916   0.9833
   1.5    .5 -1.00   2.64    0.1144   0.1144   0.4096   0.8916   0.9833
   2.5    .5 -1.00   2.15    0.0229   0.3410   0.8229   0.9302   1.0000
   3.5    .5 -1.00   1.69    0.0229   0.3410   0.8229   0.9302   1.0000
   4.5    .5 -1.00    .51    0.4430   0.7657   0.9612   1.0000   1.0000
   5.5    .5 -1.00    .27    0.8959   0.9563   1.0000   1.0000   1.0000
   6.5    .5 -1.00    .16    0.8959   0.9563   1.0000   1.0000   1.0000
   7.5    .5 -1.00    .31    0.8959   0.9563   1.0000   1.0000   1.0000
   8.5    .5 -1.00    .39    0.8959   0.9563   1.0000   1.0000   1.0000
   9.5    .5 -1.00    .40    0.8959   0.9563   1.0000   1.0000   1.0000
 • • •
  40.5  49.5 -1.00   7.22    0.0000   0.1033   0.2785   0.5041   0.8463
  41.5  49.5 -1.00   8.03    0.0000   0.1033   0.2785   0.5041   0.8463
  42.5  49.5 -1.00   5.70    0.0000   0.1033   0.2785   0.5041   0.8463
  43.5  49.5 -1.00   8.36    0.0000   0.1033   0.2785   0.5041   0.8463
  44.5  49.5 -1.00   8.84    0.0000   0.1033   0.2785   0.5041   0.8463
  45.5  49.5 -1.00   8.82    0.0000   0.1033   0.2785   0.5041   0.8463
  46.5  49.5 -1.00   7.97    0.0000   0.1033   0.2785   0.5041   0.8463
  47.5  49.5 -1.00   8.62    0.0000   0.1033   0.2785   0.5041   0.8463
  48.5  49.5 -1.00   6.59    0.0000   0.1033   0.2785   0.5041   0.8463
  49.5  49.5 -1.00   5.83    0.0000   0.1033   0.2785   0.5041   0.8463
```

Figure A.79: Output from bicalib corresponding to the parameter file shown in Figure A.78.

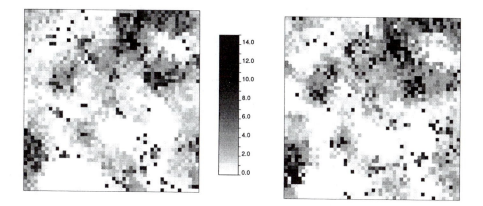

Figure A.80: Two output realizations of `sisim` accounting for secondary data with the Markov-Bayes option.

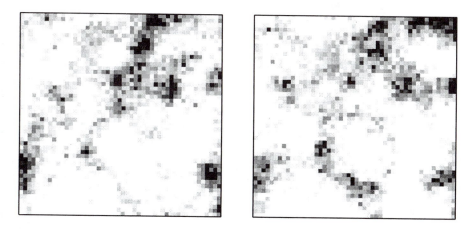

Figure A.81: Two output realizations of `sgsim` with an external drift variable. This result can be compared to that of the Markov-Bayes simulation generated by `sisim` shown in Figure A.80, the true distribution shown in Figure A.36, and the secondary variable distribution shown in Figure A.37.

```
                          Parameters for PFSIM
                          ********************

START OF PARAMETERS:
1                                   \1=continuous(cdf), 0=categorical(pdf)
1                                   \1=indicator ccdfs, 0=Gaussian (mean,var)
9                                   \   number thresholds/categories
0.159 0.278 0.515 0.918 1.211 1.708 2.325 3.329 5.384    \   thresholds
../../data/cluster.dat     \   file with global distribution
3    5      -1.0    1.0e21  \      ivr,   iwt
0.0     30.0               \   minimum and maximum Z value
1    0.0                   \   lower tail: option, parameter
1    1.0                   \   middle   : option, parameter
1    30.0                  \   upper tail: option, parameter
../prob5/ik3d.out          \file with input conditional distributions
1  2  3   4  5  6  7  8  9  \  columns for mean, var, or ccdf values
-1.0e21      1.0e21        \  trimming limits
sgsimp.out                 \file with input p-field realizations
1                          \  column number in p-field file
0                          \  0=Gaussian, 1=uniform [0,1]
pfsim.out                  \file for output realizations
5                          \number of realizations
50    50    1              \nx, ny, nz
```

Figure A.82: Input parameter file for **pfsim**.

```
PFSIM outputSGSIM Realizations
1
value
      12.9385
       3.4036
       3.1687
       3.5090
       0.0306
       0.1020
       0.0087
       0.0982
       0.0683
       0.0177
• • •
      24.1757
      22.7239
       4.9085
      27.4815
       3.1243
       3.1366
       2.7428
       4.1550
       2.3705
       2.4171
```

Figure A.83: Output of **pfsim** corresponding to the parameter file shown in Figure A.82.

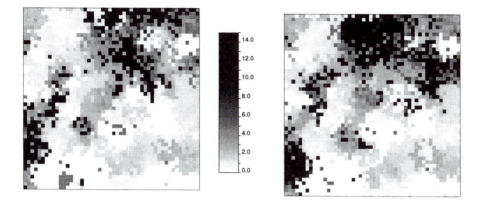

Figure A.84: First two output realizations of `pfsim`.

```
                 Parameters for ELLIPSIM
                 ************************

START OF PARAMETERS:
ellipsim.out                    \file for output realizations
2                               \number of realizations
100   0. 1.0                    \nx,xmn,xsiz
100   0. 1.0                    \ny,ymn,ysiz
1     0. 1.0                    \nz,zmn,zsiz
69069                           \random number seed
0.25                            \target proportion (in ellipsoids)
20.0 4.0 1.0 90.0 0.0 0.0 1.0   \radius[1,2,3],angle[1,2,3],weight
```

Figure A.85: Input parameter file for `ellipsim`.

```
Output from ELLIPSIM
1
in/out code
0
0
0
0
0
0
0
0
0
0
 • • •
0
0
0
0
0
0
0
0
0
1
```

Figure A.86: Output of `ellipsim` corresponding to the parameter file shown in Figure A.85.

Figure A.87: Two `ellipsim` realizations.

Figure A.88: An `ellipsim` realization with varying ellipse radii, anisotropy ratios, and direction angles.

```
                          Parameters for SISIM
                          *********************

START OF PARAMETERS:
0                                 \1=continuous(cdf), 0=categorical(pdf)
2                                 \number thresholds/categories
0.0   1.0                         \    thresholds / categories
0.75  0.25                        \    global cdf / pdf
nodata                            \file with data
1    2    0    3                  \    columns for X,Y,Z, and variable
nodata                            \file with soft indicator input
1    2    0    3 4                 \    columns for X,Y,Z, and indicators
0                                 \    Markov-Bayes simulation (0=no,1=yes)
0.61  0.54                        \      calibration B(z) values
-1.0e21      1.0e21               \trimming limits
0.0   1.0                         \minimum and maximum data value
1     1.0                         \   lower tail option and parameter
1     1.0                         \   middle      option and parameter
1     1.0                         \   upper tail option and parameter
nodata                            \   file with tabulated values
3   5                             \      columns for variable, weight
1                                 \debugging level: 0,1,2,3
sisimpdf.dbg                      \file for debugging output
sisimpdf.out                      \file for simulation output
2                                 \number of realizations
100   0.5    1.0                  \nx,xmn,xsiz
100   0.5    1.0                  \ny,ymn,ysiz
1     1.0    10.0                 \nz,zmn,zsiz
69069                             \random number seed
12                                \maximum original data  for each kriging
12                                \maximum previous nodes for each kriging
1                                 \maximum soft indicator nodes for kriging
1                                 \assign data to nodes? (0=no,1=yes)
0      3                          \multiple grid search? (0=no,1=yes),num
0                                 \maximum per octant    (0=not used)
40.0   8.0   1.0                  \maximum search radii
90.0   0.0   0.0                  \angles for search ellipsoid
1     0.0                         \0=full IK, 1=median approx. (cutoff)
0                                 \0=SK, 1=OK
1     0.00                        \One   nst, nugget effect
1     1.00 90.0  0.0     0.0      \     it,cc,ang1,ang2,ang3
            40.0  8.0     1.0      \     a_hmax, a_hmin, a_vert
1     0.00                        \Two   nst, nugget effect
1     1.00 90.0  0.0     0.0      \     it,cc,ang1,ang2,ang3
            40.0  8.0     1.0      \     a_hmax, a_hmin, a_vert
```

Figure A.89: An example parameter file for sisim (categorical variable option).

```
SISIM SIMULATIONS:
1
Simulated Value
   0.0000
   0.0000
   0.0000
   0.0000
   0.0000
   0.0000
   0.0000
   0.0000
   0.0000
   0.0000
 • • •
   0.0000
   0.0000
   0.0000
   0.0000
   0.0000
   0.0000
   0.0000
   0.0000
   0.0000
   0.0000
```

Figure A.90: Output of sisim corresponding to the parameter file shown in Figure A.89.

Figure A.91: Two sisim realizations with a variogram that approximately corresponds to the images shown in Figure A.87.

```
                         Parameters for ANNEAL
                         **********************

START OF PARAMETERS:
sisimpdf.out                        \file with input image(s)
ellipsim.out                        \file with training Image
3    10                             \debug level, reporting Interval
anneal.dbg                          \file for debuggubg output
anneal.out                          \file for output simulation
2                                   \number of realizations
100   100   1                       \nx, ny, nz
69069                               \random number seed
5    0.000001                       \maximum iterations, tolerance
4                                   \number of directions
1   0   0   1                       \ixl, iyl, izl, nlag
0   1   0   1                       \ixl, iyl, izl, nlag
1   1   0   1                       \ixl, iyl, izl, nlag
1  -1   0   1                       \ixl, iyl, izl, nlag
```

Figure A.92: Input parameter file for **anneal**.

```
ANNEAL SIMULATIONS: Output from ELLIPSIM
1
simulation
0
0
0
0
0
0
0
0
0
0
• • •
0
0
0
0
0
0
0
0
0
0
```

Figure A.93: Output of **anneal** corresponding to the parameter file shown in Figure A.92.

Figure A.94: Two `anneal` realizations starting from the `sisim` realizations shown in Figure A.91 and using control statistics taken from the left side of Figure A.87.

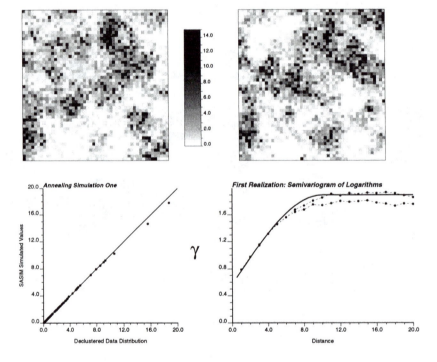

Figure A.95: Two annealing realizations generated by sasim. A Q-Q plot with the declustered histogram, experimental semivariogram (dots) in the NS and EW directions, and the model semivariogram (solid line) are shown for the first realization.

```
                    Simulated Annealing Based Simulation
                    ***********************************

START OF PARAMETERS:
1  1  0   0   0                    \components: hist,varg,ivar,corr,bivh
1  5  1   1   1                    \weight:    hist,varg,ivar,corr,bivh
1                                  \0=no transform, 1=log transform
2                                  \number of realizations
50        0.5     1.0             \grid definition: nx,xmn,xsiz
50        0.5     1.0             \                 ny,ymn,ysiz
 1        0.5     1.0             \                 nz,zmn,zsiz
69069                             \random number seed
4                                 \debugging level
sasim.dbg                         \file for debugging output
sasim.out                         \file for simulation output
1                                 \schedule (0=automatic,1=set below)
1.0   0.25   10    5    5   0.001 \   schedule: t0,redfac,ka,k,num,Omin
100.0                             \   maximum number of perturbations
 1.0                              \   reporting interval
1                                 \conditioning data:(0=no, 1=yes)
../../data/cluster.dat            \   file with data
1   2   0    3                    \   columns: x,y,z,attribute
-1.0e21     1.0e21                \   trimming limits
1                                 \file with histogram:(0=no, 1=yes)
../../data/cluster.dat            \   file with histogram
3   5                             \   column for value and weight
48                                \   number of quantiles for obj. func.
1                                 \number of indicator variograms
2.78                              \   indicator thresholds
../../data/ydata.dat              \file with gridded secondary data
4                                 \   column number
0                                 \   vertical average (0=no, 1=yes)
0.60                              \correlation coefficient
../../data/cluster.dat            \file with bivariate data
3   4   0                         \   columns for primary, secondary, wt
-0.5    100.0                     \   minimum and maximum
5                                 \   number of primary thresholds
5                                 \   number of secondary thresholds
51                                \Variograms: number of lags
1                                 \   standardize sill (0=no,1=yes)
1   0.3                           \   nst, nugget effect
1   0.7  0.0   0.0    0.0         \   it,cc,ang1,ang2,ang3
         10.0  10.0   10.0        \   a_hmax, a_hmin, a_vert
```

Figure A.96: An example parameter file for sasim.

```
SASIM Realizations
1
value
   18.660
    1.814
    0.611
    2.051
    0.110
    0.090
    0.160
    0.090
    0.090
    0.203
 ● ● ●
    2.740
    9.080
    9.080
    0.781
    1.259
    4.819
    6.096
    3.683
   18.364
    2.999
```

Figure A.97: Output of **sasim** corresponding to the parameter file shown in Figure A.96.

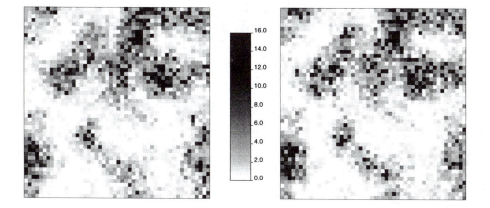

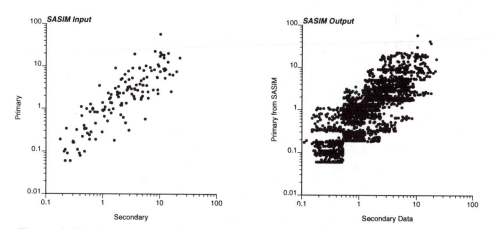

Figure A.98: Two realizations conditioned to local data, histogram, variogram, and conditional distributions from secondary data. The scatterplot on the left is from the reference 140 data and the scatterplot on the right is from the first realization.

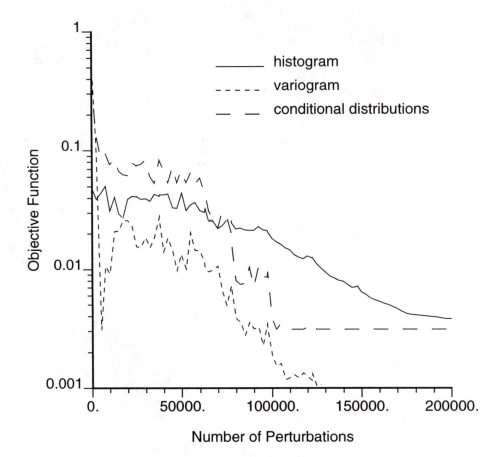

Figure A.99: Component objective function versus the number of perturbations (time).

Appendix B

Software Installation

GSLIB does not provide executable programs and tutorials. All the programs must be compiled prior to running them. This appendix provides some information and hints about loading and compiling the GSLIB programs.

The programs in this version of GSLIB have been developed and are used at Stanford primarily in the UNIX environment; however, there are no restrictions on the type of operating system. These programs have also been tested in the DOS environment and are distributed on 3.5 inch DOS diskettes. The following information is necessarily general because of the many different possible operating systems.

As mentioned in Chapter I, the GSLIB programs are distributed without any support or warranty. The disclaimer shown on Figure B.1 is included in all source code files.

B.1 Installation

Choose a target computer with a Fortran compiler. The first step is to choose a specific computer to load GSLIB. In many cases this is evident; nevertheless, keep the intended applications in mind. Workstations or larger machines would be appropriate if large 3D simulations and kriging runs are anticipated. A compiler for ANSI standard Fortran 77 (or any later release) must be available on the target computer.

Load the distribution software. The second step is to copy all the files from the distribution diskettes to the target computer. Ensure that the directory structure is unchanged; there are files with the same name in different directories. The files may be copied directly if the target computer reads IBM-type high-density diskettes; otherwise, they will have to be transferred through an IBM-type PC to the target computer via some file transfer protocol.

The type of transfer protocol will be accessible from the local system

```
C%%%%%%%%%%%%%%%%%%%%%%%%%%%%%%%%%%%%%%%%%%%%%%%%%%%%%%%%%%%%%%%%%%%%%%%%%%
C                                                                      %
C Copyright (C) 1996, The Board of Trustees of the Leland Stanford     %
C Junior University.  All rights reserved.                            %
C                                                                      %
C The programs in GSLIB are distributed in the hope that they will be  %
C useful, but WITHOUT ANY WARRANTY.  No author or distributor accepts  %
C responsibility to anyone for the consequences of using them or for   %
C whether they serve any particular purpose or work at all, unless he  %
C says so in writing.  Everyone is granted permission to copy, modify  %
C and redistribute the programs in GSLIB, but only under the condition %
C that this notice and the above copyright notice remain intact.       %
C                                                                      %
C%%%%%%%%%%%%%%%%%%%%%%%%%%%%%%%%%%%%%%%%%%%%%%%%%%%%%%%%%%%%%%%%%%%%%%%%%%
```

Figure B.1: The 16 line disclaimer included in all source code files.

expert or administrator. Some possibilities include a modem connection to the target computer, an ethernet card, or some other network connection.

File naming conventions. Different Fortran compilers require different file naming conventions; e.g., most UNIX implementations require Fortran source code files to be names with a .f extension, most DOS implementations require a .for extension, and some compilers require a .ftn extension. All of the Fortran code (named with a .for extension on the distribution diskettes) should be systematically renamed if appropriate. A UNIX shell script mvsuff is provided to change file suffixes; e.g., entering mvsuff for f will cause all files with a .for extension in the current working directory to be renamed with a .f extension.

Choose compiler options. The next step is to choose the compiler options and to document the changes that are required for the programs to be compiled. Compiler options vary considerably; the following should be kept in mind:

- The code can be optimized unless the use of a symbolic debugger is being considered. A high level of optimization on certain computers can introduce problems. For example, the vargplt program has failed to work properly when compiled with a high level of optimization on certain compilers.

- The source code is in mixed upper and lower case. In some cases a compiler option must be set or the source code may have to be systematically changed to upper case (the ANSI standard).

- Some compilers force each do loop to be executed at least once (i.e., the test for completion is at the end of the loop). This will cause problems and must be avoided by additional coding (adding an if statement before entering a loop) or setting the appropriate option when compiling the programs.

Modifications required in one program will surely be required in other programs. It is good practice to modify systematically all programs at one time rather than when needed.

On UNIX machines a `makefile` can be used to facilitate source code management. If extensive modifications and upgrades are anticipated, some type of source code control system could be considered.

Compile the GSLIB subroutine library. The 37 programs in GSLIB version 2 require a compiled GSLIB subroutine library. The subroutines are in the GSLIB subdirectory on the diskettes.

Compile the programs. Each program is in its own subdirectory and can be compiled separately once the GSLIB library has been compiled.

Testing. A necessary but not sufficient check that the programs are working correctly is to reproduce the example run of each program.

B.2 Troubleshooting

This section describes some known installation pitfalls. When attempting to isolate a problem, it would be a good idea to scan all the following points:

MAC users note that the quotes around the file name in the "include" statement must be removed.

VAX/VMS users note that the file opening convention will have to be modified in some cases. The VMS compiler does not accept the `status='UNKNOWN'` option when opening files.

Lahey Fortran users note that a variable that is in a common block cannot be initialized by a "data" statement; a "block data" statement must be used.

File not found errors are often due to users forgetting that all file names are read as 40 characters. For example, accidentally removing some of the blank spaces before the start of a comment (e.g., `\data file`) will cause the program to fail.

Nonexistent file names are often used to inform the program that a particular operation is not needed; e.g., if the file specifying conditioning data, in the simulation programs, is missing then an unconditional simulation is constructed. It is good practice explicitly to specify a name that is never used, e.g., $\boxed{\text{nodata}}$.

Difficulty reproducing simulation results may be due to differences of numerical precision (machine representation of numbers) and a different sequence of pseudorandom numbers.

A DOS "end of line" is marked by two characters. Some file transfer pro-
tocols automatically remove the extra character or a special command
is available on most computers (`dos2unix` on SUN computers, `to_unix`
on SGI machines,...).

End-of-record code. The programs in GSLIB read in the data in free format, so each line should contain the correct number of data values
separated by white space. Some of the programs and editors, such as
EMACS, that are used to create data files for GSLIB may terminate
the very last line with an end-of-file (EOF) code without an explicit
end-of-record code; if this is the case, GSLIB may not be able to read
the very last record in the file. If users notice that the number of data
being read is one fewer than the number that they intended, a blank
line at the end of their data file may solve the problem.

Appendix C

Programming Conventions

C.1   General

Many of these programs have been written and revised over the course of 12 years by numerous geostatisticians, most of whom were Stanford graduate students. Although they have been standardized as much as possible, some quirks from various programmers remain. Care has been taken to keep the code as streamlined and efficient as possible. Nevertheless, the clarity and simplicity of the code has taken precedence over tricks for saving speed and storage.

Some other programming considerations:

- The names of frequently used variables have been standardized. For example, the number of variables is typically referred to as *nvar* and the variable values as an array named *vr*. Section C.2 contains a dictionary of commonly used variables.

- Most dimensioning parameters are specified in an include file to make it easy to compile versions with different storage capabilities. Note that the "include" statement is not standard Fortran, but is supported on virtually all compilers.

- Common blocks have been used to declare the variables used in a number of the more complex subroutines. This avoids potential errors with long argument lists and mismatched array sizes.

- Common blocks are typically defined in include files to facilitate quick modifications and avoid typographical errors.

- Maximum dimensioning parameters have been standardized as follows:

  MAXCTX: maximum x size of the covariance table

  MAXCTY: maximum y size of the covariance table

329

MAXCTZ: maximum z size of the covariance table

MAXCUT: maximum number of cutoffs for an indicator-based method

MAXDAT: maximum number of data

MAXDIR: maximum number of directions in variogram programs

MAXDIS: maximum number of discretization points for a block

MAXLAG: maximum number of variogram lags

MAXNOD: maximum number of previously simulated nodes to use

MAXNST: maximum number of structures defining a variogram model

MAXSAM: maximum number of samples for kriging or simulation

MAXSBX: maximum number of nodes in x direction of super block network

MAXSBY: maximum number of nodes in y direction of super block network

MAXSBZ: maximum number of nodes in z direction of super block network

MAXVAR: maximum number of variables per datum

MAXSOF: maximum number of soft data

MAXX: maximum number of nodes in x direction

MAXY: maximum number of nodes in y direction

MAXZ: maximum number of nodes in z direction

MXJACK: maximum number of points to jackknife

MXVARG: maximum number of variograms to compute

Other less common dimensioning parameters, e.g., **MAXDT** maximum number of drift terms, are encountered in different include files.

- Most low-level routines such as the sorting subroutine and matrix solvers have been collected in a GSLIB subroutine library. To extract a single program and all its associated source code, it will be necessary to include some of the code from this subroutine library.

C.2   Dictionary of Variable Names

a: left-hand-side matrix of covariance values

aa: range of a variogram nested structure

anis1, anis2: the anisotropy factors that apply after rotations using the angles $ang1, ang2, ang3$ (see Figure II.4 and the discussion in Section II.3)

ang: azimuth angle measured in degrees clockwise from north.

ang1, ang2, ang3: angles defining the orientation of an ellipsoid in 3D (see Figure II.4 and the discussion in Section II.3)

atol: angular tolerance from *azm*

avg: average of the input data

azm: azimuth angle measured in degrees clockwise from north

bandw: the maximum acceptable distance perpendicular to the direction vector

c0: isotropic nugget effect

cbb: average covariance within a block

cc: variance contribution of a nested structure

ccdf: array of ccdf values for an indicator-based method

cdf: array of cdf values for an indicator-based method

close: array of pointers to the close data

cmax: maximum covariance value (used to convert semivariogram values to pseudocovariance values)

covtab: table lookup array of covariance values

cut: an array of cutoff values for indicator kriging or simulation

datafl: the name of the input data file

dbgfl: the name of the debugging output file

dip: the dip angle measured in negative degrees down from horizontal

dis: an array used in the variogram programs to store the average distance for a particular lag

dtol: angular tolerance from *dip*

EPSLON: a small number used as a test to avoid division by zero

gam: an array used to store the computed semivariogram measures

gcdf: an array of global cdf values for indicator-based methods

gcut: an array of global cutoff values for indicator-based methods

h2: squared distance between two data

hm: an array used to store the mean of the "head" values in the variogram subroutines

hv: an array used to store the variance of the "head" values in the variogram subroutines

idbg: integer debugging level (0=none, 1=normal, 2=intermediate, 3=serious debugging)

idrif: integer flags specifying whether or not to estimate a drift term in KT

inoct: array containing the number of data in each octant

isim: a particular simulated realization of interest

isrot: pointer to the location of the rotation matrix for the anisotropic search

isk: flag specifying SK or not (0=no, 1=yes)

it: integer flag specifying type of variogram model

itrend: integer flag specifying whether to estimate the trend of the variable in KT

ivhead: the variable number for the head of a variogram pair (see Section III.1)

ivtail: the variable number for the tail of a variogram pair (see Section III.1)

ivtype: the integer code specifying the type of variogram to be computed (see Section III.1)

ixd, iyd, izd: the node offset in each coordinate direction as used to define a direction in gam2, gam3, and anneal

ixinc, iyinc, izinc: the particular node offset currently being considered

ixnode, iynode, iznode: arrays containing the offsets for nodes to consider in a spiral search

KORDEI: order of ACORN random number generator; see [189]

ktype: kriging type (0=simple kriging, 1=ordinary kriging)

ldbg: Fortran logical unit number for debugging output

lin: Fortran logical unit number for input

lint: Fortran logical unit number for intermediate results

lout: Fortran logical unit number for output

ltail: indicator flag of lower tail option

ltpar: parameter for lower tail option

middle: indicator flag specifying option for interpolation within middle classes of a cdf

mik: indicator flag specifying median IK or not (0=no, 1=yes)

minoct: the minimum number of informed octants before estimating or simulating a point

nclose: the number of close data

nctx: the number x nodes in the covariance lookup table

ncty: the number y nodes in the covariance lookup table

nctz: the number z nodes in the covariance lookup table

nd: the number of data

ndir: the number of directions to be being considered in a variogram or simulated annealing simulation program

ndmax: the maximum number of data used to estimate or simulate a point

ndmin: the minimum number of data required before estimating or simulating a point

nlag: the number of lags to be computed in a variogram program or the number of lags to be matched in the simulated annealing simulation

nict: an array with the cumulative number of data in the super block and all other super blocks with a lower index

nlooku: the number of nodes to be searched in a spiral search

noct: the maximum number of data to retain from any one octant

np: an array used to store the number of pairs for a particular lag

nsbx: number of super blocks in x direction

nsby: number of super blocks in y direction

nsbz: number of super blocks in z direction

nsim: number of simulations to generate

nst: number of nested variogram structures (not including nugget)

nsxsb: number of super block nodes in x direction to search

nsysb: number of super block nodes in y direction to search

nszsb: number of super block nodes in z direction to search

nv: number of variables being considered

nvar: number of variables in cokriging or variogram subroutines

nvari: number of variables in an input data file

nvarg: number of variograms to compute

nx: number of blocks/nodes in x direction

nxdis: number of points to discretize a block in the x direction

ny: number of blocks/nodes in y direction

nydis: number of points to discretize a block in the y direction

nz: number of blocks/nodes in z direction

nzdis: number of points to discretize a block in the z direction

ortho: array containing the orthogonal decomposition of a covariance matrix

orthot: transpose of ortho

outfl: the name of the output file

r: right-hand-side vector of covariance values

rr: copy of right-hand-side vector of covariance values

radius: maximum radius for data searching

radsqd: maximum radius squared

rotmat: array with the rotation matrix for all variogram structures and the anisotropic search

s: vector of kriging weights

sang1,2,3: angles (see ang1,2,3) for anisotropic search

sanis1,2: anisotropy (see anis1,2) for anisotropic search

seed: integer seed to the pseudorandom number generator

sill: array of sill values in the cokriging subroutine

skmean: mean used if performing SK

ssq: sum of squares of the input data (used to compute the variance)

sstrat: indicator flag of search strategy (0=two part, 1=assign data to the nearest nodes)

title: a title for the current plot or output file

tm: an array used to store the mean of the "tail" values in the variogram subroutines

tmax: maximum allowable data value (values $\geq tmax$ are trimmed)

tmin: minimum data value (values<tmin are accepted)

tv: an array used to store the variance of the "tail" values in the variogram subroutines

UNEST: parameter used to flag unestimated nodes

unbias: constant used in unbiasedness constraint

utail: indicator flag of upper tail option

utpar: parameter for upper tail option

vr: an array name used for data. There may be 2 or more dimensions for cokriging or some indicator implementations

vra: an array name used for data close to the point being considered

x: array of x locations of data

xa: array of x locations of data close to the point being considered

xlag: the lag distance used to partition the experimental pairs in variogram calculation

xloc: x location of current node

xltol: the half-window tolerance for the lag distance xlag

xmn: location of first x node (origin of x axis)

xmx: location of last x node

xsbmn: x origin of super block search network

xsbsiz: x size of super blocks

xsiz: spacing of nodes or block size in x direction

y: array of y locations of data

ya: array of y locations of data close to the point being considered

yloc: y location of current node

ymn: location of first y node (origin of y axis)

ymx: location of last y node

ysbmn: y origin of super block search network

ysbsiz: y size of super blocks

ysiz: spacing of nodes or block size in y direction

z: array of z locations of data

za: array of z locations of data close to the point being considered

zloc: z location of current node

zmn: location of first z node (origin of z axis)

zmx: location of last z node

zsbmn: z origin of super block search network

zsbsiz: z size of super blocks

zsiz: spacing of nodes or block size in z direction

C.3   Reference Lists of Parameter Codes

The following codes are used in the variogram programs to indicate the type
of variogram being computed; see Section III.1:

ivtype	Type of variogram
1	traditional semivariogram
2	traditional cross semivariogram
3	covariance
4	correlogram
5	general relative semivariogram
6	pairwise relative semivariogram
7	semivariogram of logarithms
8	semimadogram
9	indicator semivariogram - continuous variable
10	indicator semivariogram - categorical variable

The following codes are used in the kriging, simulation, and utility programs
to specify the type of variogram model being considered:

it	Type of variogram model
1	spherical model (a=range, c=variance contribution)
2	exponential model (a=practical range, c=variance contribution)
3	Gaussian model (a=practical range, c=variance contribution)
4	power model (a=power, c=slope)
5	hole effect model (a=distance to first peak, c=variance contribution)

The following codes are used in the simulation and utility programs to specify the type of interpolation/extrapolation used for going beyond a discrete cdf:

ltail, middle, utail	option
1	linear interpolation between bounds
2	ω-power model interpolation between bounds
3	linear interpolation between tabulated quantiles
4	hyperbolic model extrapolation in upper tail

Appendix D

Alphabetical Listing of Programs

`addcoord`: add coordinates to a grid file

`anneal`: postprocessing of simulated realizations by annealing

`backtr`: normal scores back transformation

`bicalib`: Markov-Bayes calibration

`bigaus`: computes indicator semivariograms for a Gaussian model

`bivplt`: PostScript gray/color-scale map of bivariate probability density

`cokb3d`: cokriging

`declus`: cell declustering

`draw`: cell declustering

`ellipsim`: Boolean simulation of ellipses

`gam`: computes variograms of regularly spaced data

`gamv`: computes variograms of irregularly spaced data

`gtsim`: Gaussian truncated simulation

`histplt`: PostScript plot of histogram with statistics

`histsmth`: histogram smoothing program

`ik3d`: indicator kriging

`kb2d`: 2D kriging

`kt3d`: 2D kriging

`locmap`: PostScript plot of a location map

`lusim`: LU simulation

`nscore`: normal scores transformation

`pfsim`: probability field simulation

`pixelplt`: PostScript gray/color scale map

`postik`: IK post processing

`postsim`: postprocessing of multiple realizations

`probplt`: PostScript probability paper plot

`qpplt`: PostScript P-P or Q-Q plot

`rotcoord`: change coordinate systems

`sasim`: simulated annealing simulation

`scatplt`: PostScript plot of scattergram with statistics

`scatsmth`: scattergram smoothing program

`sgsim`: sequential Gaussian simulation

`sisim`: sequential indicator simulation

`sisim_gs`: sequential indicator simulation - gridded secondary variable

`sisim_lm`: sequential indicator simulation - locally varying mean

`trans`: general transformation program

`vargplt`: PostScript plot of variogram

`varmap`: calculation of variogram map

`vmodel`: variogram file from model

There are 39 programs (compared to 37 in the first edition). Most programs have been changed from the first edition; there are 12 new programs.

Appendix E

List of Acronyms and Notations

E.1   Acronyms

BLUE: best linear unbiased estimator

ccdf: conditional cumulative distribution function

cdf: cumulative distribution function

coIK: indicator cokriging

COK: cokriging

DK: disjunctive kriging

E-type: conditional expectation estimate

IK: indicator kriging

iid: independent, indentically distributed

IPCK: indicator principal components kriging

IRF-k: intrinsic random functions of order k

Geo-EAS: geostatistical software released by the U.S. Environmental Protection Agency [58]

GSLIB: geostatistical software library (the software released with this guide)

KT: kriging with a trend model, also known as universal kriging

LS: least squares

M-type: conditional median estimate

Median IK: indicator kriging with the same indicator correlogram for all cutoff values

MG: multi-Gaussian (algorithm or model)

OK: ordinary kriging

pdf: probability density function

p-field: probability field simulation

P-P plot: probability-probability plot

PK: probability kriging

Q-Q plot: quantile-quantile plot

RF: random function

RV: random variable denoted by capital letter Z or Y

SK: simple kriging

sGs: sequential Gaussian simulation

sis: sequential indicator simulation

E.2   Common Notation

$\forall$: whatever

a: range parameter

a_k: coefficient of component number k of the trend model

$B(z)$: Markov-Bayes calibration parameter

$C(\mathbf{0})$: covariance value at separation vector $\mathbf{h} = 0$. It is also the stationary variance of random variable $Z(\mathbf{u})$

$C(\mathbf{h})$: stationary covariance between any two random variables $Z(\mathbf{u})$, $Z(\mathbf{u} + \mathbf{h})$ separated by vector $\mathbf{h}$

$C(\mathbf{u}, \mathbf{u}')$: nonstationary covariance of two random variables $Z(\mathbf{u})$, $Z(\mathbf{u}')$

$C_l(\mathbf{h})$: nested covariance structure in the linear covariance model
$$C(\mathbf{h}) = \sum_{l=1}^{L} C_l(\mathbf{h})$$

$C_I(\mathbf{h}; z_k, z_{k'})$: stationary indicator cross covariance for cutoffs z_k, $z_{k'}$; it is the cross covariance between the two indicator random variables $I(\mathbf{u}; z_k)$ and $I(\mathbf{u} + \mathbf{h}; z_{k'})$

$C_I(\mathbf{h}; z_k)$: stationary indicator covariance for cutoff z_k

$C_I(\mathbf{h}; s_k)$: stationary indicator covariance for category s_k

$C_{ZY}(\mathbf{h})$: stationary cross covariance between the two random variables $Z(\mathbf{u})$ and $Y(\mathbf{u} + \mathbf{h})$ separated by lag vector $\mathbf{h}$

$d\mathbf{u}$: denotes an infinitesimal area (volume) centered at location $\mathbf{u}$

$E\{\cdot\}$: expected value

$E\{Z(\mathbf{u})|(n)\}$: conditional expectation of the random variable $Z(\mathbf{u})$ given the realizations of n other neighboring random variables (called data)

$Exp(\mathbf{h})$: exponential semivariogram model, a function of separation vector $\mathbf{h}$

$F(\mathbf{u}; z)$: nonstationary cumulative distribution function of random variable $Z(\mathbf{u})$

$F(\mathbf{u}; z|(n))$: nonstationary conditional cumulative distribution function of the continuous random variable $Z(\mathbf{u})$ conditioned by the realizations of n other neighboring random variables (called data)

$F(\mathbf{u}; k|(n))$: nonstationary conditional probability distribution function of the categorical variable $Z(\mathbf{u})$

$F(\mathbf{u}_1, \ldots, \mathbf{u}_K; z_1, \ldots, z_K)$: K variate cumulative distribution function of the K random variables $Z(\mathbf{u}_1), \ldots, Z(\mathbf{u}_K)$

$F(z)$: cumulative distribution function of a random variable Z, or stationary cumulative distribution function of a random function $Z(\mathbf{u})$

$F^{-1}(p)$: inverse cumulative distribution function or quantile function for the probability value $p \in [0, 1]$

$f_k(\cdot)$: function of the coordinates used in a trend model

$\gamma(\mathbf{h})$: stationary semivariogram between any two random variables $Z(\mathbf{u})$, $Z(\mathbf{u} + \mathbf{h})$ separated by lag vector $\mathbf{h}$

$\gamma_I(\mathbf{h}; z)$: stationary indicator semivariogram for lag vector $\mathbf{h}$ and cutoff z: it is the semivariogram of the binary indicator random function $I(\mathbf{u}; z)$

$\gamma_I(\mathbf{h}; p)$: same as above, but the cutoff z is expressed in terms of p quantile with $p = F(z)$

$\gamma_{ZY}(\mathbf{h})$: stationary cross semivariogram between the two random variables $Z(\mathbf{u})$ and $Y(\mathbf{u} + \mathbf{h})$ separated by lag vector $\mathbf{h}$

$G(y)$: standard normal cumulative distribution function

$G^{-1}(p)$: standard normal quantile function such that $G(G^{-1}(p)) = p \in [0, 1]$

h: separation vector

$I(\mathbf{u}; z)$: binary indicator random function at location $\mathbf{u}$ and for cutoff z

$i(\mathbf{u}; z)$: binary indicator value at location $\mathbf{u}$ and for cutoff z

$I(\mathbf{u}; s_k)$: binary indicator random function at location $\mathbf{u}$ and for category s_k

$i(\mathbf{u}; s_k)$: binary indicator value at location $\mathbf{u}$ and for category s_k

$j(\mathbf{u}; z)$: binary indicator transform arising from constraint interval

λ_α, $\lambda_\alpha(\mathbf{u})$: kriging weight associated to datum α for estimation at location $\mathbf{u}$. The superscripts (OK), (KT), (m) are used when necessary to differentiate between the various types of kriging

M: stationary median of the distribution function $F(z)$

m: stationary mean of the random variable $Z(\mathbf{u})$

$m(\mathbf{u})$: mean at location $\mathbf{u}$, expected value of random variable $Z(\mathbf{u})$; trend component model in the decomposition $Z(\mathbf{u}) = m(\mathbf{u}) + R(\mathbf{u})$, where $R(\mathbf{u})$ represents the residual component model

$m_{KT}^*(\mathbf{u})$: estimate of the trend component at location $\mathbf{u}$

μ, $\mu(\mathbf{u})$: Lagrange parameter for kriging at location $\mathbf{u}$

$N(\mathbf{h})$: number of pairs of data values available at lag vector $\mathbf{h}$

$\prod_{i=1}^{n} y_i = y_1 \cdot y_2 \cdots y_n$: product

$p_k = E\{I(\mathbf{u}; s_k)\}$: stationary mean of the indicator of category k

$q(p) = F^{-1}(p)$: quantile function, i.e., inverse cumulative distribution function for the probability value, $p \in [0, 1]$

$R(\mathbf{u})$: residual random function model in the decomposition $Z(\mathbf{u}) = m(\mathbf{u}) + R(\mathbf{u})$, where $m(\mathbf{u})$ represents the trend component model

ρ: correlation coefficient $\in [-1, +1]$

$\rho(\mathbf{h})$: stationary correlogram function $\in [-1, +1]$

$\sum_{i=1}^{n} y_i = y_1 + y_2 + \ldots + y_n$: summation

$\Sigma(\mathbf{h})$: matrix of stationary covariances and cross covariances

σ^2: variance

$\sigma_{OK}^2(\mathbf{u})$: ordinary kriging variance of $Z(\mathbf{u})$

$\sigma_{SK}^2(\mathbf{u})$: simple kriging variance of $Z(\mathbf{u})$

$Sph(\mathbf{h})$: spherical semivariogram function of separation vector $\mathbf{h}$

s_k: kth category

$\mathbf{u}$: coordinates vector

$Var\{\cdot\}$: variance

$Y = \varphi(Z)$: transform function $\varphi(\cdot)$ relating two random variables Y and Z

$Z = \varphi^{-1}(Y)$: inverse transform function $\varphi(\cdot)$ relating random variables Z and Y

Z: generic random variable

$Z(\mathbf{u})$: generic random variable at location $\mathbf{u}$, or a generic random function of location $\mathbf{u}$

$Z^*_{COK}(\mathbf{u})$: cokriging estimator of $Z(\mathbf{u})$

$Z^*_{KT}(\mathbf{u})$: estimator of $Z(\mathbf{u})$ using some form of prior trend model

$Z^*_{OK}(\mathbf{u})$: ordinary kriging estimator of $Z(\mathbf{u})$

$Z^*_{SK}(\mathbf{u})$: simple kriging estimator of $Z(\mathbf{u})$

$\{Z(\mathbf{u}), \mathbf{u} \in A\}$: set of random variables $Z(\mathbf{u})$ defined at each location $\mathbf{u}$ of a zone A

$z(\mathbf{u})$: generic variable function of location $\mathbf{u}$

$z(\mathbf{u}_\alpha)$: z datum value at location $\mathbf{u}$

z_k: k th cutoff value

$z^{(l)}(\mathbf{u})$: l th realization of the random function $Z(\mathbf{u})$ at location $\mathbf{u}$

$z_c^{(l)}(\mathbf{u})$: l th realization conditional to some neighboring data

$z^*(\mathbf{u})$: an estimate of value $z(\mathbf{u})$

$[z(\mathbf{u})]^*_E$: E type estimate of value $z(\mathbf{u})$, obtained as an arithmetic average of multiple simulated realizations $z^{(l)}(\mathbf{u})$ of the random function $Z(\mathbf{u})$

Bibliography

[1] E. Aarts and J. Korst. *Simulated Annealing and Boltzmann Machines.* John Wiley & Sons, New York, 1989.

[2] M. Abramovitz and I. Stegun, editors. *Handbook of Mathematical Functions: with Formulas, Graphs, and Mathematical Tables.* Dover, New York, 1972. 9th (revised) printing.

[3] Adobe Systems Incorporated. *PostScript Language Reference Manual.* Addison-Wesley, Menlo Park, CA, 1985.

[4] Adobe Systems Incorporated. *PostScript Language Tutorial and Cookbook.* Addison-Wesley, Menlo Park, CA, 1985.

[5] F. Alabert. The practice of fast conditional simulations through the LU decomposition of the covariance matrix. *Math Geology*, 19(5):369–86, 1987.

[6] F. Alabert. Stochastic imaging of spatial distributions using hard and soft information. Master's thesis, Stanford University, Stanford, CA, 1987.

[7] F. Alabert and G. J. Massonnat. Heterogeneity in a complex turbiditic reservoir: Stochastic modelling of facies and petrophysical variability. In *65th Annual Technical Conference and Exhibition*, pages 775–90. Society of Petroleum Engineers, September 1990. SPE Paper 20604.

[8] A. Almeida, F. Guardiano, and L. Cosentino. Generation of the Stanford 1 turbiditic reservoir. In *Report 5*, Stanford, CA, May 1992. Stanford Center for Reservoir Forecasting.

[9] A. Almeida and A. Journel. Joint simulation of multiple variables with a Markov-type coregionalization model. *Math Geology*, 26(5):565–88, July 1994.

[10] T. Anderson. *An Introduction to Multivariate Statistical Analysis.* John Wiley & Sons, New York, 1958.

[11] M. Armstrong. Improving the estimation and modeling of the variogram. In G. Verly et al., editors, *Geostatistics for Natural Resources Characterization*, pages 1–20. Reidel, Dordrecht, Holland, 1984.

[12] P. Ballin, A. Journel, and K. Aziz. Prediction of uncertainty in reservoir performance forecasting. *JCPT*, 31(4):52–62, April 1992.

[13] R. Barnes and T. Johnson. Positive kriging. In G. Verly et al., editors, *Geostatistics for Natural Resources Characterization*, pages 231–44. Reidel, Dordrecht, Holland, 1984.

[14] S. Begg, E. Gustason, and M. Deacon. Characterization of a fluvial-dominated delta, zone 1 of the Prudhoe bay field. SPE paper # 24698, 1992.

[15] J. Berger. *Statistical Decision Theory and Bayesian Analysis*. Springer Verlag, New York, 1980.

[16] J. Besag. On the statistical analysis of dirty pictures. *J. R. Statistical Society B*, 48(3):259–302, 1986.

[17] L. Borgman. New advances in methodology for statistical tests useful in geostatistical studies. *Math Geology*, 20(4):383–403, 1988.

[18] L. Borgman, M. Taheri, and R. Hagan. Three-dimensional frequency-domain simulations of geological variables. In G. Verly et al., editors, *Geostatistics for Natural Resources Characterization*, pages 517–41. Reidel, Dordrecht, Holland, 1984.

[19] S. Bozic. *Digital and Kalman Filtering*. E. Arnold, London, 1979.

[20] R. Bratvold, L. Holden, T. Svanes, and K. Tyler. STORM: Integrated 3D stochastic reservoir modeling tool for geologists and reservoir engineers. SPE paper # 27563, 1994.

[21] J. Bridge and M. Leeder. A simulation model of alluvial stratigraphy. *Sedimentology*, 26:617–44, 1979.

[22] P. Brooker. Kriging. *Engineering and Mining Journal*, 180(9):148–53, 1979.

[23] P. Brooker. Two-dimensional simulations by turning bands. *Math Geology*, 17(1):81–90, 1985.

[24] J. Carr and N. Mao. A general form of probability kriging for estimation of the indicator and uniform transforms. *Math Geology*, 25(4):425–38, 1993.

[25] J. Carr and D. Myers. COSIM: A Fortran IV program for co-conditional simulation. *Computers & Geosciences*, 11(6):675–705, 1985.

[26] G. Christakos. On the problem of permissible covariance and variogram models. *Water Resources Research*, 20(2):251–65, 1984.

[27] G. Christakos. *Random Field Models in Earth Sciences*. Academic Press, San Diego, CA, 1992.

[28] J. Chu. *Conditional Fractal Simulation, Sequential Indicator Simulation and Interactive Variogram Modeling*. PhD thesis, Stanford University, Stanford, CA, 1993.

[29] J. Chu and A. Journel. Conditional fBm simulation with dual kriging. In R. Dimitrakopoulos, editor, *Geostatistics for the Next Century*, pages 407–21. Kluwer, Dordrecht, Holland, 1993.

[30] D. Cox and V. Isham. *Point Processes*. Chapman and Hall, New York, 1980.

[31] N. Cressie. *Statistics for Spatial Data*. John Wiley & Sons, New York, 1991.

[32] M. Dagbert, M. David, D. Crozel, and A. Desbarats. Computing variograms in folded strata-controlled deposits. In G. Verly et al., editors, *Geostatistics for Natural Resources Characterization*, pages 71–89. Reidel, Dordrecht, Holland, 1984.

[33] E. Damsleth, C. Tjolsen, K. Omre, and H. Haldorsen. A two-stage stochastic model applied to a north sea reservoir. In *65th Annual Technical Conference and Exhibition*, pages 791–802. Society of Petroleum Engineers, September 1990.

[34] M. David. *Geostatistical Ore Reserve Estimation*. Elsevier, Amsterdam, 1977.

[35] B. Davis. Indicator kriging as applied to an alluvial gold deposit. In G. Verly et al., editors, *Geostatistics for Natural Resources Characterization*, pages 337–48. Reidel, Dordrecht, Holland, 1984.

[36] B. Davis. Uses and abuses of cross validation in geostatistics. *Math Geology*, 19(3):241–48, 1987.

[37] J. Davis. *Statistics and Data Analysis in Geology*. John Wiley & Sons, New York, 2nd edition, 1986.

[38] M. W. Davis. Production of conditional simulations via the LU decomposition of the covariance matrix. *Math Geology*, 19(2):91–98, 1987.

[39] P. Delfiner. Linear estimation of non-stationary spatial phenomena. In M. Guarascio et al., editors, *Advanced Geostatistics in the Mining Industry*, pages 49–68, Reidel, Dordrecht, Holland, 1976.

[40] C. Deutsch. A probabilistic approach to estimate effective absolute permeability. Master's thesis, Stanford University, Stanford, CA, 1987.

[41] C. Deutsch. DECLUS: A Fortran 77 program for determining optimum spatial declustering weights. *Computers & Geosciences*, 15(3):325–332, 1989.

[42] C. Deutsch. *Annealing Techniques Applied to Reservoir Modeling and the Integration of Geological and Engineering (Well Test) Data*. PhD thesis, Stanford University, Stanford, CA, 1992.

[43] C. Deutsch. Constrained modeling of histograms and cross plots with simulated annealing. *Technometrics*, 38(3):266–274, August 1996.

[44] C. Deutsch. Correcting for negative weights in ordinary kriging. *Computers & Geosciences*, 22(7):765–773, 1996.

[45] C. Deutsch and P. Cockerham. Practical considerations in the application of simulated annealing to stochastic simulation. *Math Geology*, 26(1):67–82, 1994.

[46] C. Deutsch and A. Journel. The application of simulated annealing to stochastic reservoir modeling. *SPE Advanced Techology Series*, 2(2), April 1994.

[47] C. Deutsch and R. Lewis. Advances in the practical implementation of indicator geostatistics. In *Proceedings of the 23rd International APCOM Symposium*, pages 133–48, Tucson, AZ, April 1992. Society of Mining Engineers.

[48] C. V. Deutsch and L. Wang. Hierarchical object-based stochastic modeling of fluvial reservoirs. *Math Geology*, 28(7):857–880, 1996.

[49] L. Devroye. *Non-Uniform Random Variate Generation*. Springer Verlag, New York, 1986.

[50] J. Doob. *Stochastic Processes*. John Wiley & Sons, New York, 1953.

[51] P. Doyen. Porosity from seismic data: A geostatistical approach. *Geophysics*, 53(10):1263–75, 1988.

[52] P. Doyen and T. Guidish. Seismic discrimination of lithology in sand/shale reservoirs: A Bayesian approach. Expanded Abstract, SEG 59th Annual Meeting., 1989, Dallas, TX.

[53] P. Doyen, T. Guidish, and M. de Buyl. Monte Carlo simulation of lithology from seismic data in a channel sand reservoir. SPE paper # 19588, 1989.

[54] O. Dubrule. A review of stochastic models for petroleum reservoirs. In M. Armstrong, editor, *Geostatistics*, pages 493–506. Kluwer, Dordrecht, Holland, 1989.

[55] O. Dubrule and C. Kostov. An interpolation method taking into account inequality constraints. *Math Geology*, 18(1):33–51, 1986.

[56] B. Efron. *The Jackknife, the Bootstrap, and Other Resampling Plans.* Soc. for Industrial and Applied Math, Philadelphia, 1982.

[57] A. Emanuel, G. Alameda, R. Behrens, and T. Hewett. Reservoir performance prediction methods based on fractal geostatistics. *SPE Reservoir Engineering*, pages 311–18, August 1989.

[58] E. Englund and A. Sparks. *Geo-EAS 1.2.1 User's Guide, EPA Report # 60018-91/008.* EPA-EMSL, Las Vegas, NV, 1988.

[59] C. Farmer. The generation of stochastic fields of reservoir parameters with specified geostatistical distributions. In S. Edwards and P. King, editors, *Mathematics in Oil Production*, pages 235–52. Clarendon Press, Oxford, 1988.

[60] C. Farmer. Numerical rocks. In P. King, editor, *The Mathematical Generation of Reservoir Geology*, Oxford, 1992. Clarendon Press. (Proceedings of a conference held at Robinson College, Cambridge, 1989).

[61] G. Fogg, F. Lucia, and R. Sengen. Stochastic simulation of inter-well scale heterogeneity for improved prediction of sweep efficiency in a carbonate reservoir. In L. Lake, H. Caroll, and P. Wesson, editors, *Reservoir Characterization II*, pages 355–81. Academic Press, 1991.

[62] R. Froidevaux. *Geostatistical Toolbox Primer, version 1.30.* FSS International, Troinex, Switzerland, 1990.

[63] I. Gelfand. Generalized random processes. volume 100, pages 853–56. Dokl. Acad. Nauk., USSR, 1955.

[64] S. Geman and D. Geman. Stochastic relaxation, Gibbs distributions, and the Bayesian restoration of images. *IEEE Transactions on Pattern Analysis and Machine Intelligence*, PAMI-6(6):721–41, November 1984.

[65] D. J. Gendzwill and M. Stauffer. Analysis of triaxial ellipsoids: Their shapes, plane sections, and plane projections. *Math Geology*, 13(2):135–52, 1981.

[66] C. Gerald. *Applied Numerical Analysis.* Addison-Wesley, Menlo Park, CA, 1978.

[67] I. M. Glacken. Change of support by direct conditional block simulation. In *Fifth International Geostatistics Congress*, Wollongong, September 1996.

[68] A. Goldberger. Best linear unbiased prediction in the generalized linear regression model. *JASA*, 57:369–75, 1962.

[69] G. Golub and C. Van Loan. *Matrix Computations*. The Johns Hopkins University Press, Baltimore, 1990.

[70] J. Gómez-Hernández. *A Stochastic Approach to the Simulation of Block Conductivity Fields Conditioned upon Data Measured at a Smaller Scale*. PhD thesis, Stanford University, Stanford, CA, 1991.

[71] J. Gómez-Hernández and A. Journel. Joint sequential simulation of multiGaussian fields. In A. Soares, editor, *Geostatistics-Troia*, pages 85–94. Kluwer, Dordrecht, Holland, 1993.

[72] J. Gómez-Hernández and R. Srivastava. ISIM3D: An ANSI-C three dimensional multiple indicator conditional simulation program. *Computers & Geosciences*, 16(4):395–410, 1990.

[73] P. Goovaerts. Comparative performance of indicator algorithms for modeling conditional probability distribution functions. *Math Geology*, 26(3):385–410, 1994.

[74] P. Goovaerts. *Geostatistics for Natural Resources Evaluation*. Oxford University Press, New York, 1997. In Press.

[75] F. Guardiano and R. M. Srivastava. Multivariate geostatistics: Beyond bivariate moments. In A. Soares, editor, *Geostatistics-Troia*, pages 133–44. Kluwer, Dordrecht, Holland, 1993.

[76] A. Gutjahr. Fast Fourier transforms for random fields. Technical Report No. 4-R58-2690R, Los Alamos, NM, 1989.

[77] H. Haldorsen, P. Brand, and C. Macdonald. Review of the stochastic nature of reservoirs. In S. Edwards and P. King, editors, *Mathematics in Oil Production*, pages 109–209. Clarendon Press, Oxford, 1988.

[78] H. Haldorsen and E. Damsleth. Stochastic modeling. *J. of Pet. Technology*, pages 404–12, April 1990.

[79] R. L. Hardy. Theory and application of the multiquadric-biharmonic method: 20 years of discovery: 1968-1988. *Computers Math. Applic*, 19(8-9):163–208, 1990.

[80] T. Hewett. Fractal distributions of reservoir heterogeneity and their influence on fluid transport. SPE paper # 15386, 1986.

[81] T. Hewett and R. Behrens. Conditional simulation of reservoir heterogeneity with fractals. *Formation Evaluation*, pages 300–10, August 1990.

[82] M. Hohn. *Geostatistics and Petroleum Geology*. Van Nostrand, New York, 1988.

[83] A. Householder. *Principles of Numerical Analysis*. McGraw-Hill, New York, 1953.

[84] P. Huber. *Robust Statistics*. John Wiley & Sons, New York, 1981.

[85] C. Huijbregts and G. Matheron. Universal kriging - An optimal approach to trend surface analysis. In *Decision Making in the Mineral Industry*, pages 159–69. Canadian Institute of Mining and Metallurgy, 1971. Special Volume 12.

[86] E. Isaaks. Risk qualified mappings for hazardous waste sites: A case study in distribution-free geostatistics. Master's thesis, Stanford University, Stanford, CA, 1984.

[87] E. Isaaks. *The Application of Monte Carlo Methods to the Analysis of Spatially Correlated Data*. PhD thesis, Stanford University, Stanford, CA, 1990.

[88] E. Isaaks and R. Srivastava. Spatial continuity measures for probabilistic and deterministic geostatistics. *Math Geology*, 20(4):313–41, 1988.

[89] E. Isaaks and R. Srivastava. *An Introduction to Applied Geostatistics*. Oxford University Press, New York, 1989.

[90] M. Johnson. *Multivariate Statistical Simulation*. John Wiley & Sons, New York, 1987.

[91] R. Johnson and D. Wichern. *Applied Multivariate Statistical Analysis*. Prentice Hall, Englewood Cliffs, NJ, 1982.

[92] T. Jones, D. Hamilton, and C. Johnson. *Contouring Geologic Surfaces with the Computer*. Van Nostrand Reinhold, New York, 1986.

[93] A. Journel. Geostatistics for conditional simulation of orebodies. *Economic Geology*, 69:673–80, 1974.

[94] A. Journel. The lognormal approach to predicting local distributions of selective mining unit grades. *Math Geology*, 12(4):285–303, 1980.

[95] A. Journel. Non-parametric estimation of spatial distributions. *Math Geology*, 15(3):445–68, 1983.

[96] A. Journel. The place of non-parametric geostatistics. In G. Verly et al., editors, *Geostatistics for Natural Resources Characterization*, pages 307–55. Reidel, Dordrecht, Holland, 1984.

[97] A. Journel. Recoverable reserves estimation - the geostatistical approach. *Mining Engineering*, pages 563–68, June 1985.

[98] A. Journel. Constrained interpolation and qualitative information. *Math Geology*, 18(3):269–86, 1986.

[99] A. Journel. Geostatistics: Models and tools for the earth sciences. *Math Geology*, 18(1):119–40, 1986.

[100] A. Journel. Geostatistics for the environmental sciences, EPA project no. cr 811893. Technical report, U.S. EPA, EMS Lab, Las Vegas, NV, 1987.

[101] A. Journel. New distance measures: The route towards truly non-Gaussian geostatistics. *Math Geology*, 20(4):459–75, 1988.

[102] A. Journel. *Fundamentals of Geostatistics in Five Lessons*. Volume 8 Short Course in Geology. American Geophysical Union, Washington, D.C., 1989.

[103] A. Journel. Modeling uncertainty: Some conceptual thoughts. In R. Dimitrakopoulos, editor, *Geostatistics for the Next Century*, pages 30–43. Kluwer, Dordrecht, Holland, 1993.

[104] A. Journel. Resampling from stochastic simulations. *Environmental and Ecological Statistics*, 1:63–84, 1994.

[105] A. Journel and F. Alabert. Non-Gaussian data expansion in the earth sciences. *Terra Nova*, 1:123–34, 1989.

[106] A. Journel and F. Alabert. New method for reservoir mapping. *J. of Pet. Technology*, pages 212–18, February 1990.

[107] A. Journel and C. Deutsch. Entropy and spatial disorder. *Math Geology*, 25(3):329–55, 1993.

[108] A. Journel and J. Gómez-Hernández. Stochastic imaging of the Wilmington clastic sequence. *SPEFE*, pages 33–40, March 1993. SPE paper # 19857.

[109] A. Journel and C. J. Huijbregts. *Mining Geostatistics*. Academic Press, New York, 1978.

[110] A. Journel and E. Isaaks. Conditional indicator simulation: Application to a Saskatchewan uranium deposit. *Math Geology*, 16(7):685–718, 1984.

[111] A. Journel and D. Posa. Characteristic behavior and order relations for indicator variograms. *Math Geology*, 22(8):1011–25, 1990.

[112] A. Journel and M. Rossi. When do we need a trend model in kriging? *Math Geology*, 21(7):715–39, 1989.

[113] A. Journel and W. Xu. Posterior identification of histograms conditional to local data. *Math Geology*, 22(8):323–359, 1994.

[114] A. G. Journel. The abuse of principles in model building and the quest for objectivity: Opening keynote address. In *Fifth International Geostatistics Congress*, Wollongong, September 1996.

[115] A. G. Journel and C. V. Deutsch. Rank order geostatistics: A proposal for a unique coding and common processing of diverse data. In *Fifth International Geostatistics Congress*, Wollongong, September 1996.

[116] W. Kennedy Jr. and J. Gentle. *Statistical Computing*. Marcel Dekker, Inc., New York, 1980.

[117] S. Kirkpatrick, C. Gelatt Jr., and M. Vecchi. Optimization by simulated annealing. *Science*, 220(4598):671–80, May 1983.

[118] A. Kolmogorov. *Grundbegriffe der Wahrschein-lichkeitrechnung. Ergebnisse der Mathematik, English Translation: Foundations of the Theory of Probability.* Chelsea Publ., New York, English translation: 1950 edition, 1933.

[119] C. E. Koltermann and S. Gorelick. Paleoclimatic signature in terrestrial flood deposits. *Science*, 256:1775–82, September 1992.

[120] C. E. Koltermann and S. Gorelick. Heterogeneity in sedimentary deposits: A review of structure-imitating, process-imitating, and descriptive approaches. *Water Resources Research*, 32(9):2617–58, September 1996.

[121] D. G. Krige. A statistical approach to some mine valuations and allied problems at the witwatersrand. Master's thesis, University of Witwatersrand, South Africa, 1951.

[122] G. Kunkel. *Graphic Design with PostScript*. Scott, Foreman and Company, Glenview, IL, 1990.

[123] I. Lemmer. Mononodal indicator variography - part 1: Theory - part 2: Application to a computer simulated deposit. *Math Geology*, 18(7):589–623, 1988.

[124] D. Luenberger. *Optimization by Vector Space Methods*. John Wiley & Sons, New York, 1969.

[125] G. Luster. *Raw Materials for Portland Cement: Applications of Conditional Simulation of Coregionalization*. PhD thesis, Stanford University, Stanford, CA, 1985.

[126] Y. Ma and J. Royer. Local geostatistical filtering application to remote sensing. In Namyslowska and Royer, editors, *Geomathematics and Geostatistics Analysis Applied to Space and Time Data*, volume 27, pages 17–36. Sciences de la Terre, Nancy, France, 1988.

[127] J. Mallet. Regression sous contraintes lineaires: Application au codage des variables aleatoires. *Rev. Stat. Appl.*, 28(1):57–68, 1980.

[128] A. Mantoglou and J. Wilson. Simulation of random fields with the turning band method. Technical Report No. 264, Department of Civil Engineering, M.I.T., 1981.

[129] D. Marcotte and M. David. The biGaussian approach, a simple method for recovery estimation. *Math Geology*, 17(6):625–44, 1985.

[130] A. Marechal. Kriging seismic data in presence of faults. In G. Verly et al., editors, *Geostatistics for Natural Resources Characterization*, pages 271–94. Reidel, Dordrecht, Holland, 1984.

[131] G. Marsaglia. The structure of linear congruential sequences. In S. Zaremba, editor, *Applications of Number Theory to Numerical Analysis*, pages 249–85. Academic Press, London, 1972.

[132] B. Matern. *Spatial Variation*, volume 36 of *Lecture Notes in Statistics*. Springer Verlag, New York, second edition, 1980. First edition published by Meddelanden fran Statens Skogsforskningsinstitut, Band 49, No. 5, 1960.

[133] G. Matheron. Traité de géostatistique appliquée. Vol. 1 (1962), Vol. 2 (1963), ed. Technip, Paris, 1962.

[134] G. Matheron. La théorie des variables régionalisées et ses applications. Fasc. 5, Ecole Nat. Sup. des Mines, Paris, 1971.

[135] G. Matheron. The intrinsic random functions and their applications. *Advances in Applied Probability*, 5:439–68, 1973.

[136] G. Matheron. A simple substitute for conditional expectation: the disjunctive kriging. In M. Guarascio et al., editors, *Advanced Geostatistics in the Mining Industry*, pages 221–36, Reidel, Dordrecht, Holland, 1976.

[137] G. Matheron. The internal consistency of models in geostatistics. In M. Armstrong, editor, *Geostatistics*, volume 1, pages 21–38. Kluwer, Dordrecht, Holland, 1989.

[138] G. Matheron, H. Beucher, H. de Fouquet, A. Galli, D. Guerillot, and C. Ravenne. Conditional simulation of the geometry of fluvio-deltaic reservoirs. SPE paper # 16753, 1987.

[139] N. Metropolis, A. Rosenbluth, M. Rosenbluth, A. Teller, and E. Teller. Equation of state calculations by fast computing machines. *J. Chem. Phys.*, 21(6):1087–92, June 1953.

[140] M. Monmonier. *Computer-Assisted Cartography: Principles and Prospects*. Prentice Hall, Englewood Cliffs, NJ, 1982.

[141] F. Mosteller and J. Tukey. *Data Analysis and Regression*. Addison-Wesley, Reading, MA, 1977.

[142] C. Murray. Indicator simulation of petrophysical rock types. In A. Soares, editor, *Geostatistics-Troia*, pages 399–411. Kluwer, Dordrecht, Holland, 1993.

[143] D. Myers. Matrix formulation of co-kriging. *Math Geology*, 14(3):249–57, 1982.

[144] D. Myers. Cokriging - new developments. In G. Verly et al., editors, *Geostatistics for Natural Resources Characterization*, pages 295–305. Reidel, Dordrecht, Holland, 1984.

[145] R. Olea, editor. *Optimum Mapping Techniques: Series on Spatial Analysis No. 2*. Kansas Geol. Survey, Lawrence, KA, 1975.

[146] R. Olea, editor. *Geostatistical Glossary and Multilingual Dictionary*. Oxford University Press, New York, 1991.

[147] H. Omre. The variogram and its estimation. In G. Verly et al., editors, *Geostatistics for Natural Resources Characterization*, volume 1, pages 107–25. Reidel, Dordrecht, Holland, 1984.

[148] H. Omre. Bayesian kriging, merging observations and qualified guesses in kriging. *Math Geology*, 19(1):25–39, 1987.

[149] H. Omre. Heterogeneity models. In *SPOR Monograph: Recent Advances in Improved Oil Recovery Methods for North Sea Sandstone Reservoirs*, Norway, 1992. Norwegian Petroleum Directorate.

[150] H. Omre, K. Solna, and H. Tjelmeland. Calcite cementation description and production consequences. SPE paper # 20607, 1990.

[151] H. Parker. The volume-variance relationship: a useful tool for mine planning. In P. Mousset-Jones, editor, *Geostatistics*, pages 61–91, New York, 1980. McGraw Hill.

[152] H. Parker, A. Journel, and W. Dixon. The use of conditional lognormal probability distributions for the estimation of open pit ore reserves in stratabound uranium deposits: A case study. In *Proceedings of the 16th International APCOM Symposium*, pages 133–48, Tucson, AZ, October 1979. Society of Mining Engineers.

[153] D. Posa. Conditioning of the stationary kriging matrices for some well-known covariance models. *Math Geology*, 21(7):755–65, 1988.

[154] W. Press, B. Flannery, S. Teukolsky, and W. Vetterling. *Numerical Recipes*. Cambridge University Press, New York, 1986.

[155] S. Rao and A. G. Journel. Deriving conditional distributions from ordinary kriging. In *Fifth International Geostatistics Congress*, Wollongong, September 1996.

[156] J.-M. M. Rendu. Normal and lognormal estimation. *Math Geology*, 11(4):407–22, 1979.

[157] B. Ripley. *Stochastic Simulation*. John Wiley & Sons, New York, 1987.

[158] B. Ripley. *Statistical Inference for Spatial Processes*. Cambridge University Press, New York, 1988.

[159] M. Rosenblatt. Remarks on a multivariate transformation. *Annals of Mathematical Statistics*, 23(3):470–472, 1952.

[160] F. Samper Calvete and J. Carrera Ramirez. *Geoestadistica: Aplicaciones a la Hidrologia Subterranea*. Centro Int. de Metodos Numericos en Ingenieria, Barcelona, Spain, 1990.

[161] L. Sandjivy. The factorial kriging analysis of regionalized data. its application to geochemical prospecting. In G. Verly et al., editors, *Geostatistics for Natural Resources Characterization*, pages 559–71. Reidel, Dordrecht, Holland, 1984.

[162] E. Schnetzler. Visualization and cleaning of pixel-based images. Master's thesis, Stanford University, Stanford, CA, 1994.

[163] H. Schwarz. *Numerical Analysis, A Comprehensive Introduction*. John Wiley & Sons, New York, 1989.

[164] H. Schwarz, H. Rutishauser, and E. Stiefel. *Numerical Analysis of Symmetric Matrices*. Prentice-Hall, Englewood Cliffs, NJ, 1973.

[165] B. Silverman. *Density Estimation for Statistics and Data Analysis*. Chapman and Hall, New York, 1986.

[166] A. Solow. *Kriging under a Mixture Model*. PhD thesis, Stanford University, Stanford, CA, 1985.

[167] R. Srivastava. Minimum variance or maximum profitability? *CIM Bulletin*, 80(901):63–68, 1987.

[168] R. Srivastava. A non-ergodic framework for variogram and covariance functions. Master's thesis, Stanford University, Stanford, CA, 1987.

[169] R. Srivastava. Reservoir characterization with probability field simulation. In *SPE Annual Conference and Exhibition, Washington, DC*, number 24753, pages 927–38, Washington, DC, October 1992. Society of Petroleum Engineers.

[170] R. Srivastava. The visualization of spatial uncertainty. In J. Yarus and R. Chambers, editors, *Stochastic modeling and geostatistics. Principles, Methods, and Case studies*, volume 3 of *AAPG Computer Applications in Geology*, pages 339–45. AAPG, 1994.

[171] R. Srivastava. An overview of stochastic methods for reservoir characterization. In J. Yarus and R. Chambers, editors, *Stochastic Modeling and Geostatistics: Principles, Methods, and Case Studies*, pages 3–16. AAPG Computer Applications in Geology, No. 3, 1995.

[172] R. Srivastava and H. Parker. Robust measures of spatial continuity. In M. Armstrong, editor, *Geostatistics*, pages 295–308. Reidel, Dordrecht, Holland, 1988.

[173] M. Stein and M. Handcock. Some asymptotic properties of kriging when the covariance is unspecified. *Math Geology*, 21(2):171–190, 1989.

[174] D. Stoyan, W. Kendall, and J. Mecke. *Stochastic Geometry and its Applications*. John Wiley & Sons, New York, 1987.

[175] J. Sullivan. Conditional recovery estimation through probability kriging: theory and practice. In G. Verly et al., editors, *Geostatistics for Natural Resources Characterization*, pages 365–84. Reidel, Dordrecht, Holland, 1984.

[176] V. Suro-Perez. *Indicator Kriging and Simulation Based on Principal Component Analysis*. PhD thesis, Stanford University, Stanford, CA, 1992.

[177] V. Suro-Perez and A. Journel. Stochastic simulation of lithofacies: an improved sequential indicator approach. In D. Guerillot and Guillon, editors, *Proceedings of 2nd European Conference on the Mathematics of Oil Recovery*, pages 3–10, Paris, September 1990. Technip Publ.

[178] V. Suro-Perez and A. Journel. Indicator principal component kriging. *Math Geology*, 23(5):759–88, 1991.

[179] D. Tetzlaff and J. Harbaugh. *Simulating Clastic Sedimentation*. Reinhold Van Nostrand, New York, 1989.

[180] T. Tran. Improving variogram reproduction on dense simulation grids. *Computers & Geosciences*, 20(7):1161–1168, 1994.

[181] J. Tukey. *Exploratory Data Analysis*. Addison-Wesley, Reading, MA, 1977.

[182] G. Verly. The multiGaussian approach and its applications to the estimation of local reserves. *Math Geology*, 15(2):259–86, 1983.

[183] G. Verly. Sequential Gaussian cosimulation: A simulation method integrating several types of information. In A. Soares, editor, *Geostatistics-Troia*, pages 543–54. Kluwer, Dordrecht, Holland, 1993.

[184] R. Voss. *The Science of Fractal Images*. Springer Verlag, New York, 1988.

[185] H. Wackernagel. *Multivariate Geostatistics: An Introduction with Applications*. Springer-Verlag, Berlin, 1995.

[186] J. Westlake. *A Handbook of Numerical Matrix Inversion and Solution of Linear Equations*. John Wiley & Sons, New York, 1968.

[187] P. Whittle. *Prediction and Regulation by Linear Least-Squares Methods*. University of Minnesota Press, MN, 1963. Revised edition 1983.

[188] N. Wiener. *Extrapolation, Interpolation, and Smoothing of Stationary Time Series*. MIT Press, Cambridge, 7th edition, 1966.

[189] R. Wikramaratna. ACORN - a new method for generating sequences of uniformly distributed pseudo-random numbers. *Journal of Computational Physics*, 83:16–31, 1989.

[190] Y. Wu, A. Journel, L. Abramson, and P. Nair. *Uncertainty Evaluation Methods for Waste Package Performance Assessment*. U.S. Nuclear Reg. Commission, NUREG/CR-5639, 1991.

[191] H. Xiao. A description of the behavior of indicator variograms for a bivariate normal distribution. Master's thesis, Stanford University, Stanford, CA, 1985.

[192] W. Xu and A. Journel. Gtsim: Gaussian truncated simulations of reservoir units in a West Texas carbonate field. SPE paper # 27412, 1993.

[193] W. Xu and A. Journel. Histogram and scattergram smoothing using convex quadratic programming. *Math Geology*, 27(1):83–103, January 1994.

[194] W. Xu, T. Tran, R. Srivastava, and A. Journel. Integrating seismic data in reservoir modeling: the collocated cokriging alternative. SPE paper # 24742, 1992.

[195] A. Yaglom and M. Pinsker. Random processes with stationary increments of order n. volume 90, pages 731–34. Dokl. Acad. Nauk., USSR, 1953.

[196] H. Zhu. *Modeling Mixture of Spatial Distributions with Integration of Soft Data.* PhD thesis, Stanford University, Stanford, CA, 1991.

[197] H. Zhu and A. Journel. Mixture of populations. *Math Geology,* 23(4):647–71, 1991.

[198] H. Zhu and A. Journel. Formatting and integrating soft data: Stochastic imaging via the Markov-Bayes algorithm. In A. Soares, editor, *Geostatistics-Troia,* pages 1–12. Kluwer, Dordrecht, Holland, 1993.

Index